Building Blocks

Volume **1**

Teacher's Edition

PreK

Authors:

Douglas H. Clements

Julie Sarama

Bothell, WA • Chicago, IL • Columbus, OH • New York, NY

Authors

Douglas H. Clements
Professor of Early Childhood and Mathematics Education
University at Buffalo
State University of New York, NY

Julie Sarama
Professor of Mathematics Education
University at Buffalo
State University of New York, NY

www.mheonline.com

Send all inquiries to:
McGraw-Hill Education
8787 Orion Place
Columbus, OH 43240

ISBN: 978-0-02-125485-9
MHID: 0-02-125485-0

Printed in the United States of America.

3 4 5 6 7 8 9 WEB 17 16 15

This curriculum was supported in part by the National Science Foundation under Grant No. ESI-9730804, “Building Blocks-Foundations for Mathematical Thinking, Pre-Kindergarten to Grade 2: Research-based Materials Development” to Douglas H. Clements and Julie Sarama. The curriculum was also based partly upon work supported in part by the Institute of Educational Sciences (U.S. Dept. of Education, under the Interagency Educational Research Initiative, or IERI, a collaboration of the IES, NSF, and NICHHD) under Grant No. R305K05157, “Scaling Up TRIAD: Teaching Early Mathematics for Understanding with Trajectories and Technologies” and by the IERI through a National Science Foundation NSF Grant No. REC-0228440, “Scaling Up the Implementation of a Pre-Kindergarten Mathematics Curricula: Teaching for Understanding with Trajectories and Technologies.” Any opinions, findings, and conclusions or recommendations expressed in this material are those of the authors and do not necessarily reflect the views of the funding agencies.

Table of Contents

WEEKS 1-3

Volume 1

Week 1

Week 2

Week 3

Table of Contents

WEEKS 4-6

Week 4

Week 5

Week 6

Table of Contents

WEEKS 7-9

Week 7

Week 8

Week 9

WEEKS 10-12

Table of Contents

Week 10

Week 11

Week 12

Table of Contents

WEEKS 13-15

Week 13

Week 14

Week 15

Table of Contents

WEEKS 16-18

Volume 2

Week 16

Week 17

Week 18

Week 19

Week 20

Week 21

Table of Contents

WEEKS 22-24

Week 22

Week 23

Week 24

Table of Contents

WEEKS 25-27

Week 25

Week 26

Week 27

Table of Contents

Week 28

Week 29

Week 30

Building Blocks

Overview

Good early mathematics is broader and deeper than early practice on "school skills." Quality mathematics is a joy, not a pressure. It emerges from children's play, their curiosity, and their natural ability to think. ***Building Blocks*** *builds on children's love of patterns, counting and shape to develop foundational understandings and skills.*

Building Blocks *is one of a small number of projects that the National Science Foundation funded to create mathematics curriculum materials for young children. Its basic approach is to* find the mathematics in, and develop mathematics from children's experiences and interests. *The materials are intended to help children extend and "mathematize" their everyday activities.*

Eclipse Studios

***Building Blocks** is a program that acknowledges the critical role teachers play in math education. The program is designed to provide thorough background, teaching strategies, and resources to support teacher delivery of a coherent and effective mathematics curriculum.*

The program is designed to

- Build upon young children's experiences with mathematics with activities that integrate ways to explore and represent mathematics: with children's bodies, manipulatives, computers, books, and children's drawings
- Involve children in "doing mathematics"
- Establish a solid foundation for future study of mathematics
- Develop a strong conceptual framework that provides anchoring for skill acquisition
- Emphasize the development of children's mathematical thinking and reasoning abilities
- Develop the big ideas for early childhood mathematics learning in line with state and national standards
  - Number and Operations
  - Geometry
  - Measurement
  - Patterns and Algebra
  - Data Analysis and Classification
- Make appropriate and ongoing use of technology
- Develop teachers' understanding of mathematics so that they direct and understand mathematical topics, integrate math into the curriculum at large, and provide appropriate pedagogy. Successful teachers interpret what a child is doing and thinking and attempt to see the situation from the child's point of view. From their interpretations, these teachers speculate about what concepts the child might be able to learn or abstract from his or her experiences.
- Incorporate assessment as an integral part of learning events

Building Blocks develops the power of young children's mathematical thinking. Using their bodies, manipulatives, paper, and computers, children engage in activities that guide them through fine-tuned research-based learning trajectories. These activities connect children's informal knowledge to more formal school mathematics. The materials include research-based computer tools, with activities and a management system that guides children through research-based learning trajectories. These activities-through-trajectories connect children's informal knowledge to more formal school mathematics.The result is a curriculum that is not only motivating for children but also comprehensive.

The Research Behind Building Blocks

Building Blocks was designed upon research conducted in a well-defined, rigorous, and complete fashion. Results indicate strong positive effects with achievement gains near or exceeding those recorded for individual tutoring.

Building Blocks' development was supported by a grant from the National Science Foundation Research. Phases of the research included the following:

- Drafting curriculum goals
- Building an explicit model of the learning trajectories of children's knowledge and learning for each goal
- Creating initial activities and software
- Assessing prototypes and curriculum with one-on-one interviews with students and teachers
- Conducting pilot tests in several classrooms
- Conducting field tests in numerous classrooms

Results

Three different studies were conducted.

1. ***Building Blocks Summative Evaluation.*** This tested the effectiveness in a small number of classrooms. In this study, ***Building Blocks*** was shown to increase knowledge of multiple essential mathematical concepts and skills.

The results are illustrated in two graphs. We computed effect sizes using the accepted benchmarks of .25 as indicating practical significance (for example, educationally meaningful), .5 as indicating moderate strength, and .8 as indicating a large effect. The effect sizes comparing ***Building Blocks*** children's posttest to the control children's posttest were .85 and 1.44 for number and geometry, respectively, and the effect sizes comparing ***Building Blocks*** children's posttest to their pretest (measuring achievement gains) were 1.71 and 2.12. Therefore, *all effects were positive and large. Achievement gains were comparable to the Bloom's coveted "2-sigma," or two standard deviations, effect of excellent individual tutoring.*

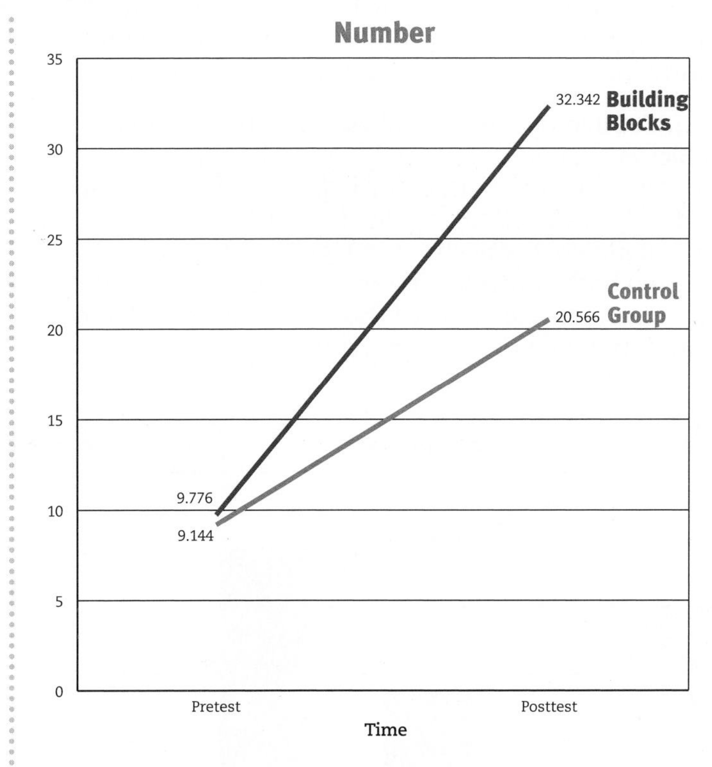

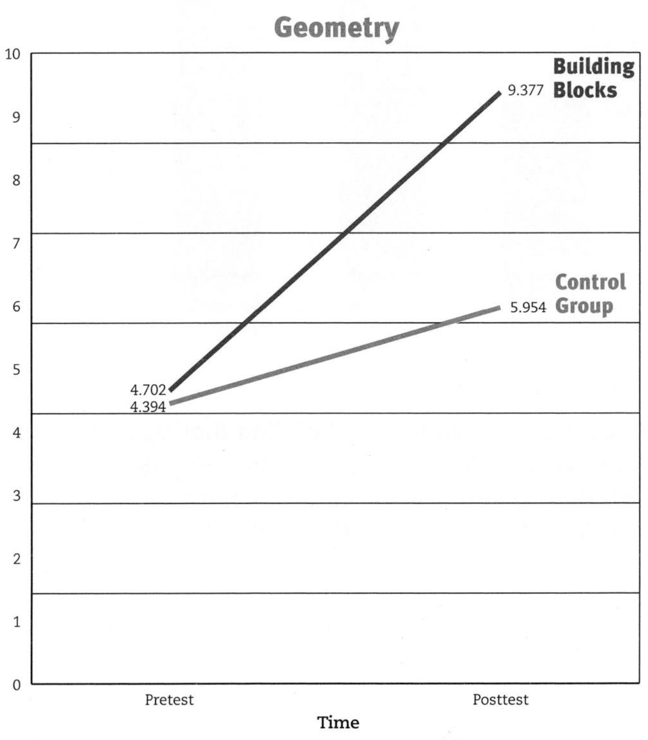

Results indicate strong positive effects of the ***Building Blocks*** materials, with achievement gains near or exceeding those recorded for individual tutoring. This is the result of implementing a curriculum built on comprehensive research-based principles.

2. ***Preschool Curriculum Evaluation Research.*** In this study, Building Blocks was used in 40 classrooms with no additional support or training. Mathematics achievement significantly increased in these classrooms as a result.

3. ***The TRIAD/Building Blocks Studies.*** This study tested ***Building Blocks*** against a comparable preschool math program and a no-treatment control group. All classrooms were randomly assigned, the "gold standard" of scientific evaluation. ***Building Blocks*** children significantly outperformed both control children and the comparison group. Again, effect sizes *doubled* those usually considered "strong" and matched those of individual tutoring.

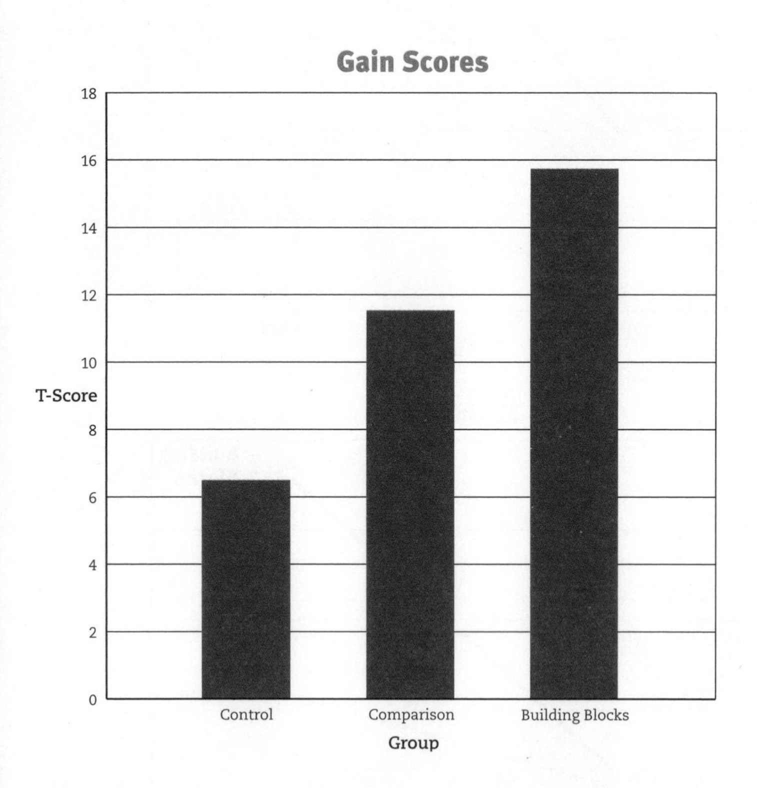

As a result of the positive results ***Building Blocks,*** under a grant from the U.S. Department of Education, is now being tested in a rigorous scale-up study with random assignment in more than a hundred classrooms.

"Basing the curriculum on learning trajectories is even more important than we originally assumed. They helped sequence activities and were critical for allowing our software to provide correlated, individualized activities. In addition, we found that teachers who understood the learning trajectories were more effective in teaching small groups and encouraging informal, incidental mathematics at an appropriate and deep level....We have found this the most powerful way to help our teachers understand children's development, conduct observational assessment, teach, and appreciate the worth of a curriculum."

Sarama, Julie and Clements, Douglas. "***Building Blocks*** for Early Childhood Mathematics" *Early Childhood Research Quarterly* 19 (2004) 181–189.

Eclipse Studios

Building Blocks *activities are based on the developmental levels of mathematics learning trajectories.*

Learning Trajectories

Learning trajectories are the observable, natural developmental progressions in learning. Curriculum research has revealed sequences of activities that are effective in guiding children through these levels of thinking. These developmental paths are the basis for ***Building Blocks*** learning trajectories.

Learning trajectories have three parts: a mathematical goal, a developmental path along which children develop to reach that goal, and a set of activities matched to each of the levels of thinking in that path that help children develop the next higher level of thinking. Thus, each learning trajectory has levels of understanding and skill, each more sophisticated than the last, with tasks that promote growth from one level to the next. The ***Building Blocks*** learning trajectories give simple labels, descriptions, and examples of each level. Complete learning trajectories describe the goals of learning, the thinking and learning processes of children at various levels, and the learning activities in which they might engage.

Building Blocks activities are carefully designed and sequenced to address each level of the learning trajectories in the following areas of mathematics:

Number

- Counting
- Comparing and Ordering Numbers
- Recognizing Number and Subitizing (Instantly Recognizing)
- Composing (Knowing Combinations of) Numbers
- Adding and Subtracting
- Multiplying and Dividing

Measuring

Patterning and Early Algebra

Classifying and Analyzing Data

Geometry

- Recognizing Geometric Shapes
- Composing Geometric Shapes
- Comparing Geometric Shapes
- Spatial Sense and Motions

As children successfully complete activities, they are presented with the challenge of the next developmental level.

For more information about Learning Trajectories, see Appendix B.

The Big Ideas In Early Childhood Learning

The specific topics **Building Blocks** teaches are children's mathematical **Building Blocks**—ways of knowing the world mathematically. They are organized into two areas: (1) number and simple arithmetic and (2) geometry, measurement, and spatial sense. These are the two emphases of NCTM's *preschool standards*. Three mathematical subthemes, (1) patterns; (2) data and graphing, and (3) classifying, sorting, and ordering, are woven through both main areas. These are not elementary school topics "pushed down" to younger ages, but developmentally appropriate areas that are meaningful and interesting to children.

The program sequences these topics based on the considerable research identifying specific "developmental continua" or "learning trajectories" that young children follow.

Number and Operations

- Numbers can be used to tell us how many, describe order, and measure; they involve numerous relations, and can be represented in various ways.
- Operations with numbers can be used to model a variety of real-world situations and to solve problems; they can be carried out in various ways.

Geometry

- Geometry can be used to understand and to represent the objects, directions, locations in our world, and the relationships between them.
- Geometric shapes can be described, analyzed, transformed, and composed and decomposed into other shapes.

Measurement

- Comparing and measuring can be used to specify "how much" of an attribute (for example, length) objects possess.
- Measures can be determined by repeating a unit or using a tool.

Patterns and Algebra

Patterns can be used to recognize relationships and can be extended to make generalizations.

Data Analysis and Classification

Objects can be sorted and classified in a variety of ways. Data analysis can be used to classify, represent, and use information to ask and answer questions.

Building Blocks mathematics is distinct in several ways.

- It connects children's informal and school mathematics. Research tells us this is early childhood mathematics education's "missing link."
- It includes everyday activities and objects, and also mathematical objects specifically designed to facilitate mathematical thinking.
- It helps children "mathematize" key activities from everyday life, such as setting a table.
- It encourages children to explore special mathematical objects and actions or processes, especially in the **Building Blocks** software.

In this way, the **Building Blocks** mathematics program offers the best of natural everyday life, as well as low-tech (manipulatives) and high-tech support for children's mathematical thinking.

Much of our world can be better understood with mathematics. Preschool is a good time for children to become interested in counting, sorting, building shapes, finding patterns, measuring, and estimating. Quality preschool mathematics is not elementary arithmetic pushed onto younger children. Instead, it invites children to experience mathematics as they play in, describe, and think about their world.

Why Preschool Mathematics?

We need preschool mathematics for four reasons:

1. Preschoolers already experience curricula. Currently most preschool curricula is limited in mathematics. Effective preschool mathematics curriculum can strengthen this area.

2. Many preschoolers especially from low-income groups later experience considerable difficulty in school mathematics. Effective preschool mathematics curriculum can narrow the gap between these children and others.

3. Preschoolers possess informal mathematical abilities and enjoy using them. They are self-motivated to investigate patterns, shapes, measurement, the meaning of numbers, and how numbers work, but they need assistance to bring these ideas to an explicit level of awareness. Effective preschool mathematics curriculum can nurture these interests.

4. Preschoolers' brains undergo significant development and grow most as the result of complex activities, not from simple skill learning. Effective preschool mathematics curriculum can support this development.

Preschoolers can and should engage in mathematical thinking. All young children possess informal mathematics and can learn more. Teachers should build on and extend the mathematics that arises in children's daily activities, interests, and questions. They should struggle to see children's points of view and use their interpretations to plan their interactions with children and the curriculum. A combination of an environment that is conducive to mathematical explorations, appropriate observations and interventions, and specific mathematical activities helps preschoolers build premathematical and explicit mathematical knowledge.

Eclipse Studios

Building Blocks software is an essential element of the Building Blocks curriculum.

Why Use Computers?

While opinions may vary, research is clear: Used in a developmentally-appropriate way, computers are interesting and beneficial for preschoolers. They help children "mathematize" and learn in a variety of ways. Research shows that when used wisely, computers can be developmentally appropriate, fun, and beneficial for young children. The computer offers many practical advantages. Children enjoy that the blocks "snap" to each other and stay together accurately. They like saving and returning to, as well as printing, their work. Children often learn more by using the computer's tools to perform actions on the shapes. Because they have to figure out how to choose a motion such as slide, flip, or turn, they become more conscious of these geometric motions. They think ahead and talk to one another about which shape and action to choose next.

Most importantly the ***Building Blocks*** software provides activities at the appropriate level of the learning trajectory for each student.

Very young children have shown comfort and confidence in using software. They can follow pictorial directions and use situational and visual cues to understand and think about their activities.

Building Blocks software has these advantages:

- It combines visual displays, animated graphics and speech.
- It provides feedback.

- It provides opportunities to explore.
- It focuses children's attention and increases their motivation.
- It individualizes—gives children tasks at children's own ability levels.
- It keeps a variety of records.
- It provides more manageable manipulatives (for example, manipulatives "snap" into position).
- It offers more flexible and extensible manipulatives (for example, manipulatives can be cut apart).
- It provides more manipulatives (you never run out!).
- It stores and retrieves children's work, so they can work on it again and again, which facilitates reflection and long-term projects.
- It records and replays children's actions.
- It links "concrete" (graphical) and symbolic (for example, numerals or spoken words) representations, which builds understanding and provides valuable feedback.
- It presents clearer mathematics (for example, using tools, such as a "turn tool," helps children become aware of mathematical processes).

Building Blocks software stores records of how children are doing on every activity. It assigns them to just the right difficulty level. It's like each child having a personal tutor! Also, you can view records of how the whole group or any individual is doing at any time. It's like having a personal aide!

Finally, computers make a special contribution to special education. These advantages lead to significant learning improvements for children with special needs:

- It is patient and non-judgmental.
- It provides undivided attention, proceeding at the child's pace.
- It provides immediate reinforcement.

For more information on the software and research on its advantages see Appendix A.

Matt Meadows

Getting Started with

Building Blocks

*This section provides an overview of classroom management issues and explanations of the **Building Blocks** program elements and how to use them.*

Program Materials

A variety of program materials are designed to help teachers provide a quality mathematics curriculum. The first step in getting started is to familiarize yourself with the program resources.

Teacher's Edition

The ***Teacher's Edition*** is the heart of the Building Blocks curriculum. It provides background for teachers and complete lesson plans with explicit suggestions on how to develop math concepts. It explains when and how to use the program resources.

Teacher's Resource Guide

The ***Teacher's Resource Guide*** offers key resources that help in delivering the curriculum. These include:

- Family Letters for each week
- English Learner support for each week
- Counting Cards
- Puzzles and Patterns
- Shape Sets
- Shape Flip Book

Building Blocks Software

The engaging software activities are essential to the curriculum. Each activity addresses a specific developmental level of the math learning trajectories.

Assessment

Building Blocks Assessment is a comprehensive research-based guide to assessing and preschool children's math proficiencies.

Manipulatives

The ***PreKindergarten Manipulative Kit*** includes key manipulatives and props for hands-on activity.

Big Books

Four big books provide excellent math related literature children will want to experience again and again.

Building Shapes

Makayla's Magnificent Machine

Victor Diego Seahawk's Big Red Wagon

Where's One?

Technology Resources

Math resources designed to facilitate instruction and record keeping expand student learning

For Teachers		For Students	
Online Planner	A tool to help teachers plan daily lessons and plot year-long goals	Building Blocks	Engaging research-based activities designed to reinforce levels of mathematical development in different strands of mathematics
Online Assessment	An assessment tool to grade, track and report electronic versions of all assessments		
Interactive Whiteboard Activities	Resources for teachers and students to explore numbers, shapes, and patterns		

Program Organization

The curriculum is organized into 30 weeks of activities and concept development.

Teaching for Understanding provides information about how children learn the key concepts.

Big Ideas outline the key concepts that will be developed throughout the week.

Math Throughout the Year overviews math strategies and props that teachers can use throughout the day to build math understanding.

Literature Connections identifies specific trade books that can enhance mathematics.

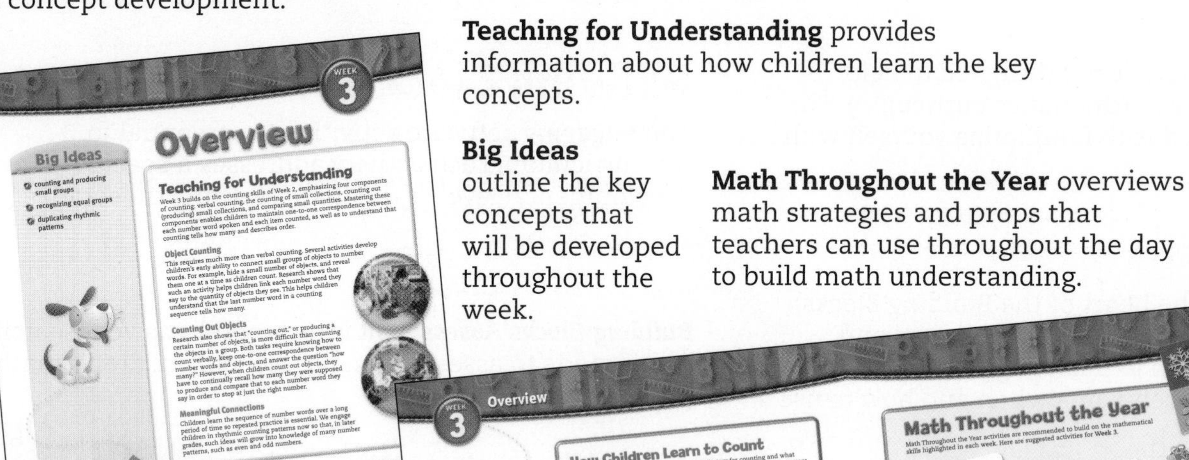

What's Ahead outlines where students are headed and how teachers can facilitate their learning.

Overview provides information about how children learn.

Center Preview helps teachers prepare for the week's Computer and Hand's On Math Centers.

For more detailed information about Learning Trajectories and Building Blocks software, see Appendix B.

(tl)Eclipse Studios; (br)(t)Matt Meadows; (b)Eclipse Studios

Weekly Planner

Provides objectives, learning trajectories, correlating activities, materials, and program-specific resources to prepare for each week.

The **Wrap-Up** for each week includes **Assess** and **Differentiate** strategies for teachers based on where students are in a week's key learning trajectories for math.

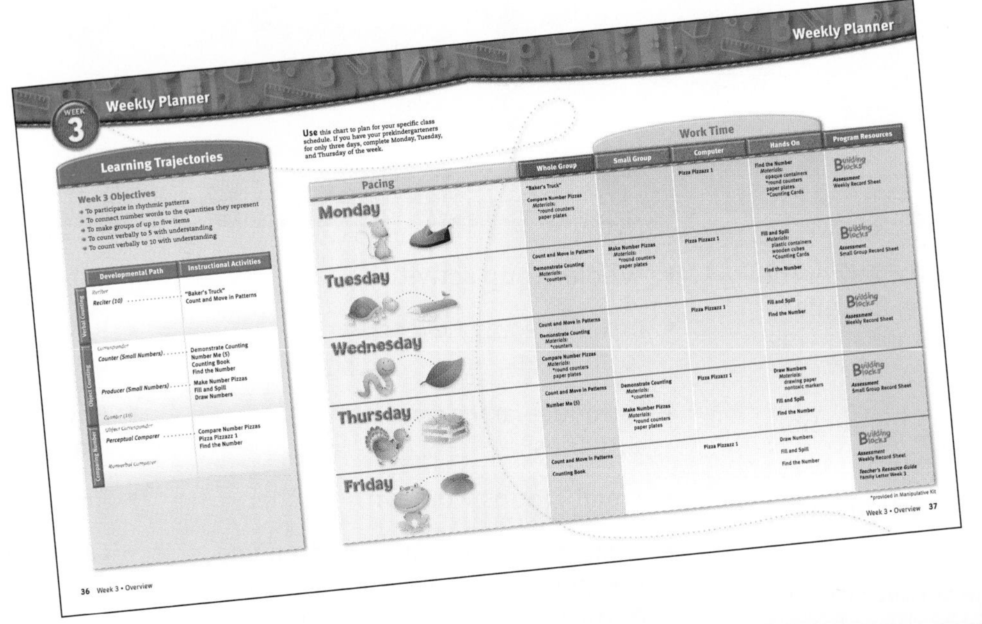

Daily Lessons

Each daily lesson follows a consistent plan.

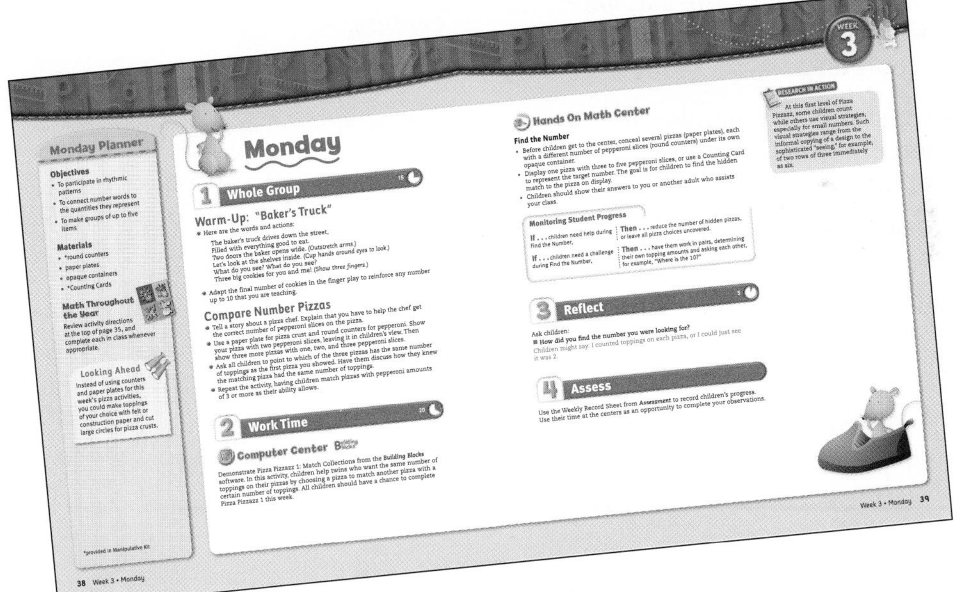

1. **Whole Group** includes Warm-Up activity to get children ready for math.

2. **Work Time** outlines the Computer Center, the Hands On Math Center, and Small Group on Tuesday and Thursday.

3. **Reflect** engages children in summarizing and analyzing their mathematical thinking.

4. **Assess** reminds teachers of their informal assessment opportunities each day.

Family Letters from the ***Teacher's Resource Guide*** for each week communicate what children are doing in school and provide an opportunity for children to demonstrate their knowledge.

The following schedule describes one way to engage children in all the activities effectively.

Week by Week

Monday

- *Before school,* set the computer for the week's activity(ies) (do this Friday or anytime before starting school on Monday).
- *Before school,* read the instructions in the ***Teacher's Edition.***
- Whole Group introduction and activities
- In addition to the Whole Group activities, introduce the software to the children, along with any new centers.
- Demonstrate new Hands On centers (Assistant monitors other children).
- Have children sign-up on list—two at a time work on each of 2 computers. Make sure an adult is available, especially at the beginning of the year, to help at the computer as needed. They can help children know it's their turn, sign on, and provide scaffolding as children work.

Tuesday

- Whole group activities
- Introduce any new software to children.
- Introduce any new center activities.
- Small groups work with teacher.
- Assistant helps other children use the sign-up lists to rotate their use of the computer and monitors all centers.

Wednesday

- Whole group activities
- Review any Work Time activities that need discussion.
- Engage in Computer and Hands On Math Centers.
- Planning time: Management
- Check the computer management system: Are all children completing activities?
- Check all assessment notes. Who needs extra help? Challenge?

Thursday

- Whole group activities
- Small Group
- Repeat Tuesday's Work time as needed, guiding children you have not yet worked with or any who need extra experience.
- Other children engage in Computer and Hands On Math Centers.

Friday

- Whole Group activities
- Engage in computer and Hands On Math Centers.
- During any Whole Group time, children show what they have done on the computer, or the teacher can run the activity again and discuss how children solved the problems ("Computer Show").
- Ask the "Reflect/Assess" questions at Whole Group.
- Complete all recording sheets.
- Send home Family Letter.
- After dismissal, set the computer to do the activity(ies) for the following week. Read the instructions in the ***Teacher's Edition,*** and try the activity yourself.
- Review all activities in preparation for next week.
- Planning time: Management
- Check the computer management system: Are all children completing activities?
- Check any observation and assessment notes: Who needs extra help? Challenge?
- Plan to gather any materials needed for next week.

Weekly Routines

Every Week Day

1. Try to do each day as written in the curriculum.

2. Continue using any Centers from previous weeks that children are still learning from and interested in.

3. In addition to the Whole Group activities, introduce any new off-computer centers during Whole Group.

4. Ask the Reflect questions each day it is possible, but especially on Friday.

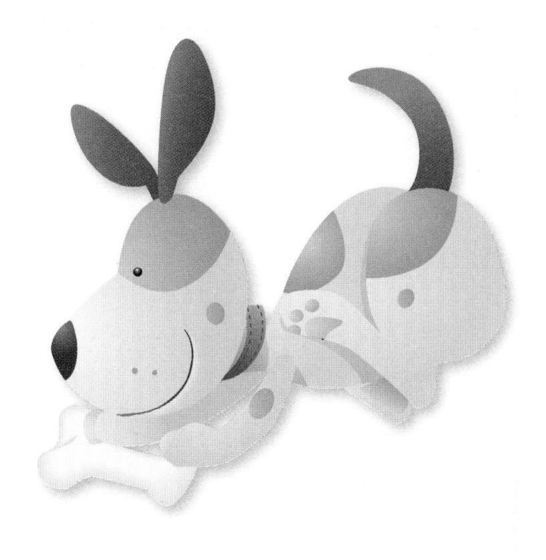

Effective whole group and small group activities develop more than mathematics knowledge. They develop children's ability to maintain attention and persist, their receptive (listening) and expressive (talking) language, their general cognitive and problem solving skills, and their social skills. The following teaching strategies help children develop in all these areas.

Conducting Successful Whole and Small Group Activities

- Be prepared ahead of time and organized so there is little or no "wait time" for the children.
- Keep the activity "moving forward" and vary the tone and pitch of your voice so that the activity is interesting to and motivating for children.
- Avoid interruptions of any kind, especially those that could come from focusing on a few misbehaving children. Instead, "draw them into" the activity.
- Use a song or finger play to increase children's attention whenever it wanes.
- In small groups especially, develop children's language and vocabulary by describing what individual children are doing and asking them to do so, developing their abilities to connect language to their activity and encouraging them to build stronger mental representations of their work by enriching them with language.

Eclipse Studios

Whole Group Activities

Whole Group activities, conducted every day, should be well-paced and engaging.

- Daily whole group activities should be completed each time they appear in the week's plan. Do the activities whenever your class meets as a group. This might be story time, morning circle, or any other whole group time.
- Repeat finger plays as often as you like. If a new finger play is introduced, and you have a better one that accomplishes the same goal, substitute yours or a previous one that children would benefit from repeating.
- Have children respond or act together to keep everyone active and motivated.

Small Group Activities

Small groups, conducted at least two days per week, should be intimate and interactive. Work with a small group of children, often four at a time, to complete the activity. Then repeat the activity with another small group. Some teachers work with all the children on the first small group day, then repeat the activity the next small group day with any children who would benefit from extra work with you on the activity. Others repeat the activity on the first day until at least half of the children have had a chance. They then complete the small group activity with the remainder of the children on the next small group day. These teachers find other times to repeat the activity with children who need extra work.

- Work with a small group of four children work to solve the problems posed in the ***Teacher's Edition.*** All children should participate at least once each week.
- Record each child's name as they work with you, along with their performance (see the following section). You may also wish to set up a pocket chart with names and the days children are scheduled to work with you.
- Repeat the activity that week or a future week with any children who need more experience.

Eclipse Studios

Computers are a way to help children learning mathematics. This section emphasizes guidelines for using computers. However, many of the teaching strategies apply to all centers, whether that center involves computers, manipulatives, books, or any other materials.

Computer Center

Arranging and Managing Computer Centers

Even preschool children can work on computers cooperatively with minimal instruction and supervision, if they have adult support initially. However, adults play a significant role in successful computer use. So, consider the following suggestions.

- Place your computer where adults can supervise and assist children as they need it.
- Place the computer so there is no glare on the screen.
- Ideally, use the computer as a "learning center." If you have 3 to 5 computers, you can cycle all children through the computer activity. If you have 1 or 2 computers, you will need to use the computer throughout the day to ensure that all children have a chance.
- Place one to two seats in front of the computer and one at the side for an adult to encourage positive social interaction. Placing computers close to each other can facilitate the sharing of ideas among children. Computers that are centrally located in the classroom invite other children to pause and participate in the computer activity. Such an arrangement also helps keep teacher participation at an optimum level. They are nearby to provide supervision and assistance as needed.
- Expect independent work from children gradually. Prepare them for independence, and increase the degree of such work slowly. Provide substantial support and guidance initially, even sitting with children at the computer to encourage turn taking. Then gradually foster self-directed and cooperative learning.
- Set up the computer sign-up sheet that has the children's icons on it.
- Monitor student interactions to ensure active participation of all. It is critical to make sure special education children are accepted, supported, and given equal access.

Very First Time with Children

Providing a lot of guidance and help at the beginning will pay handsomely for the rest of the year.

- Show the sign-in sheet and "act out" how to use it.
- Show how the sign-in screen has their names listed.
- Discuss the pictures that go with their names.
- Choose a child and click his or her name and password.
- Demonstrate the first assigned activity.
- Demonstrate how to return to the sign-in screen.
- Re-introduce this in small groups.
- At the beginning of each week, demonstrate and discuss the new computer activity: Describe the mathematical problem and how you solve it. Demonstrate every step of the activity. Tell children exactly what you are doing, moving the mouse to point here, clicking, and so on. Ask children to show you where to click as soon as they are able.

Teaching Strategies

- Make sure all children work on the assigned computer activities individually at least twice per week for about 15 minutes each time. See the next session, "Assessing and Recording," for suggestions on how to rotate children through the computer activities.
- Once the children have finished the assigned activities, they should always get the chance to play and learn with the *"free explore"* activity. This might be individual, but is also an excellent opportunity for *children to explore cooperatively,* posing problems for each other, solving problems together, or just learning through play.
- Demonstrate these activities to children on the computer every Monday and demonstrate and discuss them again every Tuesday.
- Children should spend about 15–20 minutes at the computer at a time, at least two times per week.
- Remember that preparation and follow-up are as necessary for computer activities as they are for any other. Do not omit critical whole-group discussion sessions following computer work. Consider using a single computer with a large screen or with overhead projection equipment.
- Use Friday's *"Computer Shows"* to lead discussions with the whole group. Help children communicate about their solution strategies and reflect on what they've learned.
- Research shows that the introduction of a computer often places many additional demands on the teacher. Be kind to yourself! Get help!
- Integrate mathematics on and off the computer throughout your day. The following section provides a wealth of suggestions.
- Ultimately, go *at the children's pace*. If the first few weeks are too slow, move on! Or, if you get near the end of the year and children need more time, take more time. Learning with understanding is the point.
- Make sure children make sense of the mathematics.

Matt Meadows

Hands On Math Centers give children concrete experiences with math concepts.

Hands On Math Center

- An effective way to manage centers is to keep all materials for any center in one box or container, and keep all these boxes on a special shelf. If you mark each box with a picture or symbol, children can help you get them out and put them away; for example, if you have a square blue rug, mark a box with a blue square, and then children will know where that box is placed in the room.
- Introduce each center when you are putting out the materials for the first time.
- Remember to have an adult—you or an aide—visit the center frequently. Sometimes, having your aide sit at the center is a productive way to introduce and engage children in the center. After that, visiting the center and discussing what children are doing helps children build and communicate about mathematical ideas.
- Children do these activities with some adult supervision.
- Explain or demonstrate the centers when you talk to the class about that day's options on the day they are introduced. Encourage children to visit the center and try the activity at least once, preferably more, during the week.

Effective Teaching Strategies at Centers

The following suggestions help adults teach effectively at centers with computer or manipulatives.

- Stay active! Closely guide children's work in and learning from the activities and encourage experimentation with the open-ended, free explore activities. Always encourage, question, and prompt children; also, demonstrate as need. Help children reflect on their strategies for solving problems.
- Once children are working independently, provide enough guidance, but not too much. However, observe each child at least once per week, so you know how they are doing and can provide appropriate help.
- Avoid quizzing or offering help before children request it. Instead, prompt children to teach each other by physically placing one child in a teaching role or verbally reminding a child to explain his or her actions and respond to specific requests for help.

Children should spend about 15 minutes at the computer at a time.

Engaging children in reflection is as important as assessing.

Reflect

Reflect is a critical part of every Building Blocks lesson. When children talk about their thinking, using their own words, they engage in mathematical generalizing and communicating. Allowing children to discuss what they did during an activity helps build mathematical reasoning, but also develops social skills such as turn taking, listening, and speaking.

A powerful reflection question is, "How do you know?" Early in the year children may or may not answer you and often cannot provide reasons for their answers. They may shrug their shoulders and say, "I don't know," "Because," or, "Because I'm smart." As the year progresses, children become accustomed to explaining their ideas. Their answers give more insight into their mathematical thinking. For example, one child picked an open shape—a "V"—as a triangle.

The key to reflect is asking questions.

Teacher: ***How do you know that's a triangle?***

Child: *It has three sides. (She runs her finger along the two sides and runs her finger along the open space, as if there is a third side.)*

Teacher: ***But, there's no line here (pointing to the open space).***

Child: *Because they forgot to draw it.*

Without asking questions, this teacher may have assumed the child could not recognize shapes. Asking the question showed that the child had a budding understanding of triangles but did not yet understand that shapes need to be closed, and that the third line must be visible. Young children who have such discussions with teachers and with each other begin to question and correct each other. For these reasons, we recommend that you incorporate time for reflection into your classroom mathematics activities. The following are good questions and challenges.

- How do you know?
- Why?
- Show me how . . .
- Tell me about . . .
- How is that the same?
- How is that different?

Assessment is a crucial part of making informed decisions. Effective assessment is a continual process and should involve many different types of data for a complete picture of each student's abilities.

Goals of Assessment

1. Improve instruction by informing teachers about the effectiveness of their lessons
2. Promote growth of students by identifying where they need additional instruction and support
3. Recognize accomplishments

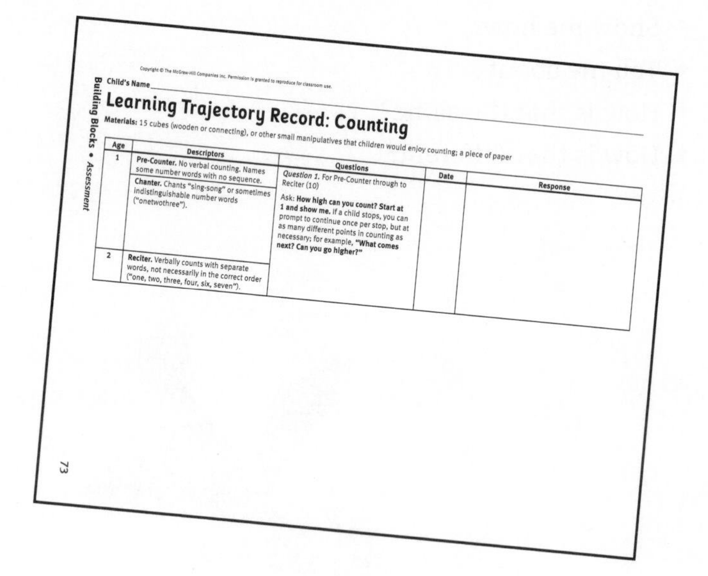

Building Blocks • Assessment

Child's Name ____________

Learning Trajectory Record: Counting

Materials: 15 cubes (wooden or connecting), or other small manipulatives that children would enjoy counting; a piece of paper

Age	Descriptors	Questions	Date	Response
1	**Pre-Counter.** No verbal counting. Names some number words with no sequence. **Chanter.** Chants "sing-song" or sometimes indistinguishable number words ("onetwothree").	*Question 1.* For Pre-Counter through to Reciter (10) Ask: **How high can you count? Start at 1 and show me.** If a child stops, you can prompt to continue once per stop, but at as many different points in counting as necessary; for example, **"What comes next? Can you go higher?"**		
2	**Reciter.** Verbally counts with separate words, not necessarily in the correct order ("one, two, three, four, six, seven").			

T3

Phases of Assessment

1. *Planning* As you develop lesson plans, you can consider how you might assess the instruction, determining how you will tell whether students have grasped the material.
2. *Gather Evidence* Throughout the instructional phase, you can informally and formally gather evidence of student understanding. Student Assessment Records (in print and online) and Anecdotal Assessment Reports (online) are provided to help you record data. The end of every lesson is designed to help in conducting meaningful assessments.
3. *Summarize Findings* Taking time to reflect on the assessments to summarize findings and make plans for follow-up is a critical part of any lesson.
4. *Use Results* Use the results of your findings to differentiate instruction or to adjust or confirm future lessons.

Building Blocks is rich in opportunities to monitor student progress to accomplish these goals.

See Appendix A for more information about Assessment.

The following assessment and recording items guide assessment and help ensure that every child has the experiences they need (for example, rotating through computers centers).

- *Small Group Activities:* Perhaps the most important assessments are the observations of children, and the most insightful observations often occur during the small group activities. Observe and interact with children during these activities and complete the assessment/record sheet for each small group activity.
- *Weekly Records:* Complete the Weekly Record Sheet for all whole group, centers, computer, small group, and everyday activities.
- *Computer Activities:* For the computer activities, use the ***ConnectEd Password Card***. This sheet has a printout of the icons from the software. Make copies of that sheet. Fill out and place a new one next to the computer each week. Have children mark (if they can, their name or initials) next to their name, filling in a new column each time they use the computer.
- *Your Assessments:* Of course, your own assessment for teaching purposes might include observation, anecdotal records, samples of children's work, interviews (preplanned or spontaneous), rating scales, photographs, audiotapes, videotapes, time sampling, and running records.

Small Group Record Sheet

Week 3 Activity—Demonstrate Counting

5 Corresponder
6 Counter (Small Numbers)
7 Producer (Small Numbers)

Child's name:	Names numbers:	Strategies/Trajectory Level:	Comments:

20 Building Blocks • *Assessment*

2 **Weekly Record Sheet**

Directions:
- Under the week number, write the date you started that week's activities.
- In the ✔ columns, put a ✔ when an activity is completed.
- For Whole Group and Math Throughout the Year, put a ✔ for each time the class does the activity.
- For the other columns, put a ✔ when you start each activity.

Week	✔	Whole Group	✔	Hands On Math Center	✔	Computer Center	✔	Small Group	✔	Home Connection	✔	Math Throughout the Year
3 ____ Date		"Baker's Truck" Compare Number Pizzas Count and Move in Patterns Demonstrate Counting Number Me (5) Counting Book		Find the Number Fill and Spill Draw Numbers		Pizza Pizzazz 1		Make Number Pizzas Demonstrate Counting		Family Letter		Counting Wand (Count All) Snack Time Simon Says Numbers
4 ____ Date		Count and Move in Patterns Match and Name Shapes Match Blocks Circle Time! ***Building Shapes*** Circles and Cans "Wake Up, Jack-in-the-Box" "Circle" Circle or Not?		Explore Shape Sets Match Shape Sets Circles and Cans		Mystery Pictures 1		Match and Name Shapes		Family Letter		Counting Wand Dough Shapes Foam Puzzles

Building Blocks • *Assessment*

Differentiated Instruction

Instruction can be differentiated in three key ways:

- *Content*—what the teacher wants students to learn and the materials or mechanisms through which that is accomplished. Differentiating the content may be teaching prerequisite concepts to students who need intervention.
- *Process*—how or what activities will the students do to ensure that they use key skills to make sense out of the content. Differentiating the process may include moving through the lesson more quickly or dwelling on a particular practice activity.
- *Product*—how the student will demonstrate what he or she has come to know. Differentiating the product may include assigning Enrichment, Practice, or Reteaching activities to complete.

Building Blocks offers a wealth of support for differentiating instruction.

- Every lesson includes a variety of small group, whole group, and individual activities that address differentiating the process of learning.
- ***Building Blocks*** software activities are tailored to individual needs to address differentiating the learning content. The software activities help develop students math proficiencies along the learning trajectories. Activities are supported by drills and instruction based on student performance.
- Assessments, including the ***Building Blocks*** software management system, Small Group Record Sheets, Trajectory Assessments, and informal weekly assessments, provide teachers with reliable data on which to gauge children's proficiency and inform their instruction.

See Appendix A for more information about Differentiating Instruction.

Eclipse Studios

When English learners work in the core content areas, they face the challenge of both the new concept load and the new language load.

English Language Learners

Sheltered instruction designs lesson delivery to take advantage of student language strengths and techniques that do not rely solely on 'telling' but on doing so that children see, hear, and experience the new concept along with the new language. ***Building Blocks*** employs the following sheltered instruction strategies throughout the program.

Accessing Prior Knowledge Sheltered instruction begins with **accessing children's prior knowledge** to determine what they already know and how much background building may be necessary. It activates student thinking and draws on experience to enhance the new learning. Children are encouraged to think about and share what they already know, allowing the teacher to diagnose where any gaps may exist. Accessing prior knowledge enhances the lessons by connecting to children's real-life experience which reduces abstractions.

Modeling the new math skills and concepts is a huge part of ***Building Blocks,*** and it is one of the key strategies involved with Sheltered Instruction. When we introduce new skills and concepts they are built on prior learning and children see teachers demonstrate using manipulatives, hear the appropriate language terminology modeled and practice in a guided setting using these tools.

Checking for Understanding Sheltered Instruction involves interaction with the teacher, including lots of **checking for understanding** along the way. When children are in the early stages of English language acquisition, they may not always be the first to answer a question or be able to fully express their ideas in speech or in writing. The wonderful step-by-step instruction mapped out in ***Building Blocks*** builds in many modes of checking comprehension throughout the lesson so that all children can show what they know. Children are asked to respond to questions in ways including pointing, raising hands and fingers, and moving objects. In this way, English learners access the core concepts but do not have to rely only on English expression to show what they know or where they need additional help.

Explicit Instruction The step-by-step instruction in ***Building Blocks*** is carefully structured to lead children to understanding the concepts. Built into the teaching are explanations of new terms and skills. It provides for plenty of repetition and revisiting of these concepts using the new language so that children have plenty of modeling before they are expected to use these terms themselves.

Math Games English learners need to practice their new language in authentic settings rather than drills that do not capture the language tasks chikldren need to perform in school. Language learning can be a high-anxiety experience because the nature of the learning means that the speaker will make many mistakes. This can be profoundly difficult in front of one's peers or in answering individual questions in front of the whole class. Working in pairs and groups allows English learners a way to understand in a low-anxiety environment and allows more learning to take place.

A big part of the practice built into ***Building Block*** is in the form of learning activities. In addition to being fun and motivating, these math activities provide English learners ample opportunity to practice their English while practicing their math. The games provide plenty of meaningful practice. For example, children learning their numbers get not just rote counting practice, but counting while moving on a game board. Vocabulary is woven into appropriate language repetition and revisiting skills and strategies without boredom or work in isolation. In turn, children bring these games home which draws the family into the practice of English and extends the learning of mathematics beyond the school day.

In addition, these specially designed games model strategic thinking so that even children at beginning levels of proficiency learn strategy and can demonstrate strategic, higher-level thinking.

It is important to keep parents informed about what their children are doing in mathematics so they can support their mathematical understanding and development at home. Families can play a critical role in student's success in mathematics if they understand how to help.

Family Involvement

Building Blocks has several elements built into the program that can enable family school communications.

Teacher Resource Guide Home Connections

This book includes ready-made Parent Surveys, Family Letters, and Games that teachers can use to communicate with student families. Family Letters are available for every week.

Assessment Resources

The Assessment book includes several resources to communicate with families.

- The Student Assessment Record is a convenient form to record all student assessments on a daily or weekly basis. These forms are handy to use at parent-teacher conferences.
- Parent Teacher Conference Checklist provides a helpful way to organize thoughts about students in preparation for parent teacher conferences.

Parent Aides

Often parents are willing to volunteer to help out in the classroom. There are many ways they can help.

- Computer Management—Make sure computers are on and loaded with appropriate software. Be available to trouble shoot and answer questions while students use the computers.
- English Learner Aide—Use the English Learner Support Guide ideas to work with English Learners to preview and review lesson concepts.
- Manipulatives Manager—Make sure manipulatives are available for student use.

Family Participation

There are many ways that families can assist students in learning math.

1. Establish daily family routines.

a. Provide a time and place to study.
b. Assign responsibility for household chores.
c. Have a regular bedtime.
d. Have family meals.

2. Monitor out-of-school activities.

a. Set limits on television watching and computer use.
b. Know what children are watching on television.
c. Approve of what children are doing on the computer.
d. Check up on children when parents are not home.
e. Arrange after-school activities.
f. Arrange for supervised care.

3. Model the value of learning, self-discipline, and hard work.

a. Ask about school work.
b. Discuss what children are learning.
c. Communicate what everyone is learning.
d. Reward hard work.
e. Demonstrate achievement that comes from hard work.
f. Demonstrate interest in learning.
g. Express enjoyment of mathematics.

4. Express high but realistic expectations for achievement.

a. Set goals and standards appropriate for children's age and maturity.
b. Recognize and encourage talents.
c. Inform friends and family about successes.
d. Set high expectations for math achievement for both girls and boys.

5. Encourage children's progress in school.

a. Show interest in children's progress in school.
b. Help with homework, but don't do homework for children.
c. Discuss the value of a good education.
d. Communicate with teachers and school staff.

6. Encourage reading, writing, playing board and card games, and discussion among family members.

a. Read books, magazines, and newspapers, and discuss what you have read.
b. Listen to children read and talk about what they are reading.
c. Identify and discuss the different roles that numbers and math concepts play in everyday life.

See Appendix A for more information about Parent Involvement

Preschoolers do not see the world as if it were divided into separate subjects. Successful preschool teachers help children develop mathematical knowledge throughout the day.

Math Throughout the Year Activities

Classrooms are filled with opportunities to include mathematics. Use children's spontaneous questions and observations as catalysts for mathematical exploration. By encouraging children's ideas and sharing of math stories, you can support their efforts to solve problems in different ways.

Throughout the day, week, months, and year there are simple ways of incorporating math ideas and concepts. Some routines can be set up early in the year and continue throughout the year. Some of these will be done nearly every day. Others are done when they "fit" but not as a separate unit of instruction. By incorporating and emphasizing the math in these activities, you can develop and reinforce concepts. By connecting daily events to mathematical concepts, children may come to see math as an integral part of life and the classroom environment.

Incorporate these ideas to "mathematize" everyday classroom routines whenever appropriate. Use these or other similar ideas to integrate mathematics into your classroom day.

- Attendance Routines
- Meals
- Cleanup
- Lining Up
- Hello Games
- Goodbye Games
- Physical Activities, including counting motions and spatial relations
- Voting and Graphing
- Daily Calendar
- Daily Weather
- Time

For more complete descriptions of these ideas, see Appendix A.

Teachers who have used this curriculum successfully with their children have shared many ideas that you may wish to use.

Teacher Tips

- Enlarge the materials, especially for whole group activities. Reproduce the existing sheets but make them bigger. Or modify the activities; for example, make shapes shadows on walls, cut out large shapes and tape them to the floor. Also consider making some shapes more colorful.
- Look ahead to what *new materials* are "coming up" in the next week or two and lay them out, encouraging children to play with them and explore what they can do with them *before* you use them in specific activities.
- For some activities, vary the size of the group as needed. Some whole group activities could be done with only half your class if that helps. Some small group activities, typically done with 4 children at a time, can be conducted with 2, 6, or 8 children at a time, depending on the activity and your children.
- Use the "extra help" suggestions when needed. Also, use your own ideas. For example, use only *some* of the shape sets to make an activity easier for children who need it.
- Keep active helping children with the computer! Review the suggestions on pages T28 and T29.
- *Use your aide* wisely, monitoring the children at the computer center and the manipulative (off-computer) centers.

Overview

Big Ideas

- Math is numbers, shapes, and patterns.
- Counting tells how many.
- Math can be explored through materials.
- Groups can be named with numbers.

Teaching for Understanding

Teaching early childhood mathematics for understanding involves engaging the complete child in mathematics. To do so, we will create an environment in which children see, touch, and build mathematical ideas and skills at every opportunity. This week is a gradual introduction to mathematics in the classroom, as well as your introduction to the purpose of daily and weekly features and sections found in the program.

Getting Started

Children will learn about centers and explore general materials beneficial to math, such as using blocks to discover mathematics through their own play. Children participate in Whole and Small Group activities, learning how to focus their attention during such activities. Children begin to realize that math is all around us and that they already know a lot about it, which makes it even more fun to learn more.

Reciting Numbers

Children learn to recite number words in the same way ABCs are learned, which is why rhythms, songs, and other daily links to their lives are utilized. Additionally, Small Group activities encourage children to attach corresponding number words to groups of one, two, and three items. Such activities build meaning for number words and help support counting with understanding.

Meaningful Connections

Relating numbers to children's experiences in their everyday world provides a context for learning, which expands the meaningful connection between math and other areas of study. Being able to identify and eventually understand the mathematics in a common task, such as counting money to pay for something, can open a child's eyes and mind to math in less obvious scenarios.

What's Ahead?

In the weeks to come, children will name small groups ("That's three!") and learn to construct mental images of the small numbers or quantities in such groups. Children will learn new ideas and skills in counting. Counting is not just saying "1, 2, 3..." It involves accurately counting objects, understanding the quantities counted, and linking that mathematical knowledge to situations in daily settings.

Eclipse Studios

How Children Learn to Count

This page of the Overview introduces each week's key learning trajectories, serving as a guide and point of assessment for your observations of children in various settings.

Knowing where children are on the learning trajectory for counting and what the next steps are helps facilitate their development. Children who easily surpass these trajectory levels might be challenged by larger numbers and encouraged to assist other children.

Verbal Counting

What to Look For Can the child recite number words from 1 to 5 or 10?

Pre-Counter No verbal counting; may name some number words but does not put them into a sequence.

Chanter Chants "sing-song" or sometimes runs number words together. For example, the child might say "onetwothree" similar to "J, K, LMNOP...."

Reciter Verbally counts with separate words, not necessarily in the correct order. For example, the child might say "one, two, three, four, six, seven."

Reciter (10) Verbally counts to 10 with some correspondence to objects. For example, the child says: 1 (points to it), 2 (points to it), 3 (starts to point), 4 (finishes pointing but is pointing to third object), 5, and so on.

Object Counting

What to Look For Can the child accurately count small groups?

Corresponder Keeps one-to-one correspondence between number words and objects (one word for each object), at least for small groups of objects in a row; may answer the question "how many?" by counting the objects again.

Counter (Small Numbers) Accurately counts objects in a line to 5, and answers the question "how many?" with the last number counted. When objects are visible, especially with small numbers, begins to understand cardinality.

Subitizing

What to Look For Can the child name the number of objects in a group of 1, 2, or 3?

Small Collection Namer Uses number words to name small collections of 1, 2, and sometimes 3.

English Learner

Before each lesson, preview with English learners the following vocabulary words and phrases: *counting aloud, stack,* and *counters.* Refer to the ***Teacher's Resource Guide.***

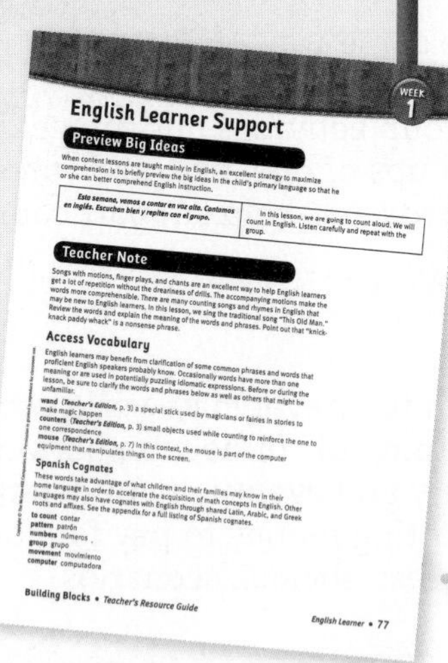

WEEK 1

English Learner Support

Preview Big Ideas

When content lessons are taught mainly in English, an excellent strategy to maximize comprehension is to briefly preview the big ideas in the child's primary language so that he or she can better comprehend English instruction.

Esta semana, vamos a contar en voz alta. Contamos en inglés. Escuchan bien y repiten con el grupo.	In this lesson, we are going to count aloud. We will count in English. Listen carefully and repeat with the group.

Teacher Note

Songs with motions, finger plays, and chants are an excellent way to help English learners get a lot of repetition without the dreariness of drills. The accompanying motions make the words more comprehensible. There are many counting songs and rhymes in English that may be new to English learners. In this lesson, we sing the traditional song "This Old Man." Review the words and explain the meaning of the words and phrases. Point out that "knick-knack paddy whack" is a nonsense phrase.

Access Vocabulary

English learners may benefit from clarification of some common phrases and words that proficient English speakers probably know. Occasionally words have more than one meaning or are used in potentially puzzling idiomatic expressions. Before or during the lesson, be sure to clarify the words and phrases below as well as others that might be unfamiliar.

wand (***Teacher's Edition***, p. 3) a special stick used by magicians or fairies in stories to make magic happen
counters (***Teacher's Edition***, p. 3) small objects used while counting to reinforce the one to one correspondence
mouse (***Teacher's Edition***, p. 7) In this context, the mouse is part of the computer equipment that manipulates things on the screen.

Spanish Cognates

These words take advantage of what children and their families may know in their home language in order to accelerate the acquisition of math concepts in English. Other languages may also have cognates with English through shared Latin, Arabic, and Greek roots and affixes. See the appendix for a full listing of Spanish cognates.

to count contar
pattern patrón
numbers números
group grupo
movement movimiento
computer computadora

Building Blocks • *Teacher's Resource Guide*

English Learner • 77

Technology Project

Technology projects will use computer tools to explore math. In these introductory weeks, however, focus on improving children's comfort level using the computer. As ability allows, challenge children to find just how high this week's Count and Race will go (how many cars they can get into the race).

Math Throughout the Year

This feature provides descriptions of Math Throughout the Year activities, which are recommended to build on the mathematical skills highlighted in each week. Most activities will be repeated from week to week, many of them with slight variations. Here are the suggested activities for **Week 1.**

Counting Wand

A counting wand draws children's attention to items that are being counted and can be made from several available materials, such as a regular pointer, a decorated ruler or wrapping paper tube, or a baton. A soft or light wand material is preferable because children enjoy using the wand themselves. Use the counting wand for counting things at every opportunity.

Find the Math

Observe children's play to "find the math" in their actions. For example, comparing: Who has more crayons? Whose building is taller?; shapes: What happens when clay is rolled out? How do pieces fit in this puzzle?; number: How old are you? Who is older, you or me?; classification: Put all red blocks here and the rest over there; and spatial position: Where did you put the truck?

I See Numbers

Help children see groups of one, two, and three everywhere, every chance you get, such as three trees, not just a group of trees. Encouraging children to see the amount of something as opposed to only seeing the "something" helps them form the habit of quantifying small groups or collections. This habit will give them lots of experience recognizing groups and numbers.

Center Preview

This feature summarizes what to expect from each week's Computer and Hands On Math Centers, which are completed with some adult supervision. Directions are provided during the week. Demonstrate centers when you tell your class about the day's options. Children visit centers to try an activity at least once, preferably more, during the week.

Computer Center

Get your classroom Computer Center ready for Count and Race from the *Building Blocks* software. After you introduce Count and Race, each child should complete the activity individually as you (or an assistant) monitor and guide him or her periodically. It is important to have an adult help during the first few weeks, especially to ensure that children know how to use a computer mouse, recognize icons, key passwords, and navigate activities. Ideally, each child will have at least ten minutes of computer time at least twice during the week. Assist children through the rotation of the Computer, Hands On Math, and any other classroom centers. Use their center time as an opportunity for assessment.

Hands On Math Center

This week's Hands On Math Center activities are explore manipulatives and Make Buildings. Supply wooden inch cubes, other safe stackable materials, and counters.

Literature Connections

This feature links literature to math. Many titles recur throughout the year. These books help develop counting.

- *Little Rabbits' First Number Book* by Alan Baker
- *The Very Hungry Caterpillar* by Eric Carle
- *One Was Johnny: A Counting Book* by Maurice Sendak
- *I Can Count the Petals of a Flower* by John and Stacey Wahl

(t)Matt Meadows (b)Eclipse Studios

Weekly Planner

This chart lists weekly objectives and correlations between Learning Trajectory levels and the week's activities. For example, you may use it to find an activity that exemplifies a particular level.

Use this chart to plan for your specific class schedule. If you have your prekindergarteners for only three days, complete Monday, Tuesday, and Thursday of the week.

Learning Trajectories

Week 1 Objectives

- To count verbally
- To explore the mathematics in manipulatives and materials
- To recognize and make groups of 2 or more
- To count verbally groups of 2 or more
- To quickly recognize the number of objects in small groups (subitize)

	Developmental Path	Instructional Activities
Verbal Counting	*Pre-Counter*	
	Chanter	
	Reciter	"This Old Man"
	Reciter (10)	Count and Move Count and Race Counting Wand
Object Counting	*Corresponder*	Count and Race Counting Wand
	Counter (Small Numbers)	Explore manipulatives Make Buildings *Where's One?*
	Producer (Small Numbers)	
Subitizing	*Small Collection Namer*	Explore manipulatives Make Buildings "When I Was One"
	Nonverbal Subitizer	

Work Time

Whole Group	Small Group	Computer	Hands On	Program Resources
"This Old Man" Count and Move		Count and Race	Explore manipulatives *Materials:* *inch cubes *counters	Building Blocks ***Assessment*** Weekly Record Sheet
"This Old Man" Counting Wand	Explore manipulatives *Materials:* *counters	Count and Race	Explore manipulatives	Building Blocks ***Assessment*** Small Group Record Sheet
"When I Was One" *Where's One?*		Count and Race	Make Buildings *Materials:* *inch cubes blocks	***Big Book Where's One?*** Building Blocks ***Assessment*** Weekly Record Sheet
"When I Was One" Count and Move	Explore manipulatives *Materials:* *counters	Count and Race	Make Buildings	Building Blocks ***Assessment*** Small Group Record Sheet
"This Old Man" *Where's One?*		Count and Race	Make Buildings	***Big Book Where's One?*** Building Blocks ***Teacher's Resource Guide*** Family Letter Week 1 ***Assessment*** Weekly Record Sheet

The Whole Group, Small Group, and Hands On columns list activity names and their materials (note that Hands On materials are only listed on the day an activity is introduced). The Computer column always refers to ***Building Blocks*** software, and the Program Resources column lists components of this program other than the ***Teacher's Edition*** needed for the week.

*provided in Manipulative Kit

Monday

Monday Planner

Objectives

- To count verbally
- To explore the mathematics in manipulatives and materials

Materials

- *inch cubes
- *counters

Vocabulary

Group or *collection* describes one or more items placed or organized together.

Math Throughout the Year

This feature is a reminder of daily, mathematical activities you can complete with children.

Review activity directions at the top of page 3, and complete each in class whenever appropriate.

Looking Ahead

When applicable, this class-preparation feature communicates any special needs of an activity, such as needing an uncommon material or sending something home to children's families.

Make a Counting Wand for tomorrow, and enlist an adult helper for this week's Small Group.

*provided in Manipulative Kit

1 Whole Group 10 

Warm-Up: "This Old Man"

- Sing as much of the traditional song as you would like. Here are initial verses and actions:

  This old man, he played one, *(Show one finger.)*
  He played knick-knack on his thumb. *(Wiggle thumb.)*
  Knick-knack paddy whack,
  Give the dog a bone. *(Clap hands.)*
  This old man came rolling home. *(Roll arms.)*
  This old man, he played two, *(Show two fingers.)*
  He played knick-knack on his shoe. *(Wiggle fingers near shoe.)*
  Knick-knack paddy whack,
  Give the dog a bone. *(Clap hands.)*
  This old man came rolling home. *(Roll arms.)*

- Following the same verse pattern, here are the remaining numeral rhymes: three/on his knee; four/on the floor; five/on his side; six/with some sticks; seven/under heaven; eight/on his plate; nine/all the time; ten/once again.

Count and Move

- Have all children count from 1 to 10, or an appropriate number, clapping their hands as they say each number.
- Repeat as needed to ensure all children have participated, and then say "We all clapped ten times!"
- You may repeat this throughout the day using various motions, such as hopping and marching.

Monitoring Student Progress

If . . . children listen for others to say each number word before they say it or cannot clap and count simultaneously,

Then . . . have children nod their heads instead of clapping or other motion and/or count slowly or to a smaller number.

2 Work Time 25 

Work Time encompasses Small Group, occurring Tuesday and Thursday of each week, Computer Center, and Hands On Math Center activities.

Based on your judgment of children's needs and abilities, allot time for each Work Time activity based on the overall time recommendation. Show children corresponding centers and work areas, as well as how they will work there.

Computer Center Building Blocks

- With children gathered around the computer, discuss its parts, focusing especially on the monitor, keyboard, and mouse. Review how to use a computer, modeling how to be gentle with it and its parts. Tell children how you would like them to get help when they need it.
- Demonstrate how to choose an icon, name, and password in the ***Building Blocks*** software. Then show how to complete Count and Race. As you do so, have children count aloud.
- Tell children they will all have at least two chances to work on the computer this week even if it is not today. Explain how they will know it is their turn (see page T42 of this guide).

Count and Race

Hands On Math Center

To familiarize themselves through building and other play, children explore math manipulatives including, though not limited to, cubes and counters.

3 Reflect

5

Reflect provides one or more questions to initiate and encourage math-related discussions with the entire class. Each Friday, a very brief summary Reflect reviews some skills from the week.

Ask children:

■ **Why do we count?**

Children might say: Because it is fun; I know how to count; or One, two, three...!

Accept all responses, but encourage children to realize that counting tells how many, and name situational examples for them, such as placing items in order, measuring items, and labeling groups.

4 Assess

Use the Weekly Record Sheet from ***Assessment*** to record children's progress. Use their time at the centers as an opportunity to complete your observations.

Tuesday Planner

Objectives

- To recognize and make groups of 2 or more
- To count verbally groups of 2 or more
- To explore the mathematics in manipulatives and materials

Materials

*counters

Math Throughout the Year

Review activity directions at the top of page 3, and complete each in class whenever appropriate.

Looking Ahead

Preview ***Big Book Where's One*** for tomorrow's Whole Group.

*provided in Manipulative Kit

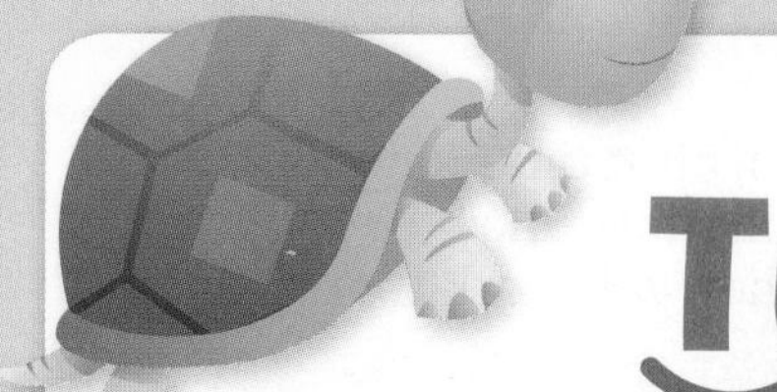

Tuesday

Whole Group

10

Warm-Up: "This Old Man"

Refer to yesterday's Warm-Up to sing as much of the traditional song as you would like with the class.

Counting Wand

- Explain to children that numbers are an important part of mathematics because they tell us how many. Tell children you wonder how many children are in the room today.
- Using the counting wand, count each child with a gentle tap on the shoulder for each number. As children listen to you count, they may begin saying the numbers with you, which provides an informal assessment of counting and number language skills.
- Once you have counted the last child, repeat the last number word, emphasizing that there are that many children. If children seem to be able to join in, invite them to do so, and repeat the count.
- This is also a Math Throughout the Year activity and will gradually occur in other variations. For now, repeat this basic version daily to establish the ideas of attendance and the amount you need to account for, acknowledging when a child is absent.

2 Work Time

25

Before starting today's Small Group, explain the concept if children are not familiar with it. You will work with four children during Small Group while remaining children complete other activities and, perhaps, centers if an adult helper is available. Assure children that they all will have a chance to work with you in Small Group.

Small Group

Explore manipulatives

- Invite children to play with counters, and observe what they do with them.
- Then give each pair of children three counters of a different color or type, and ask them whether they can talk about how many they have. Vary the amount each pair is given based on their abilities.

- Repeat the entire activity with a different group of four children until at least half the class has participated (the rest can continue Thursday). However, if all children were able to participate today, repeat the activity on Thursday with children who need extra help.

Monitoring Student Progress

If . . . children struggle naming how many counters in a group,	**Then . . .** use fewer counters and/or, instead of discussion, ask children to show how many with their fingers.
If . . . children excel at naming how many counters in a group,	**Then . . .** increase the number of counters.

Computer Center

When it is a child's turn to use the computer today, he or she will work with a partner. Review with children how to log on the computer, and start Count and Race for them. Demonstrate the activity again if needed.

Hands On Math Center

Children continue to explore math manipulatives including, though not limited to, cubes and counters.

3 Reflect

5

Ask children:

- **How high did you count on the computer with Count and Race?**

Children might say: To ten, or they may just count to the appropriate number.

4 Assess

During Small Group activities, use the Small Group Record Sheet from ***Assessment*** to observe and record children's progress.

Wednesday Planner

Objectives
- To recognize and make groups of 2 or more
- To count verbally groups of 2 or more
- To explore the mathematics in manipulatives and materials

Materials
- *inch cubes
- blocks

Math Throughout the Year

Review activity directions at the top of page 3, and complete each in class whenever appropriate.

*provided in Manipulative Kit

Wednesday

1 Whole Group 20

Warm-Up: "When I Was One"
- Here are the words and actions:

  When I was one, I was so small, *(Show one finger.)*
  I could not speak a word at all. *(Move head left to right indicating "no.")*
  When I was two, I learned to talk, *(Show two fingers.)*
  I learned to sing, I learned to walk. *(Point to mouth and feet.)*
  When I was three, I grew and grew, *(Show three fingers.)*
  Now I am four and so are you! *(Show four fingers.)*

- If most children in class are three years old, change the last line to "Soon I'll be four and so will you!" or, if there is a mixture, say "Soon we'll all be four, it's true!"

Where's One?
- Read aloud the first few pages of ***Big Book Where's One?***
- Ask children how many of a certain thing appears on those pages.

2 Work Time 15

Computer Center

Continue to provide each child with a chance to complete Count and Race.

Hands On Math Center

Make Buildings
- Children make a small stack of wooden inch cubes (or other stackable material that is small or slippery) so that buildings have few items and are safe when they fall.
- Observe children: What do they do with the materials? Do they talk to others about their buildings? Do they ask questions about the task?
- Children then count and tell you or an assistant how many cubes (or other material) are in their building.

WEEK 1

Monitoring Student Progress

If . . . children need encouragement to discuss their work,

Then . . . ask questions to extend their thinking and problem solving, such as "How many cubes are in (child's name) building? Does your building have more or fewer? Make a building that looks like (same child's name)'s." Repeat so the child in question makes his or her own building for another child to mimic.

Reflect

5 

Ask children:

- **What did you do with the cubes (or other material)?**

Children might say: I built a tower; or I counted; and there were 5.

Assess

Use the Weekly Record Sheet from ***Assessment*** to record children's progress. Use their time at the centers as an opportunity to complete your observations.

Thursday Planner

Objectives

- To recognize and make groups of 2 or more
- To count verbally groups of 2 or more
- To explore the mathematics in manipulatives and materials
- To quickly recognize the number of objects in small groups (subitize)

Materials

*counters

Math Throughout the Year

Review activity directions at the top of page 3, and complete each in class whenever appropriate.

Looking Ahead

If you have not already, make copies of Family Letter Week 1 from the ***Teacher's Resource Guide*** for Friday.

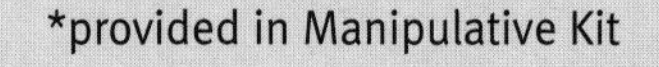

*provided in Manipulative Kit

Thursday

1 Whole Group

10 

Warm-Up: "When I Was One"

- Here are the words and actions:

  When I was one, I was so small, *(Show one finger.)*
  I could not speak a word at all. *(Move head left to right indicating "no.")*
  When I was two, I learned to talk, *(Show two fingers.)*
  I learned to sing, I learned to walk. *(Point to mouth and feet.)*
  When I was three, I grew and grew, *(Show three fingers.)*
  Now I am four and so are you! *(Show four fingers.)*

- If most children in class are three years old, change the last line to "Soon I'll be four and so will you!" or, if there is a mixture, say "Soon we'll all be four, it's true!"

Count and Move

- Have all children count from 1 to 10, or an appropriate number, clapping their hands as they say each number.
- Repeat as needed to ensure all children have participated, and then say "We all clapped ten times!"
- You may repeat this throughout the day using various motions, such as hopping and marching.

2 Work Time

25

Small Group

Explore manipulatives

- Remind children that you will be working with four of them at a time in Small Group, and remember to start with any children who did not get an opportunity to participate Tuesday.
- Invite children to play with counters, and observe what they do with them.
- Then give each pair of children three counters of a different color or type, and ask them whether they can talk about how many they have. Vary the amount each pair is given based on their abilities.
- If there is time leftover and all children have participated, repeat the activity with children who need extra help.

Monitoring Student Progress

If . . . children struggle naming how many counters in a group,	**Then . . .** use fewer counters and/or, instead of discussion, ask children to show how many with their fingers.
If . . . children excel at naming how many counters in a group,	**Then . . .** increase the number of counters.

Computer Center Building Blocks

Continue to provide each child with a chance to complete Count and Race.

Hands On Math Center

Continue to allow children to Make Buildings from yesterday's Hands On Math Center.

Reflect

5 

Ask children:

- **How many did you count when you worked in Small Group with me?**
Children might respond with various numbers or actually count aloud.

Assess

During Small Group activities, use the Small Group Record Sheet from ***Assessment*** to observe and record children's progress.

Friday Planner

Objectives

- To recognize and make groups of 2 or more
- To count verbally groups of 2 or more
- To explore the mathematics in manipulatives and materials
- To quickly recognize the number of objects in small groups (subitize)

Materials

no new materials

Math Throughout the Year

Review activity directions at the top of page 3, and complete each in class whenever appropriate.

Looking Ahead

For next week, familiarize yourself with Kitchen Counter from the ***Building Blocks*** software.

Friday

1 Whole Group 20

Warm-Up: "This Old Man"

- Sing as much of the traditional song as you would like. Here are initial verses and actions:

  This old man, he played one, *(Show one finger.)*
  He played knick-knack on his thumb. *(Wiggle thumb.)*
  Knick-knack paddy whack,
  Give the dog a bone. *(Clap hands.)*
  This old man came rolling home. *(Roll arms.)*
  This old man, he played two, *(Show two fingers.)*
  He played knick-knack on his shoe. *(Wiggle fingers near shoe.)*
  Knick-knack paddy whack,
  Give the dog a bone. *(Clap hands.)*
  This old man came rolling home. *(Roll arms.)*

- Following the same verse pattern, here are the remaining numeral rhymes: three/on his knee; four/on the floor; five/on his side; six/with some sticks; seven/under heaven; eight/on his plate; nine/all the time; ten/once again.

Where's One?

- Review ***Big Book Where'e One?*** with children.
- Return to pages other than the ones you chose Wednesday, still asking children how many of a certain thing appears on the pages. Lead children in counting aloud to check, as needed.

2 Work Time 15

Computer Center Building Blocks

Continue to provide each child with a chance to complete Count and Race.

Monitoring Student Progress

If . . . a child counts easily during Count and Race,

Then . . . partner the child with one who is challenged by the activity to allow for a tutorial to develop.

Hands On Math Center

- Children continue to stack wooden inch cubes (or other small, stackable material) to make buildings.
- Observe children: What do they do with the materials? Do they talk to others about their buildings? Do they ask questions about the task?
- Children then count and tell you (or an assistant) how many are in their building.

Reflect

5 

Ask children:

- **What have you done in class this week?**

Children might say: I counted; I worked on the computer; or I clapped and sang.

Briefly summarize other things children have done, such as working at the centers and playing Count and Race.

Assess

Use the Weekly Record Sheet from ***Assessment*** to record children's progress. Use their time at the centers as an opportunity to complete your observations.

Home Connection

From the ***Teacher's Resource Guide,*** distribute to children copies of Family Letter Week 1 to share with their family. Each letter has an area for children to show families an example of what they have been doing in class.

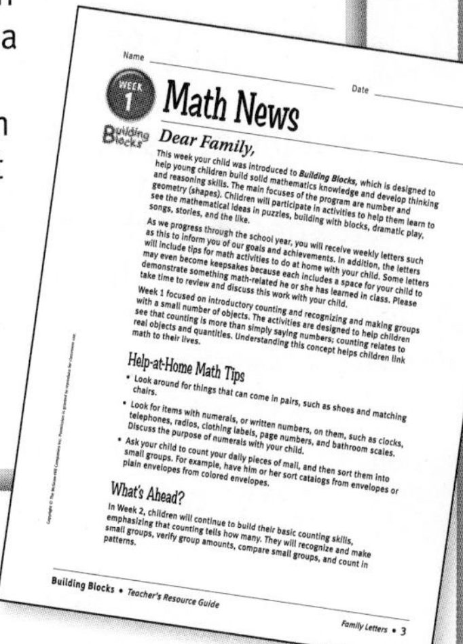

Name ______ Date ______

WEEK 1

Math News

Building Blocks

Dear Family,

This week your child was introduced to ***Building Blocks***, which is designed to help young children build solid mathematics knowledge and develop thinking and reasoning skills. The main focuses of the program are number and geometry (shapes). Children will participate in activities to help them learn to see the mathematical ideas in puzzles, building with blocks, dramatic play, songs, stories, and the like.

As we progress through the school year, you will receive weekly letters such as this to inform you of our goals and achievements. In addition, the letters will include tips for math activities to do at home with your child. Some letters may even become keepsakes because each includes a space for your child to demonstrate something math-related he or she has learned in class. Please take time to review and discuss this work with your child.

Week 1 focused on introductory counting and recognizing and making groups with a small number of objects. The activities are designed to help children see that counting is more than simply saying numbers; counting relates to real objects and quantities. Understanding this concept helps children link math to their lives.

Help-at-Home Math Tips

- Look around for things that can come in pairs, such as shoes and matching chairs.
- Look for items with numerals, or written numbers, on them, such as clocks, telephones, radios, clothing labels, page numbers, and bathroom scales. Discuss the purpose of numerals with your child.
- Ask your child to count your daily pieces of mail, and then sort them into small groups. For example, have him or her sort catalogs from envelopes or plain envelopes from colored envelopes.

What's Ahead?

In Week 2, children will continue to build their basic counting skills, emphasizing that counting tells how many. They will recognize and make small groups, verify group amounts, compare small groups, and count in patterns.

Building Blocks • *Teacher's Resource Guide*   *Family Letters* • 3

Assess and Differentiate

The Wrap-Up provides another quick reference of each week's key learning trajectories, which should help you evaluate and anticipate children's progress.

A Gather Evidence

Review children's progress in mathematics by looking at the Weekly Record Sheets (Monday, Wednesday, Friday) and the Small Group Record Sheets (Tuesday, Thursday) from this past week.

B Summarize Findings

Using ***Online Assessment,*** summarize and analyze assessment data for each child based on your weekly observations and Record Sheets. Such information helps to determine where each child is on the math trajectory for counting and subitizing. See ***Assessment*** for the print companion to each Learning Trajectory Record.

C Differentiate Instruction

Once you have seen a child exhibit specific levels of the trajectory, begin to encourage and work with that child toward the next level. Refer to Appendix A for individualized instruction opportunities, including Special Education concerns.

Verbal Counting

If...	Then...
If . . . the child says some numbers, even in incorrect order,	**Then . . .** *Pre-Counter* No verbal counting; may name some number words but does not put them into a sequence.
If . . . the child sings number words to 10,	**Then . . .** *Chanter* Chants "sing-song" or sometimes runs number words together. For example, the child might say "onetwothree" similar to "J, K, LMNOP...."
If . . . the child can count using separate words that are mostly in order,	**Then . . .** *Reciter* Verbally counts with separate words, not necessarily in the correct order. For example, the child might say "one, two, three, four, six, seven."
If . . . the child can accurately recite number words up to 10,	**Then . . .** *Reciter (10)* Verbally counts to 10 with some correspondence to objects. For example, the child says: 1 (points to it), 2 (points to it), 3 (starts to point), 4 (finishes pointing but is pointing to third object), 5, and so on.

Object Counting

If...	Then...
If . . . the child can maintain one-to-one correspondence between number words and objects,	**Then . . .** *Corresponder* Keeps one-to-one correspondence between number words and objects (one word for each object), at least for small groups of objects in a row; may answer the question "how many?" by counting the objects again.
If . . . the child can count up to five objects with understanding,	**Then . . .** *Counter (Small Numbers)* Accurately counts objects in a line to 5, and answers the question "how many?" with the last number counted. When objects are visible, especially with small numbers, begins to understand cardinality.

Subitizing

If...	Then...
If . . . the child uses numbers to name groups up to 3,	**Then . . .** *Small Collection Namer* Uses number words to name small collections of 1, 2, and sometimes 3.

Overview

Big Ideas

- introductory counting
- recognizing and making small groups
- exploring materials

Teaching for Understanding

Surprisingly, even children under one year old have some number knowledge. Number sense can develop considerably during the preschool years, particularly if children have exposure to numbers in a variety of play and everyday activities.

Grouping Activities

Primary grouping activities encourage children to attach number words to groups of one, two, or three items. These activities build meaning for number words and help support counting with understanding. The easiest grouping for preschoolers to count typically consists of a few objects arranged in a straight line that can be touched as they count.

Meaningful Connections

Relating numbers to children's everyday world expands meaningful connections. All children can learn to name small groups and begin to construct mental images of numbers by recognizing small groupings shown for a short time (subitizing).

Counting

The next step for children is connecting each number word to each object as they count. To develop meaningful counting, encourage rhythmic coordination even in simple verbal counting activities, such as clapping once as a number word is spoken.

What's Ahead?

In the weeks to come, children will learn not only about larger numbers but new ideas and skills in counting. Counting is not just saying "1, 2, 3...." It involves counting objects accurately and knowing that the last number counted tells how many. Children will also become familiar with shapes, such as knowing a triangle has three sides, and eventually link counting, number, and shape.

Eclipse Studios

WEEK 2

Overview

How Children Learn to Count

Knowing where children are on the learning trajectory for counting and what the next steps are helps facilitate their development. Children who easily surpass these trajectory levels might be challenged by larger numbers and encouraged to assist other children.

Verbal Counting

What to Look For Can the child count in correct sequence up to 5?

Reciter (10) Verbally counts to 10 with some correspondence to objects.

Object Counting

What to Look For Does the child understand one-to-one correspondence?

Corresponder Keeps one-to-one correspondence between number words and objects (one word for each object), at least for small groups of objects in a row; may answer the question "how many?" by counting the objects again.

Counter (Small Numbers) Accurately counts objects in a line to 5, and answers "how many?" with the last number counted. When objects are visible, especially with small numbers, begins to understand cardinality.

Recognition of Number and Subitizing

What to Look For Can the child recognize groups up to three or four items?

Small Collection Namer Names groups up to 2, sometimes 3.

Maker of Small Collections Nonverbally makes a small collection (no more than 4, usually up to 3) using the number of another collection, such as a mental or verbal model, not necessarily by matching.

Perceptual Subitizer to 4 Instantly recognizes collections of up to 4, and verbally names the number of items.

English Learner

Use gestures when speaking to show what you mean when you are giving directions or sharing a story with English learners. Review *snapshot, hide,* and *to match.* Refer to the ***Teacher's Resource Guide.***

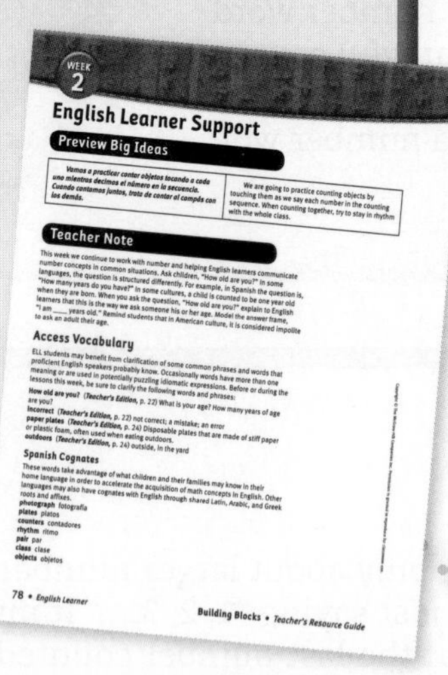

WEEK 2

English Learner Support

Preview Big Ideas

Vamos a practicar contar objetos tocando a cada uno mientras decimos el número en la secuencia. Cuando contamos juntos, trata de contar al compás con los demás.	We are going to practice counting objects by touching them as we say each number in the counting sequence. When counting together, try to stay in rhythm with the whole class.

Teacher Note

This week we continue to work with number and helping English learners communicate number concepts in common situations. Ask children, "How old are you?" In some languages, the question is structured differently. For example, in Spanish the question is, "How many years do you have?" In some cultures, a child is counted to be one year old when they are born. When you ask the question, "How old are you?" explain to English learners that this is the way we ask someone his or her age. Model the answer frame, "I am ____ years old." Remind students that in American culture, it is considered impolite to ask an adult their age.

Access Vocabulary

ELL students may benefit from clarification of some common phrases and words that proficient English speakers probably know. Occasionally words have more than one meaning or are used in potentially puzzling idiomatic expressions. Before or during the lessons this week, be sure to clarify the following words and phrases:

How old are you? (*Teacher's Edition,* p. 22) What is your age? How many years of age are you?

incorrect (*Teacher's Edition,* p. 22) not correct; a mistake; an error

paper plates (*Teacher's Edition,* p. 24) Disposable plates that are made of stiff paper or plastic foam, often used when eating outdoors.

outdoors (*Teacher's Edition,* p. 24) outside, in the yard

Spanish Cognates

These words take advantage of what children and their families may know in their home language in order to accelerate the acquisition of math concepts in English. Other languages may also have cognates with English through shared Latin, Arabic, and Greek roots and affixes.

photograph fotografía
plates platos
counters contadores
rhythm ritmo
pair par
class clase
objects objetos

78 • *English Learner* Building Blocks • *Teacher's Resource Guide*

Technology Project

Technology Projects will use computer tools to explore math. In these introductory weeks, however, focus on improving children's comfort level using the computer. As ability allows, challenge children to find just how high this week's Count and Race will go.

Math Throughout the Year

Math Throughout the Year activities are recommended to build on the mathematical skills highlighted in each week. Here are suggested activities for **Week 2.**

Counting Jar

A counting jar holds a specified number of items for children to count. The jar should be a clean, empty, plastic container, such as a peanut butter jar. Initially place a small number of relatively large items in the jar, and have children spill the items to count them. Use the same jar all year, changing the amount of items weekly. Later in the year, this can become an Estimating Jar.

Counting Wand

A counting wand draws children's attention to items that are being counted and can be made from several available materials, such as a regular pointer, a decorated ruler or wrapping paper tube, or a baton. A soft or light wand material is preferable because children enjoy using the wand themselves. Use the counting wand for counting things at every opportunity.

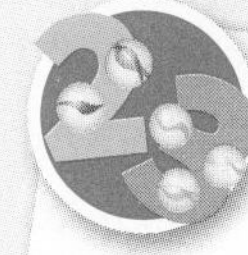

I See Numbers

Help children see groups of one, two, and three everywhere, every chance you get, such as three trees, not just a group of trees. Encouraging children to see the amount of something as opposed to only seeing the "something" helps them form the habit of quantifying small groups or collections. This habit will give them lots of experience recognizing groups and numbers.

Center Preview

Computer Center

Get your classroom Computer Center ready for Kitchen Counter and Count and Race, ***Building Blocks*** counting activities.

After you introduce each counting activity, each child should complete the activities individually as you (or an assistant) monitor and guide him or her periodically. Ideally, each child will have at least ten minutes of computer time at least twice during the week. To assess progress, observe children as they rotate through all classroom centers.

Hands On Math Center

This week's Hands On Math Center activities are exploring manipulatives, Make Buildings, Find Groups, and Make Groups. Supply the center with these materials: wooden inch cubes, other safe stackable materials, counters, blocks, paper plates, and common classroom items to make groups.

Literature Connections

These books help develop counting.

- *Little Rabbits' First Number Book* by Alan Baker
- *The Very Hungry Caterpillar* by Eric Carle
- *Three Tales of Three* by Marilyn Helmer
- *Ten Cats and Their Tales* by Martin Leman
- *Frog Counts to Ten* by John Liebler

(t)Matt Meadows (b)Eclipse Studios

WEEK 2 Weekly Planner

Learning Trajectories

Week 2 Objectives

- To name the number of objects in a group up to 3
- To count verbally groups of 2 or more with understanding
- To recognize and make groups of 1 or more
- To connect number words to the quantities they represent
- To quickly recognize the number of objects in small groups (subitize)
- To produce simple rhythmic patterns

	Developmental Path	Instructional Activities
Verbal Counting	*Chanter*	
	Reciter	
	Reciter (10)	Count and Move Count and Race
Object Counting	*Corresponder*	Kitchen Counter Count and Race
	Counter (Small Numbers)	Number Me Find Groups Make Buildings
	Producer (Small Numbers)	Counting Jar
	Counter (10)	
Recognition of Number and Subitizing	*Small Collection Namer*	"When I Was One" "Two Little Blackbirds" Make Buildings Number Me
	Maker of Small Collections	Make Groups
	Perceptual Subitizer to 4	Snapshots Find Groups
	Perceptual Subitizer to 5	

Use this chart to plan for your specific class schedule. If you have your prekindergarteners for only three days, complete Monday, Tuesday, and Thursday of the week.

Work Time

Whole Group	Small Group	Computer	Hands On	Program Resources
"When I Was One" Number Me Count and Move		• Kitchen Counter • Count and Race	Explore manipulatives *Materials:* *inch cubes *counters	Building Blocks ***Assessment*** Weekly Record Sheet
Count and Move Find Groups	Find and Make Groups *Materials:* *counters paper plates	• Kitchen Counter • Count and Race	Explore manipulatives	Building Blocks ***Assessment*** Small Group Record Sheet
Count and Move in Patterns "Two Little Blackbirds" Make Groups		• Kitchen Counter • Count and Race	Make Buildings *Materials:* *inch cubes blocks	Building Blocks ***Assessment*** Weekly Record Sheet
Count and Move in Patterns Snapshots *Materials:* dark cloth paper plates *counters	Find and Make Groups *Materials:* *counters paper plates	• Kitchen Counter • Count and Race	Find Groups *Materials:* *counters paper plates Make Buildings	Building Blocks ***Assessment*** Small Group Record Sheet
Count and Move in Patterns Make Groups Snapshots *Materials:* dark cloth paper plates *counters		• Kitchen Counter • Count and Race	Make Groups *Materials:* paper plates Make Buildings Find Groups	Building Blocks ***Assessment*** Weekly Record Sheet ***Teacher's Resource Guide*** Family Letter Week 2

*provided in Manipulative Kit

Monday Planner

Objectives

- To name the number of objects in a group up to 3
- To count verbally groups of 2 or more with understanding
- To make groups of one and two objects
- To connect number words to the quantities they represent

Materials

- *inch cubes
- *counters

Vocabulary

Group or *collection* describes one or more items placed or organized together.

Math Throughout the Year

Review activity directions at the top of page 19, and complete each in class whenever appropriate.

*provided in Manipulative Kit

Monday

1 Whole Group 15

Warm-Up: "When I Was One"

- Here are the words and actions:

  When I was one, I was so small, *(Show one finger.)*
  I could not speak a word at all. *(Shake head left to right, "no.")*
  When I was two, I learned to talk, *(Show two fingers.)*
  I learned to sing, I learned to walk. *(Point to mouth and feet.)*
  When I was three, I grew and grew. *(Show three fingers.)*
  Now I am four and so are you! *(Show four fingers.)*

- If some children are five years old, adapt the finger play's last line. Have four-year-olds say "I'm four," showing four fingers, and have five-year-olds say "I'm five," showing five fingers. Then all children say "How old are you?"

Number Me

- Ask children how old they are. They should answer by saying the number and showing that many fingers.
- Then ask children how many arms they have. Have them wave their arms.
- Repeat with hands, fingers, legs, feet, toes, head, nose, eyes, and ears. Include motion when possible, such as "Wiggle your ten fingers," and have fun!
- Make silly, incorrect statements. For example, "I have four ears… three feet… five eyes." Mix them with correct statements. Have children say whether or not you are correct. When children disagree, ask: How many do I have? How do you know?

Count and Move

- Have all children count from 1 to 10, or an appropriate number, clapping with each number.
- Repeat during the day and on different days using other motions, such as hopping or marching.

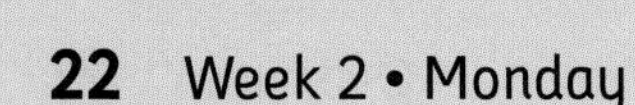

Eclipse Studios

Work Time

20

Computer Center

Building Blocks

From the ***Building Blocks*** software, introduce the Kitchen Counter activity, and demonstrate the Count and Race activity. As you do so, have children count aloud with you. In Kitchen Counter, children need to click on each item to count it; explain that they should click each item only once. Children should have a chance to complete both activities this week.

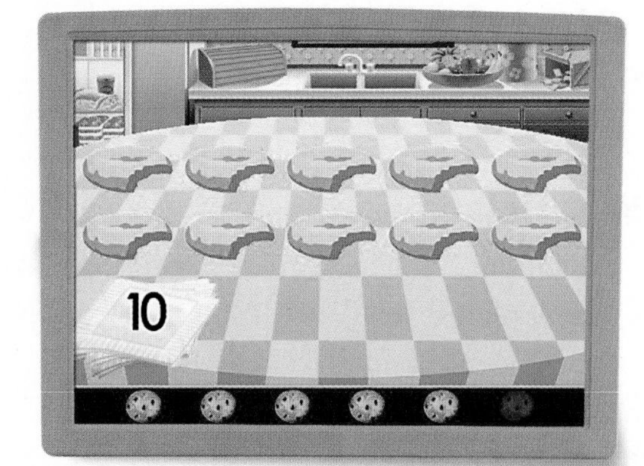

Kitchen Counter

Monitoring Student Progress

If . . . a child struggles using the computer,	**Then . . .** partner him or her with an "expert" child, one who shows computer proficiency.

Hands On Math Center

Children explore math manipulatives including, though not limited to, cubes and counters.

Reflect

5

Ask children:

- **How many doors do we have?**
- **How do you know?**

Children might say: Two doors; I just see two; or I can count them—one, two.

Encourage participation by counting items with children.

Assess

Use the Weekly Record Sheet from ***Assessment*** to record children's progress. Use their time at the centers as an opportunity to complete your observations.

Tuesday Planner

Objectives

- To recognize and make groups of 2 or more
- To count verbally groups of 2 or more with understanding

Materials

- *counters
- paper plates

Math Throughout the Year

Review activity directions at the top of page 19, and complete each in class whenever appropriate.

*provided in Manipulative Kit

Tuesday

1 Whole Group

10

Warm-Up: Count and Move

- Have all children count from 1 to 10, or an appropriate number, clapping with each number. Explain that the last number word tells how many people are in class.
- You may tap yourself for "one" using the Counting Wand and, while standing, each child counts aloud as you tap him or her on the shoulder and sits after being counted.
- Repeat on different days using various motions, such as hopping or marching.

Find Groups

- Ask children to find groups of 2 in the classroom, such as two doors, two clocks, and so on. They can look around and give their answers or walk around and show each other.
- Ask children to tell and show with fingers how many of a certain item is in the classroom.

2 Work Time

25

Small Group

Find and Make Groups

- Place six different groups of counters (or thematic items) on paper plates (or construction paper), beginning with groups of one to three items.
- Have children work in pairs to find the group of two (or another number if they are able). Partners help and check each other while you observe and assist as needed. Vary the amount pairs are to find based on their ability.
- Ask children how they knew for sure they had found a group with the correct number.
- Now model making a group of two to ensure that children understand the task, and have pairs of children make their own groups of two (or an appropriate number).
- If all children did not get a chance to participate today, they will have another chance on Thursday.
- When possible, complete this activity outdoors. Children should choose safe items they do not have to harm, such as finding two leaves on the ground, not picking them from a tree.

Monitoring Student Progress

If . . . children struggle with finding or making groups of 2,	**Then . . .** reduce the number to one and, if necessary, give children sheets of paper with two circles, and have them put an object in each circle.
If . . . children excel at finding or making groups of 2,	**Then . . .** increase the number of items they are to find or make.

Use the small group dynamic as a chance to assess how high and accurately each child can count; increase the amount for each child to discover what numbers he or she knows.

Computer Center Building Blocks

Continue to provide each child with a chance to complete Kitchen Counter and Count and Race.

Hands On Math Center

Children continue to explore math manipulatives and other items by making groups of 2. If applicable, provide items that correspond to class themes.

Count and Race

3 Reflect 5

If children made groups of both 2 and more than 2, ask:

- **Which group is larger?**
- **How do you know which group has more items?**

Children might say: That group is bigger, or I counted.

4 Assess

During Small Group activities, use the Small Group Record Sheet from ***Assessment*** to observe and record children's progress.

Wednesday Planner

Objectives

- To recognize and make groups of 2 or more
- To count verbally groups of 2 or more with understanding
- To produce simple rhythmic patterns

Materials

- *inch cubes
- blocks

Math Throughout the Year

Review activity directions at the top of page 19, and complete each in class whenever appropriate.

*provided in Manipulative Kit

Wednesday

1 Whole Group

15

Warm-Up: Count and Move in Patterns

- Spend just a few minutes counting in patterns of 2. Have all children count from 1 to 10, or an appropriate even number. For example, say 1 quietly, 2 loudly, 3 quietly, 4 loudly, and so on.
- Add to the fun by marching along with the counting: 1 (step), 2 (stomp), 3 (step), 4 (stomp), and so on.

"Two Little Blackbirds"

Here are the words and actions:

Two little blackbirds *(Show index finger on each hand.)*
Sitting on a hill.
One named Jack, *(Put right finger forward.)*
One named Jill. *(Put left finger forward.)*
Fly away, Jack. *(Wiggle right finger and place behind back.)*
Fly away, Jill. *(Wiggle left finger and place behind back.)*
Come back, Jack. *(Return right finger to the front.)*
Come back, Jill. *(Return left finger to the front.)*

Make Groups

- Place familiar items, or items based on a class theme, in easy reach of all children, and have them make groups of 2.
- Ask children how they knew they made a group of 2.
- Repeat with a larger number as long as children are successful.
- When this activity is done outdoors, check groups the next day to see whether they are intact. For example, are two rocks still where a child put them? Remind children not to harm anything to make their groups.

2 Work Time

20

Computer Center Building Blocks

Continue to provide each child with a chance to complete Kitchen Counter and Count and Race.

Hands On Math Center

Make Buildings

- Children make a short stack of wooden inch cubes (or other stackable material that is small or slippery) so that buildings have few items and are safe if they fall.
- Children then count and tell you (or an assistant) how many cubes are in their building.

Monitoring Student Progress

If . . . children need help counting,

Then . . . limit the number of materials they have to build with, or provide materials that do not stack high, such as foam cubes.

If . . . children can count higher than the number they stack,

Then . . . provide materials that can be stacked high, such as flat objects, enabling children to count higher.

3 Reflect

5

Ask children:

- **When you made buildings, how many items did you stack?**

Children should respond with a number.

Follow up by asking:

- **How did you know how many?**

Children might say: I counted, or I could just see that there were 3.

4 Assess

Use the Weekly Record Sheet from *Assessment* to record children's progress. Use their time at the centers as an opportunity to complete your observations.

Thursday Planner

Objectives
- To recognize and make groups of 2 or more
- To count verbally groups of 2 or more with understanding
- To quickly recognize the number of objects in small groups (subitize)

Materials
- dark cloth
- paper plates
- *counters

Math Throughout the Year
Read a book to children, such as *The Very Hungry Caterpillar* by Eric Carle, returning to various pages to have children tell you how many of a certain thing is shown. Count aloud with children to check.

Looking Ahead
If you have not already, make copies of Family Letter Week 2 from the ***Teacher's Resource Guide*** for Friday.

*provided in Manipulative Kit

Thursday

1 Whole Group

15

Warm-Up: Count and Move in Patterns
- Spend just a few minutes counting in patterns of 2. Have all children count from 1 to 10, or an appropriate even number. For example, say 1 quietly, 2 loudly, 3 quietly, 4 loudly, and so on.
- Add to the fun by including movement while counting.

Snapshots
- Prepare beforehand by secretly placing two counters on a paper plate (or construction paper), covering the plate with a cloth you cannot see through.
- Ask children to tell about times when their family took photographs or "snapshots" with a camera. Tell children, during Snapshots, they will use their eyes and minds like a camera by making a picture (taking a snapshot) of what they see.
- Show your covered plate, and explain that counters are hidden under the cloth. Tell children to watch carefully, quietly, and with hands in their laps as you quickly uncover the plate to expose the counters. Uncover the plate for two seconds, and cover it again.
- Have children show with their fingers how many counters they saw and, once you have seen all children's responses, ask them to tell how many counters they saw. Repeat the uncovering if needed.
- Uncover your plate indefinitely and, to check their answers, ask what children see. Allow them to say their answers, recognizing their contributions.
- Depending on children's ability, repeat the activity with up to four counters lined up.

2 Work Time

20

Small Group

Find and Make Groups
- If any children did not get a chance to complete Tuesday's Small Group, they should do so today.
- Place six different groups of counters or thematic items on paper plates, beginning with groups up to 3.
- Have children work in pairs to find the group of 2 or another number. Partners help and check each other while you observe and assist as needed. Vary the amount pairs are to find based on their ability.

- Ask how children knew they had found a group with the correct number.
- Have partners make their own groups of 2. Model this if necessary.
- If done outdoors, children should pick safe items they do not have to harm.

Monitoring Student Progress

If . . . a child struggles with making a group of 2,	**Then . . .** provide paper with two circles drawn on it so the child can put an object in each circle. Increase the number gradually.
If . . . a child excels at making a group of 2,	**Then . . .** see how large a number the child can represent accurately.

RESEARCH IN ACTION

Children who can quickly recognize numbers, taking a mental "snapshot" of them, are more able to determine the number of items in a collection, represent that number, and, later, add the numbers.

Computer Center Building Blocks

Continue to provide each child with a chance to complete Kitchen Counter and Count and Race.

Hands On Math Center

Continue to allow children to Make Buildings from yesterday's center and/or introduce the activity below, recuring from Whole and Small Groups.

Find Groups

Set different groups of counters on paper plates, and have children ask each other to find a certain number of their choice. Use this time to assess how high each child can count.

3 Reflect 5

Ask children:

- **Do we have groups in the classroom?**

Encourage them to think about storage, shelves, and so on, and ask:

- **What does each group have in it?** (Children should name actual items.)
- **Do any of the groups have more than two items?**

If children can count a group's items, have them do so.

4 Assess

During Small Group activities, use the Small Group Record Sheet from *Assessment* to observe and record children's progress.

Friday Planner

Objectives

- To recognize and make groups of 3 or more
- To count verbally groups of 2 or more with understanding
- To quickly recognize the number of objects in small groups (subitize)

Materials

- dark cloth
- paper plates
- *counters

Math Throughout the Year

Review activity directions at the top of page 19, and complete each in class whenever appropriate.

Looking Ahead

For next week, familiarize yourself with Pizza Pizzazz 1: Match Collections on the computer.

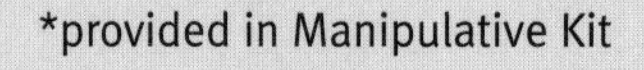

*provided in Manipulative Kit

Friday

1 Whole Group

15 

Warm-Up: Count and Move in Patterns

- Spend a few minutes counting in patterns of 3. Have all children count from 1 to an appropriate number. For example, say 1 quietly, 2 quietly, 3 loudly, 4 quietly, 5 quietly, 6 loudly, and so on.
- Add to the fun by including movement while counting: 1 (step), 2 (step), 3 (hop), 4 (step), 5 (step), 6 (hop), and so on.

Make Groups

- Place familiar or thematic items in easy reach of all children, and have them make groups of 3.
- As children work, ask how they know they made a group of 3.
- Repeat with a larger number as long as children are successful.
- When this activity is done outdoors, check groups the next day to see whether they are intact. For example, are two rocks still where a child put them? Remind children not to harm anything to make their groups.

Snapshots

- Prepare beforehand by secretly placing three counters on a paper plate, covering it with a cloth you cannot see through.
- Review with children the concept of "snapshots." They will use their eyes and minds like a camera to take a snapshot of what they see.
- Show your covered plate, and explain that counters are hidden under the cloth. Remind children to watch carefully, quietly, and with hands in their laps as you quickly uncover the plate to expose the counters. Uncover the plate for two seconds, and cover it again.
- Have children show how many counters they saw with their fingers and, once you have seen all children's responses, ask them to tell how many counters they saw. Repeat the uncovering if needed.
- Uncover your plate indefinitely and, to check their answers, ask children what they see. Allow them to say their answers and recognize their contributions.
- Depending on children's ability, repeat the activity with up to four counters lined up.

Eclipse Studios

Monitoring Student Progress

If . . . children struggle during Snapshots,	**Then . . .** arrange items in a line only, or display items for a longer time. Use fewer items as necessary.
If . . . children excel during Snapshots,	**Then . . .** gradually increase the number of items, or place items in a random arrangement rather than in a straight line.

Work Time

Computer Center Building Blocks

Continue to provide each child with a chance to complete Kitchen Counter and Count and Race.

Hands On Math Center

Continue to allow children to Find Groups and/or Make Buildings from this week's previous Hands On Math Center and/or introduce the activity below.

Make Groups

Provide numerous counters and paper plates for children to make groups of a number they choose. Suggest they start with 3, and use this time to assess counting ability.

Home Connection

From the ***Teacher's Resource Guide,*** distribute to children copies of Family Letter Week 2 to share with their family. Each letter has an area for children to show families an example of what they have been doing in class.

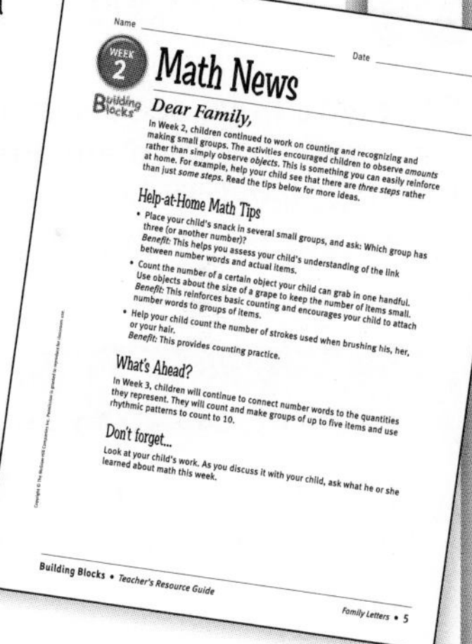

Name ______ Date ______

WEEK 2 **Math News**

Building Blocks

Dear Family,

In Week 2, children continued to work on counting and recognizing and making small groups. The activities encouraged children to observe *amounts* rather than simply observe *objects*. This is something you can easily reinforce at home. For example, help your child see that there are *three steps* rather than just *some steps*. Read the tips below for more ideas.

Help-at-Home Math Tips

- Place your child's snack in several small groups, and ask: Which group has three (or another number)?
  Benefit: This helps you assess your child's understanding of the link between number words and actual items.
- Count the number of a certain object your child can grab in one handful. Use objects about the size of a grape to keep the number of items small.
  Benefit: This reinforces basic counting and encourages your child to attach number words to groups of items.
- Help your child count the number of strokes used when brushing his, her, or your hair.
  Benefit: This provides counting practice.

What's Ahead?

In Week 3, children will continue to connect number words to the quantities they represent. They will count and make groups of up to five items and use rhythmic patterns to count to 10.

Don't forget...

Look at your child's work. As you discuss it with your child, ask what he or she learned about math this week.

Building Blocks • *Teacher's Resource Guide* *Family Letters* • 5

Reflect

Briefly discuss with children what you have done in class this week, such as making groups. Discuss how single items are combined to make groups.

Ask children:

- **What new thing did you learn about math this week?**

Encourage them to consider their counting, memory, and computer skills.

Assess

Use the Weekly Record Sheet from ***Assessment*** to record children's progress. Use their time at the centers as an opportunity to complete your observations.

Assess and Differentiate

A Gather Evidence

Review children's progress in mathematics by looking at the Weekly Record Sheets (Monday, Wednesday, Friday) and the Small Group Record Sheets (Tuesday, Thursday) from this past week.

B Summarize Findings

Using ***Online Assessment,*** summarize and analyze assessment data for each student based on your weekly observations and Record Sheets. Such information helps determine where each child is on the math trajectory for counting and subitizing. See ***Assessment*** for the print companion to each Learning Trajectory Record.

C Differentiate Instruction

Once you have seen a child exhibit specific levels of the trajectory, begin to encourage and work with that child toward the next level. Refer to Appendix A for individualized instruction opportunities, including Special Education concerns.

Verbal Counting

If . . .	Then . . .
If . . . the child can recite 1 to 5 or 10,	**Then . . .** *Reciter (10)* Verbally counts to 10 with some correspondence to objects.

Object Counting

If . . .	Then . . .
If . . . the child can understand that one number word corresponds to one object,	**Then . . .** *Corresponder* Keeps one-to-one correspondence between number words and objects (one word for each object), at least for small groups of objects in a row; may answer the question "how many?" by counting the objects again.
If . . . the child can say number words in correct sequence,	**Then . . .** *Counter (Small Numbers)* Accurately counts objects in a line to 5, and answers "how many?" with the last number counted. When objects are visible, especially with small numbers, begins to understand cardinality.

Recognition of Number and Subitizing

If . . .	Then . . .
If . . . the child can use number words to name small groups,	**Then . . .** *Small Collection Namer* groups up to 2, sometimes 3.
If . . . the child can think of a small collection and then visually represent it,	**Then . . .** *Maker of Small Collections* Nonverbally makes a small collection (no more than 4, usually up to 3) using the number of another collection, such as a mental or verbal model, not necessarily by matching.
If . . . the child can recognize a group of two or more objects when shown for only a brief time,	**Then . . .** *Perceptual Subitizer to 4* Instantly recognizes collections of up to 4 when shown briefly, and verbally names the number of items.

Overview

Big Ideas

- counting and producing small groups
- recognizing equal groups
- duplicating rhythmic patterns

Teaching for Understanding

Week 3 builds on the counting skills of Week 2, emphasizing four components of counting: verbal counting, the counting of small collections, counting out (producing) small collections, and comparing small quantities. Mastering these components enables children to maintain one-to-one correspondence between each number word spoken and each item counted, as well as to understand that counting tells how many and describes order.

Object Counting

This requires much more than verbal counting. Several activities develop children's early ability to connect small groups of objects to number words. For example, hide a small number of objects, and reveal them one at a time as children count. Research shows that such an activity helps children link each number word they say to the quantity of objects they see. This helps children understand that the last number word in a counting sequence tells how many.

Counting Out Objects

Research also shows that "counting out," or producing a certain number of objects, is more difficult than counting the objects in a group. Both tasks require knowing how to count verbally, keep one-to-one correspondence between number words and objects, and answer the question "how many?" However, when children count out objects, they have to continually recall how many they were supposed to produce and compare that to each number word they say in order to stop at just the right number.

Meaningful Connections

Children learn the sequence of number words over a long period of time so repeated practice is essential. We engage children in rhythmic counting patterns now so that, in later grades, such ideas will grow into knowledge of many number patterns, such as even and odd numbers.

What's Ahead?

Every week we continue to lay the groundwork for counting and other number abilities, while also interweaving many other mathematical topics. The next couple of weeks focus on shapes and geometric comprehension.

WEEK 3 Overview

English Learner

Reading aloud counting stories in books is an excellent strategy for building counting-related vocabulary. Review the following: *rows, hide counters,* and *empty hand.* Refer to the ***Teacher's Resource Guide.***

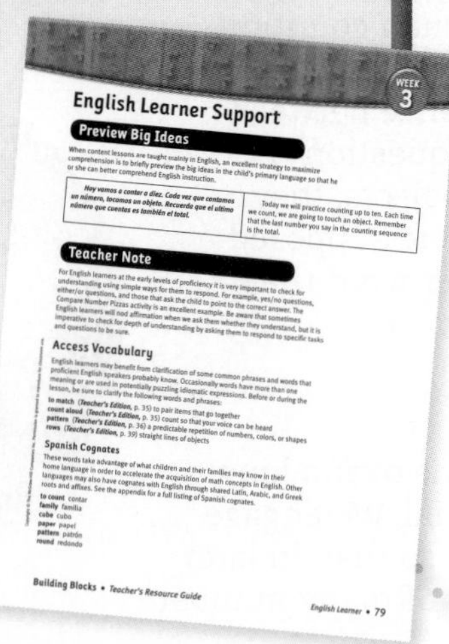

WEEK 3

English Learner Support

Preview Big Ideas

When content lessons are taught mainly in English, an excellent strategy to maximize comprehension is to briefly preview the big ideas in the child's primary language so that he or she can better comprehend English instruction.

Hoy vamos a contar a diez. Cada vez que contamos un número, tocamos un objeto. Recuerda que el último número que cuentas es también el total.	Today we will practice counting up to ten. Each time we count, we are going to touch an object. Remember that the last number you say in the counting sequence is the total.

Teacher Note

For English learners at the early levels of proficiency it is very important to check for understanding using simple ways for them to respond. For example, yes/no questions, either/or questions, and those that ask the child to point to the correct answer. The Compare Number Pizzas activity is an excellent example. Be aware that sometimes English learners will nod affirmation when we ask them whether they understand, but it is imperative to check for depth of understanding by asking them to respond to specific tasks and questions to be sure.

Access Vocabulary

English learners may benefit from clarification of some common phrases and words that proficient English speakers probably know. Occasionally words have more than one meaning or are used in potentially puzzling idiomatic expressions. Before or during the lesson, be sure to clarify the following words and phrases:

to match (*Teacher's Edition,* p. 35) to pair items that go together
count aloud (*Teacher's Edition,* p. 35) count so that your voice can be heard
pattern (*Teacher's Edition,* p. 36) a predictable repetition of numbers, colors, or shapes
rows (*Teacher's Edition,* p. 39) straight lines of objects

Spanish Cognates

These words take advantage of what children and their families may know in their home language in order to accelerate the acquisition of math concepts in English. Other languages may also have cognates with English through shared Latin, Arabic, and Greek roots and affixes. See the appendix for a full listing of Spanish cognates.

to count contar
family familia
cube cubo
paper papel
pattern patrón
round redondo

Building Blocks • *Teacher's Resource Guide*

English Learner • 79

Technology Project

Children may get additional problem-posing and -solving practice using Pizza Pizzazz Free Explore. Children can count along with the program as they place toppings on either pizza, or they may work cooperatively as one child names a number and the other puts that many toppings on a pizza.

How Children Learn to Count

Knowing where children are on the learning trajectory for counting and what the next steps are helps facilitate their development. Children who easily surpass these trajectory levels might be challenged by larger quantities and encouraged to assist other children.

Verbal Counting

What to Look For Does the child count to at least 5 with some one-to-one correspondence?

Reciter Verbally counts using separate words ("one, two, three" not "onetwothree") but not always in the correct order, such as "one, two, three, four, six, seven."

Reciter (10) Verbally counts to 10 with some correspondence to objects.

Object Counting

What to Look For Can the child accurately count and produce small groups?

Corresponder Keeps one-to-one correspondence between number words and objects (one word for each object), at least for small groups of objects in a row; may answer the question "how many?" by counting the objects again.

Counter (Small Numbers) Accurately counts objects in a line to 5, and answers "how many?" with the last number counted. When objects are visible, especially with small numbers, begins to understand cardinality.

Producer (Small Numbers) Counts out objects to 5.

Comparing Number

What to Look For Does the child accurately identify the larger of two groups?

Perceptual Comparer Compares collections that are quite different in size, such as one that is twice the size of the other (shown ten blocks and twenty-five blocks, points to the group with twenty-five as having more). If collections are similar, child can compare them accurately as well (with groups of two and four blocks, knows the group of four has more).

Math Throughout the Year

Math Throughout the Year activities are recommended to build on the mathematical skills highlighted in each week. Here are suggested activities for **Week 3.**

Counting Wand (Count All)

Count each child with a gentle tap of the wand, making sure all children count aloud with you. Emphasize that the last number word tells how many children are in class today. If children are ready to take turns tapping one another, use a very soft wand.

Simon Says Numbers

Play traditional Simon Says using only number commands, such as "Jump two times" and "Pat your head six times."

Snack Time

Children take a specified amount of a snack, as well as anything they might need to eat the snack. Demonstrate counting out the items and saying afterward how many there are. You may choose to place a Counting Card on the snack table. For example, use a five card to indicate that children take that many pretzels.

Center Preview

Computer Center Building Blocks

Get your classroom Computer Center ready for Pizza Pizzazz 1: Match Collections from the ***Building Blocks*** software.

After you introduce Pizza Pizzazz 1, each child should complete the activity individually as you (or an assistant) monitor and guide him or her periodically. Ideally, each child will have at least ten minutes of computer time at least twice during the week. To assess progress, observe children as they rotate through all classroom centers.

Hands On Math Center

This week's Hands On Math Center activities are Find the Number, Fill and Spill, and Draw Numbers. Supply the center with these materials: opaque containers, round counters, paper plates, Counting Cards, plastic containers, wooden cubes, drawing paper, crayons, and other drawing materials.

Literature Connections

These books help develop counting.

- *Little Rabbits' First Number Book* by Alan Baker
- *The Very Hungry Caterpillar* by Eric Carle
- *I Spy Two Eyes: Numbers In Art* by Lucy Micklethwait
- *One Was Johnny: A Counting Book* by Maurice Sendak
- *I Can Count the Petals of a Flower* by John and Stacey Wahl

(t)Matt Meadows (b)Eclipse Studios

WEEK 3

Weekly Planner

Learning Trajectories

Week 3 Objectives

- To participate in rhythmic patterns
- To connect number words to the quantities they represent
- To make groups of up to five items
- To count verbally to 5 with understanding
- To count verbally to 10 with understanding

	Developmental Path	Instructional Activities
Verbal Counting	*Reciter*	
	Reciter (10)	"Baker's Truck" Count and Move in Patterns
Object Counting	*Corresponder*	
	Counter (Small Numbers)	Demonstrate Counting Number Me (5) Counting Book Find the Number
	Producer (Small Numbers)	Make Number Pizzas Fill and Spill Draw Numbers
	Counter (10)	
Comparing Number	*Object Corresponder*	
	Perceptual Comparer	Compare Number Pizzas Pizza Pizzazz 1 Find the Number
	Nonverbal Comparer	

Use this chart to plan for your specific class schedule. If you have your prekindergarteners for only three days, complete Monday, Tuesday, and Thursday of the week.

Work Time

Whole Group	Small Group	Computer	Hands On	Program Resources
"Baker's Truck" **Compare Number Pizzas** *Materials:* *round counters paper plates		**Pizza Pizzazz 1**	**Find the Number** *Materials:* opaque containers *round counters paper plates *Counting Cards	Building Blocks ***Assessment*** Weekly Record Sheet
Count and Move in Patterns **Demonstrate Counting** *Materials:* *counters	**Make Number Pizzas** *Materials:* *round counters paper plates	**Pizza Pizzazz 1**	**Fill and Spill** *Materials:* plastic containers wooden cubes *Counting Cards **Find the Number**	Building Blocks ***Assessment*** Small Group Record Sheet
Count and Move in Patterns **Demonstrate Counting** *Materials:* *counters **Compare Number Pizzas** *Materials:* *round counters paper plates		**Pizza Pizzazz 1**	**Fill and Spill** **Find the Number**	Building Blocks ***Assessment*** Weekly Record Sheet
Count and Move in Patterns **Number Me (5)**	**Demonstrate Counting** *Materials:* *counters **Make Number Pizzas** *Materials:* *round counters paper plates	**Pizza Pizzazz 1**	**Draw Numbers** *Materials:* drawing paper nontoxic markers **Fill and Spill** **Find the Number**	Building Blocks ***Assessment*** Small Group Record Sheet
Count and Move in Patterns **Counting Book**		**Pizza Pizzazz 1**	Draw Numbers Fill and Spill Find the Number	Building Blocks ***Assessment*** Weekly Record Sheet ***Teacher's Resource Guide*** Family Letter Week 3

*provided in Manipulative Kit

Monday Planner

Objectives

- To participate in rhythmic patterns
- To connect number words to the quantities they represent
- To make groups of up to five items

Materials

- *round counters
- paper plates
- opaque containers
- *Counting Cards

Math Throughout the Year

Review activity directions at the top of page 35, and complete each in class whenever appropriate.

Looking Ahead

Instead of using counters and paper plates for this week's pizza activities, you could make toppings of your choice with felt or construction paper and cut large circles for pizza crusts.

*provided in Manipulative Kit

Monday

1 Whole Group

15

Warm-Up: "Baker's Truck"

- Here are the words and actions:

  The baker's truck drives down the street,
  Filled with everything good to eat.
  Two doors the baker opens wide. *(Outstretch arms.)*
  Let's look at the shelves inside. *(Cup hands around eyes to look.)*
  What do you see? What do you see?
  Three big cookies for you and me! *(Show three fingers.)*

- Adapt the final number of cookies in the finger play to reinforce any number up to 10 that you are teaching.

Compare Number Pizzas

- Tell a story about a pizza chef. Explain that you have to help the chef get the correct number of pepperoni slices on the pizza.
- Use a paper plate for pizza crust and round counters for pepperoni. Show your pizza with two pepperoni slices, leaving it in children's view. Then show three more pizzas with one, two, and three pepperoni slices.
- Ask all children to point to which of the three pizzas has the same number of toppings as the first pizza you showed. Have them discuss how they knew the matching pizza had the same number of toppings.
- Repeat the activity, having children match pizzas with pepperoni amounts of 3 or more as their ability allows.

2 Work Time

20

Computer Center Building Blocks

Demonstrate Pizza Pizzazz 1: Match Collections from the ***Building Blocks*** software. In this activity, children help twins who want the same number of toppings on their pizzas by choosing a pizza to match another pizza with a certain number of toppings. All children should have a chance to complete Pizza Pizzazz 1 this week.

Hands On Math Center

Find the Number

- Before children get to the center, conceal several pizzas (paper plates), each with a different number of pepperoni slices (round counters) under its own opaque container.
- Display one pizza with three to five pepperoni slices, or use a Counting Card to represent the target number. The goal is for children to find the hidden match to the pizza on display.
- Children should show their answers to you or another adult who assists your class.

Monitoring Student Progress

If . . . children need help during Find the Number,	**Then . . .** reduce the number of hidden pizzas, or leave all pizza choices uncovered.
If . . . children need a challenge during Find the Number,	**Then . . .** have them work in pairs, determining their own topping amounts and asking each other, for example, "Where is the 10?"

RESEARCH IN ACTION

At this first level of Pizza Pizzazz, some children count while others use visual strategies, especially for small numbers. Such visual strategies range from the informal copying of a design to the sophisticated "seeing," for example, of two rows of three immediately as six.

3 Reflect 5

Ask children:

■ **How did you find the number you were looking for?**

Children might say: I counted toppings on each pizza, or I could just see it was 2.

4 Assess

Use the Weekly Record Sheet from *Assessment* to record children's progress. Use their time at the centers as an opportunity to complete your observations.

Tuesday Planner

Objectives

- To participate in rhythmic patterns
- To count verbally to 5 with understanding
- To connect number words to the quantities they represent
- To make groups of up to five items

Materials

- *counters
- *round counters
- paper plates
- plastic containers
- wooden cubes
- *Counting Cards

Math Throughout the Year

Review activity directions at the top of page 35, and complete each in class whenever appropriate.

*provided in Manipulative Kit

Tuesday

1 Whole Group

10

Warm-Up: Count and Move in Patterns

- Spend just a few minutes counting in patterns of 2. Have all children count from 1 to 10, or an appropriate even number. For example, say 1 quietly, 2 loudly, 3 quietly, 4 loudly, and so on.
- Add to the fun by marching along with the counting: 1 (step), 2 (stomp), 3 (step), 4 (stomp), and so on. Incorporate class themes when possible.

Demonstrate Counting

- The goal of this activity is to teach object counting, emphasizing that counting tells how many. Hide four counters in your hand.
- Tell children you have some counters in your hand, and ask them to count aloud with you to find out how many.
- Remove one of the counters, and place it where children can see and focus on it. Say "one" with the children; emphasize that "one" tells how many counters there are now.
- Repeat until you have counted and displayed all four counters. Then show your empty hand.
- Ask children how many counters there are in all. If they reply "four," agree and reiterate that, together, you counted four counters.
- Repeat with a different number of counters, making sure children count aloud with you.

2 Work Time

25

Small Group

Make Number Pizzas

- In this activity, children put a given number of counters on paper plates. Place a plate with three counters to serve as their target.
- Make sure each pair of children has a paper plate and several counters. Tell each pair to make a pizza with the same number of toppings as the one you made (3).
- Ask children how they know there are three toppings on their pizzas. Discuss the various arrangements children made with their counters.
- Repeat with numbers up to 5 (or more as appropriate).

Monitoring Student Progress

If . . . children struggle with Make Number Pizzas,	**Then . . .** use fewer counters, or draw circles for children to put one counter in each until they can reliably count out 5.
If . . . children excel at Make Number Pizzas,	**Then . . .** increase the number of counters, and/or tell children the target number without showing it.

Computer Center Building Blocks

Continue to provide each child with a chance to complete Pizza Pizzazz 1.

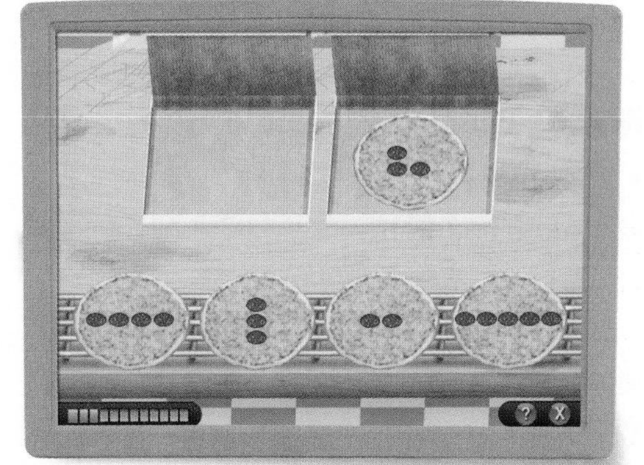

Pizza Pizzazz 1

Hands On Math Center

Continue to allow children to Find the Number from yesterday's Hands On Math Center, and introduce the following activity:

Fill and Spill

- Place a Counting Card on the table to indicate a target number, such as 3.
- Children put the target amount of cubes into a plastic container, and then spill the cubes to count to check there are still that many.
- Children may repeat the activity as they like. You may put a different Counting Card at another table.
- If scaffolding is needed, refer to Wednesday's Monitoring Student Progress.

Reflect

5

Ask children:

- **When does your family count to find out how many?**

Children might say: My parents count money; I count during games; and so on.

Assess

During Small Group activities, use the Small Group Record Sheet from ***Assessment*** to observe and record children's progress.

Wednesday Planner

Objectives

- To participate in rhythmic patterns
- To count verbally up to 10 with understanding
- To connect number words to the quantities they represent
- To make groups of up to five items

Materials

- *counters
- *round counters
- paper plates

Math Throughout the Year

Review activity directions at the top of page 35, and complete each in class whenever appropriate.

*provided in Manipulative Kit

Wednesday

1 Whole Group

Warm-Up: Count and Move in Patterns

- Spend just a few minutes counting in patterns of 4. Have all children count from 1 to 12. For example, say 1 quietly, 2 quietly, 3 quietly, 4 loudly, and so on.
- Add to the fun by marching along with the counting: 1 (step), 2 (step), 3 (step), 4 (stomp), and so on. Incorporate class themes when possible.

Demonstrate Counting

- Hide four counters in your hand. Tell children you have some counters in your hand, and ask them to count aloud with you to find out how many.
- Remove one of the counters, and place it where children can see and focus on it. Say "one" with the children; emphasize that "one" tells how many counters there are now.
- Repeat until you have counted and displayed all four counters. Then show your empty hand.
- Ask children how many counters there are in all. If they reply "four," agree and reiterate that, together, you counted four counters.
- Repeat with a different number of counters, making sure children count aloud with you. When possible, do this activity in small-group settings.

Compare Number Pizzas

- Remind children that you have to help a chef get the correct number of toppings on pizzas.
- Show your pizza (paper plate) with three pepperoni slices (round counters). Then show three more pizzas with one, two, and three pepperoni slices.
- Ask children which pizza has the same number of toppings as the first pizza you showed. Then ask how they knew the matching pizza had the same number of toppings.
- Repeat the activity, having children match pizzas with as many toppings as possible.

Eclipse Studios

2 Work Time 20

Computer Center Building Blocks

Continue to provide each child with a chance to complete Pizza Pizzazz 1.

Hands On Math Center

Provide both Fill and Spill and Find the Number for today's Hands On Math Center. If needed, consult the Weekly Planner for corresponding materials and Monday's and Tuesday's lessons for directions.

Monitoring Student Progress

If . . . children struggle with Fill and Spill,	**Then . . .** reduce the number of cubes, or place the cube collections on paper plates.
If . . . children excel at Fill and Spill,	**Then . . .** have them work in pairs, determining their own amounts and saying, for example, "Find eight."

3 Reflect 5

Ask children:

- **How did you figure out on the computer which pizza to choose?**

Children might say: I counted; I matched toppings; or I could just see how many there were.

4 Assess

Use the Weekly Record Sheet from ***Assessment*** to record children's progress. Use their time at the centers as an opportunity to complete your observations.

Thursday Planner

Objectives

- To participate in rhythmic patterns
- To count verbally up to 10 with understanding
- To make groups of up to five items
- To connect number words to the quantities they represent

Materials

- *counters
- *round counters
- paper plates
- drawing paper
- nontoxic markers

Math Throughout the Year

Review activity directions at the top of page 35, and complete each in class whenever appropriate.

Looking Ahead

You may choose to enlist an adult helper for today's Small Group. Select a counting book, such as Eric Carle's *The Very Hungry Caterpillar,* for Friday's Whole Group.

If you have not already, make copies of Family Letter Week 3 from the ***Teacher's Resource Guide*** for Friday.

*provided in Manipulative Kit

Thursday

1 Whole Group

Warm-Up: Count and Move in Patterns

- Spend just a few minutes counting in patterns of 3. Have all children count from 1 to 15. For example, say 1 loudly, 2 loudly, 3 quietly, 4 loudly, 5, loudly, 6 quietly, and so on.
- Add to the fun by marching along with the counting: 1 (stomp), 2 (stomp), 3 (step), and so on. Incorporate class themes when possible.

Number Me (5)

- Tell children to show you five of something on their body. They will most likely show their fingers on one hand. Ask how they can prove there are 5. Children might answer by actually counting to 5.
- Make silly statements, such as "I see six fingers on your hand," encouraging children to prove there are only five by counting.

2 Work Time

Small Group

Demonstrate Counting

- Hide five counters in your hand. Ask children to count aloud with you to find out how many.
- Remove one counter, and place it where children can see and focus on it.
- Repeat until you have counted and displayed all five counters. Then show your empty hand.
- Ask children how many counters there are in all. If they reply "five," agree and reiterate that, together, you counted five counters.
- Repeat with a different number of counters, making sure children count aloud with you.

Make Number Pizzas

- Show a plate with three counters on it to model children's target number.
- Make sure each pair of children has a paper plate and several counters. Tell each pair to make a pizza with the same number of toppings as yours.
- Ask children how they know there are three toppings on their pizzas. Discuss the various arrangements children made with their counters.
- Repeat with numbers up to 5 (or more as appropriate).

Monitoring Student Progress

If . . . children struggle with Make Number Pizzas,	**Then . . .** use fewer counters, or draw circles for children to put one counter in each until they can reliably count out 5.
If . . . children excel at Make Number Pizzas,	**Then . . .** increase the number of counters, and/or tell children the target number without showing it.

Computer Center Building Blocks

Continue to provide each child with a chance to complete Pizza Pizzazz 1.

Hands On Math Center

If still beneficial to children, continue to provide both Fill and Spill and Find the Number, this week's previous Hands On Math Center activities, and introduce the following activity:

Draw Numbers

- Building on a class theme, have children draw five items on paper.
- You may model this on a board or an overhead projector. It may help to have an adult nearby to remind children of the goal: to draw only five items.
- If scaffolding is needed, refer to Friday's Monitoring Student Progress.

3 Reflect

5 

Ask children:

■ **How do you make a pizza with four toppings?**

Children might say: I count; or I just know.

Have them show you how they "just know."

4 Assess

During Small Group activities, use the Small Group Record Sheet from *Assessment* to observe and record children's progress.

Friday Planner

Objectives

- To participate in rhythmic patterns
- To count verbally up to 10 with understanding
- To connect number words to the quantities they represent

Materials

no new materials

Math Throughout the Year

Review activity directions at the top of page 35, and complete each in class whenever appropriate.

Looking Ahead

For next week, familiarize yourself with Mystery Pictures 1: Match Shapes on the computer.

Friday

1 Whole Group 15

Warm-Up: Count and Move in Patterns

- Spend just a few minutes counting in patterns of 5. Have all children count from 1 to 15, for example, say 1 loudly, 2 loudly, 3 loudly, 4 loudly, 5 quietly, and so on.
- Add to the fun by marching along with the counting: 1 (stomp), 2 (stomp), 3 (stomp), 4 (stomp), 5 (step), and so on. Incorporate class themes when possible.

Counting Book

- Read aloud a counting book, such as *The Very Hungry Caterpillar,* in its entirety.
- Return to various pages, and ask children how many of a certain thing appears on those pages. Lead children in counting aloud to check.
- In a book like *The Very Hungry Caterpillar,* lead children in pretending to be the caterpillar, and ask how many berries they want to eat. For example, count five berries: 1 (bite and gulp), 2 (bite and gulp), 3 (bite and gulp), and so on. Adapt as needed for a character from another story.

2 Work Time 20

Computer Center Building Blocks

Continue to provide each child with a chance to complete Pizza Pizzazz 1.

Eclipse Studios

 Hands On Math Center

Based on what children continue to learn and benefit from most, choose from this week's Hands On Math Center activities: Draw Numbers, Fill and Spill, and Find the Number. If needed, consult Week 3's Weekly Planner for the corresponding materials, and refer to today's Monitoring Student Progress for scaffolding of Draw Numbers.

Monitoring Student Progress

If . . . children need help during Draw Numbers,	**Then . . .** provide class-themed stamps, or age-appropriate magazines, scissors, and paste for children to create five images on paper, or reduce the number children are to represent as needed.
If . . . children need a challenge during Draw Numbers,	**Then . . .** encourage them to make another drawing to start a number book.

 Reflect

5 

Briefly discuss with children what you have done in class this week, such as counting to 15.

Ask children:

■ **How did you find the number you were looking for?**

Children might say: I counted; or I could just see there were 4.

 Assess

Use the Weekly Record Sheet from *Assessment* to record children's progress. Use their time at the centers as an opportunity to complete your observations.

Home Connection

From the ***Teacher's Resource Guide,*** distribute to children copies of Family Letter Week 3 to share with their family. Each letter has an area for children to show families an example of what they have been doing in class.

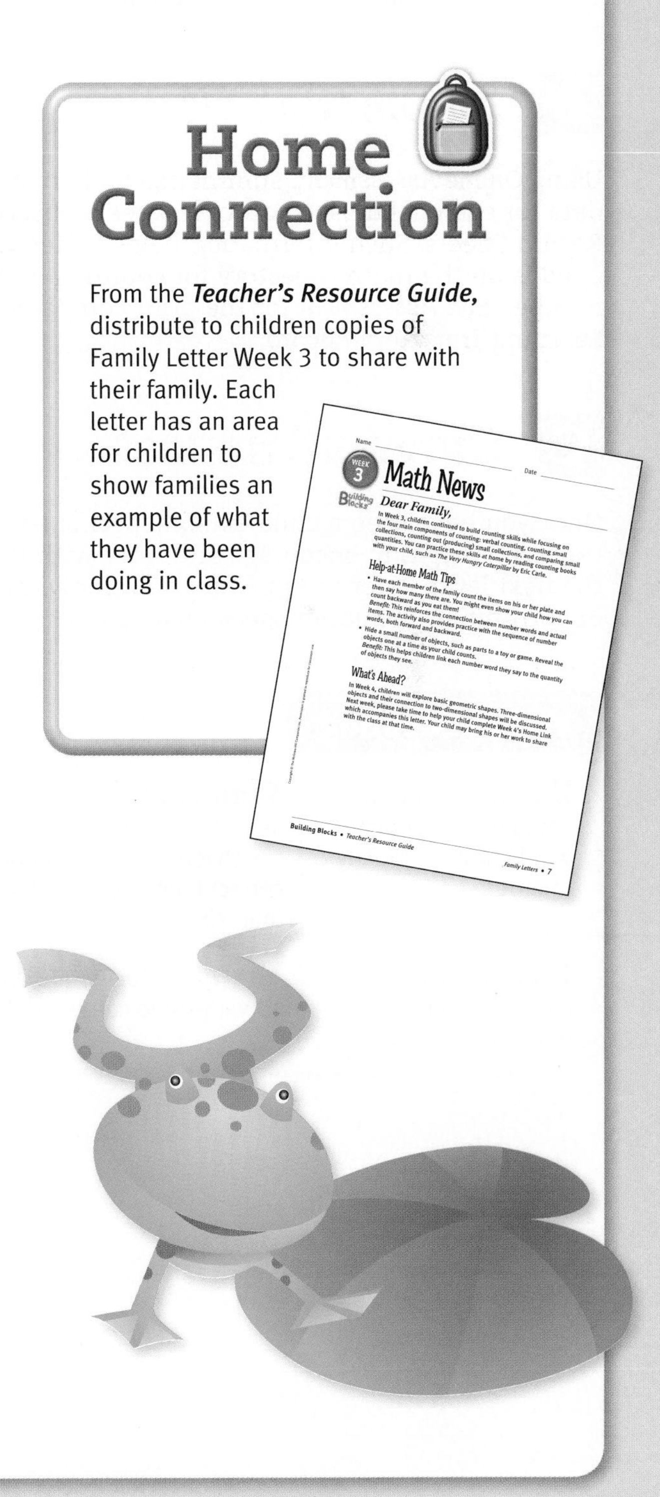

Name ______ Date ______

WEEK 3

Math News

Building Blocks

Dear Family,

In Week 3, children continued to build counting skills while focusing on the four main components of counting: verbal counting, counting small collections, counting out (producing) small collections, and comparing small quantities. You can practice these skills at home by reading counting books with your child, such as *The Very Hungry Caterpillar* by Eric Carle.

Help-at-Home Math Tips

- Have each member of the family count the items on his or her plate and then say how many there are. You might even show your child how you can count backward as you eat them! *Benefit:* This reinforces the connection between number words and actual items. The activity also provides practice with the sequence of number words, both forward and backward.
- Hide a small number of objects, such as parts to a toy or game. Reveal the objects one at a time as your child counts. *Benefit:* This helps children link each number word they say to the quantity of objects they see.

What's Ahead?

In Week 4, children will explore basic geometric shapes. Three-dimensional objects and their connection to two-dimensional shapes will be discussed. Next week, please take time to help your child complete Week 4's Home Link which accompanies this letter. Your child may bring his or her work to share with the class at that time.

Building Blocks • *Teacher's Resource Guide* *Family Letters* • 7

Assess and Differentiate

A Gather Evidence

Review children's progress in mathematics by looking at the Weekly Record Sheets (Monday, Wednesday, Friday) and the Small Group Record Sheets (Tuesday, Thursday) from this past week.

B Summarize Findings

Using ***Online Assessment***, summarize and analyze assessment data for each child based on your weekly observations and Record Sheets. Such information helps to determine where each child is on the math trajectory for counting and comparing number. See ***Assessment*** for the print companion to each Learning Trajectory Record.

C Differentiate Instruction

Once you have seen a child exhibit specific levels of the trajectory, begin to encourage and work with that child toward the next level. Refer to Appendix A for individualized instruction opportunities, including Special Education concerns.

Verbal Counting

If . . . the child can count using separate words that are mostly in order,	**Then . . .** *Reciter* Verbally counts using separate words ("one, two, three" not "onetwothree") but not always in the correct order, such as "one, two, three, four, six, seven."
If . . . the child can accurately recite number words up to 10,	**Then . . .** *Reciter (10)* Verbally counts to 10 with some correspondence to objects.

Object Counting

If . . . the child can count up to five objects with understanding,	**Then . . .** *Counter (Small Numbers)* Accurately counts objects in a line to 5, and answers the question "how many?" with the last number counted. When objects are visible, especially with small numbers, begins to understand cardinality.
If . . . the child can maintain one-to-one correspondence between number words and objects,	**Then . . .** *Corresponder* Keeps one-to-one correspondence between number words and objects (one word for each object), at least for small groups of objects in a row; may answer the question "how many?" by counting the objects again.
If . . . the child can produce up to five objects,	**Then . . .** *Producer (Small Numbers)* Counts out objects to 5.

Comparing Number

If . . . the child can accurately compare various-sized collections,	**Then . . .** *Perceptual Comparer* Compares collections that are quite different in size, such as one that is twice the size of the other (shown ten blocks and twenty-five blocks, points to the group with twenty-five as having more). If collections are similar, child can compare them accurately as well (with groups of two and four blocks, knows the group of four has more).

WEEK 4

Overview

Big Ideas

- matching shapes
- shape recognition
- counting

Teaching for Understanding

Learning about geometry and spatial concepts is crucial to young children's mathematical development. This week begins with matching, naming, and exploring shapes that are familiar to most children. The Shape Sets provided in this program will be an effective tool in children's exposure to shapes; the sets fulfill many roles in a variety of activities—from identification to comparison. Observing and engaging children in a dialogue about their ideas regarding shapes offer excellent assessment opportunities.

Geometry

Children inevitably discover that shapes combine to create their environment. By recognizing an equilateral triangle (a triangle with three equal sides) and seeing how it is related yet distinct from other triangles, children begin to find such triangles in their lives, for example, in traffic signs and structures. This curiosity leads to a broader, more accurate understanding of triangles.

Dimensions

Children are gently introduced to three-dimensional objects, such as cans or cylinders, as well as the relationship between two-dimensional (flat or plane) and three-dimensional (solid) shapes, such as two faces of a cylinder being circles. Because we live in a three-dimensional world, it is helpful for even very young children to start thinking and talking about both two- and three-dimensional shapes.

Meaningful Connections

Children who learn about shapes and spatial concepts perform better not only in geometry but in number, arithmetic, visualization, problem solving, and even language arts, especially vocabulary. Geometry and spatial thinking are important to children's math abilities because they involve understanding the space that children must live in, move in, and explore throughout their daily lives.

What's Ahead?

In the weeks to come, children will have the opportunity to build their shape knowledge by learning new shapes and shape attributes. Children develop a deeper comprehension of shape if they touch, manipulate, create with, and discuss shapes. Children begin to link shape knowledge to number knowledge.

Overview

How Children Learn about Shapes

Knowing where children are on the learning trajectory for geometry and counting and what the next steps are helps facilitate their development. Children who easily surpass these trajectory levels might be challenged by harder shapes and encouraged to assist other children.

Shapes: Matching

What to Look For What types of shapes can the child match accurately?

Shape Matcher, Identical Matches familiar shapes (circle, square, typical triangle) of same size and orientation.

Shape Matcher, Sizes Matches familiar shapes of different sizes.

Shape Matcher, Orientations Matches familiar shapes of different orientations.

Shapes: Naming

What to Look For Can the child differentiate among basic shapes of various orientations?

Shape Recognizer, Typical Recognizes and names typical circle, square, and, less often, triangle; some children correctly name different sizes, shapes, and orientations of rectangles but also call some shapes rectangles that look rectangular but are not rectangles.

Shape Recognizer, Circles, Squares, and Triangles+ Recognizes some less typical squares and triangles and may recognize some rectangles but usually not rhombuses (diamonds), often does not differentiate sides/corners.

Verbal Counting

What to Look For Can the child count with some one-to-one correspondence?

Reciter (10) Verbally counts to 10 with some correspondence to objects. For example, the child says: 1 (points to it), 2 (points to it), 3 (starts to point), 4 (finishes pointing but is pointing to third object), 5, and so on.

English Learner

Help English learners ask questions by modeling common question forms. Play games that provide practice. Review the following: *mystery, incorrect,* and *shuffle the papers.* Refer to the ***Teacher's Resource Guide.***

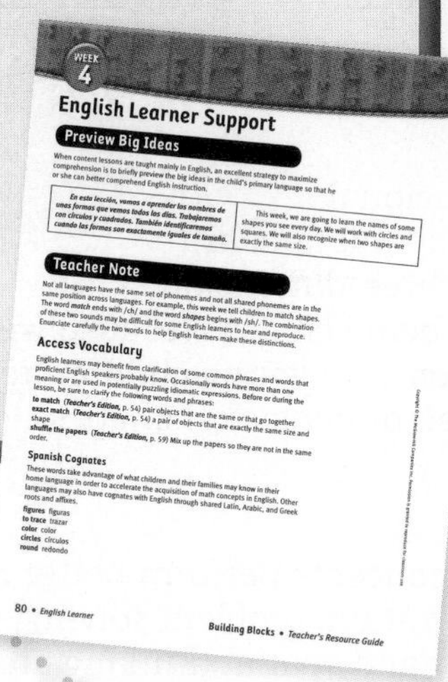

WEEK 4

English Learner Support

Preview Big Ideas

When content lessons are taught mainly in English, an excellent strategy to maximize comprehension is to briefly preview the big ideas in the child's primary language so that he or she can better comprehend English instruction.

En esta lección, vamos a aprender los nombres de unas formas que vemos todos los días. Trabajaremos con círculos y cuadrados. También identificaremos cuando las formas son exactamente iguales de tamaño.	This week, we are going to learn the names of some shapes you see every day. We will work with circles and squares. We will also recognize when two shapes are exactly the same size.

Teacher Note

Not all languages have the same set of phonemes and not all shared phonemes are in the same position across languages. For example, this week we tell children to match shapes. The word ***match*** ends with /ch/ and the word ***shapes*** begins with /sh/. The combination of these two sounds may be difficult for some English learners to hear and reproduce. Enunciate carefully the two words to help English learners make these distinctions.

Access Vocabulary

English learners may benefit from clarification of some common phrases and words that proficient English speakers probably know. Occasionally words have more than one meaning or are used in potentially puzzling idiomatic expressions. Before or during the lesson, be sure to clarify the following words and phrases:

to match (***Teacher's Edition,*** p. 54) pair objects that are the same or that go together

exact match (***Teacher's Edition,*** p. 54) a pair of objects that are exactly the same size and shape

shuffle the papers (***Teacher's Edition,*** p. 59) Mix up the papers so they are not in the same order.

Spanish Cognates

These words take advantage of what children and their families may know in their home language in order to accelerate the acquisition of math concepts in English. Other languages may also have cognates with English through shared Latin, Arabic, and Greek roots and affixes.

figures figuras
to trace trazar
color color
circles círculos
round redondo

80 • *English Learner*   **Building Blocks** • *Teacher's Resource Guide*

Technology Project

Children may get additional practice with shapes using Mystery Pictures Free Explore. Encourage children to drag as many shapes as they would like to create their own pictures. If children click the play button, they should ask a classmate to complete their picture. This is challenging because the classmate must get the correct shape in the correct orientation. Ask children to name their chosen shapes and why they chose them.

Math Throughout the Year

Math Throughout the Year activities are recommended to build on the mathematical skills highlighted in each week. Here are suggested activities for **Week 4.**

Counting Wand

Count each child with a gentle shoulder tap using the Counting Wand. Use the wand to count things at every opportunity. At this point, you should have already made your class' wand, but you might want to change its decorations periodically to correspond with class themes.

Dough Shapes

Children make inedible dough (or clay) shapes using plastic cookie cutters and other objects. Using the space from which a shape was cut, another child may replace the shape in its "hole" like a puzzle. When circles are introduced and studied, provide only various-sized circle cutters for children to use at those times.

Foam Puzzles

Children complete puzzles, observing how pieces have to be the same shape and size to fit. You may use other shape puzzles in addition to foam ones. Ideal puzzles, whether made or purchased, should include a variety of geometric shapes.

Center Preview

Computer Center Building Blocks

Get your classroom Computer Center ready for Mystery Pictures 1: Match Shapes from the ***Building Blocks*** software.

After you introduce Mystery Pictures 1, each child should complete the activity individually as you (or an assistant) monitor and guide him or her periodically. Ideally, each child will have at least ten minutes of computer time at least twice during the week. Use children's center time as an opportunity for assessment.

Hands On Math Center

This week's Hands On Math Center activities are Explore Shape Sets, Match Shape Sets, and Circles and Cans. Supply the center with these materials: Shape Sets, several food cans of various shapes and sizes, large paper, and crayons of the same color.

Literature Connections

These books help develop shape recognition.

- *Ten Black Dots* by Donald Crews
- *The Wing on a Flea* by Ed Emberley
- *What's the Shape?* by Judy Nayer
- *Round Things Everywhere* by Seymour Reit
- *Baby Bop Discovers Shapes* by Stephen White

(t)Matt Meadows (b)Eclipse Studios

Weekly Planner

Learning Trajectories

Week 4 Objectives

- To name familiar two-dimensional shapes, such as circles and squares
- To match the face of a three-dimensional object to its congruent two-dimensional outline
- To match congruent shapes
- To describe why certain figures are or are not circles
- To verbally count to at least 10

	Developmental Path	Instructional Activities
Shapes: Matching	*Shape Matcher, Identical*	Match and Name Shapes Circles and Cans
	Shape Matcher, Sizes	Match Shape Sets
	Shape Matcher, Orientations	Match and Name Shapes Match Shape Sets Mystery Pictures 1 Match Blocks
	Shape Matcher, Sizes and Orientations	
Shapes: Naming	*Shape Recognizer, Typical*	Match and Name Shapes
	Shape Recognzier, Circles, Squares, and Triangles+	Match and Name Shapes Circle Time! *Building Shapes* Circles and Cans Circle or Not?
	Shape Recognizer, All Rectangles	
Verbal Counting	*Reciter*	
	Reciter (10)	Count and Move in Patterns "Wake Up, Jack-in-the-Box"
	Counter Backward from 10	

Use this chart to plan for your specific class schedule. If you have your prekindergarteners for only three days, complete Monday, Tuesday, and Thursday of the week.

Work Time

Whole Group	Small Group	Computer	Hands On	Program Resources
Count and Move in Patterns **Match and Name Shapes** *Materials:* *Shape Sets		**Mystery Pictures 1**	**Explore Shape Sets** *Materials:* *Shape Sets	Building Blocks ***Assessment*** Weekly Record Sheet
Match Blocks *Materials:* blocks **Circle Time!** *Materials:* hula hoop	**Match and Name Shapes** *Materials:* *Shape Sets	**Mystery Pictures 1**	**Explore Shape Sets** *Materials:* *Shape Sets	Building Blocks ***Assessment*** Small Group Record Sheet
Building Shapes **Circles and Cans** *Materials:* various food cans large paper		**Mystery Pictures 1**	**Match Shape Sets** *Materials:* *Shape Sets **Circles and Cans** *Materials:* various food cans large paper	***Big Book Building Shapes*** Building Blocks ***Assessment*** Weekly Record Sheet
"Wake Up, Jack-in-the-Box" **Match Blocks** *Materials:* blocks	**Match and Name Shapes** *Materials:* *Shape Sets	**Mystery Pictures 1**	**Match Shape Sets** **Circles and Cans**	Building Blocks ***Assessment*** Small Group Record Sheet
"Circle" **Circle Time!** *Materials:* hula hoop **Circle or Not?**		**Mystery Pictures 1**	**Match Shape Sets** **Circles and Cans**	Building Blocks Family Letter Week 4 ***Assessment*** Weekly Record Sheet

*provided in Manipulative Kit

Monday Planner

Objectives

- To name familiar two-dimensional shapes, such as circles and squares
- To match congruent shapes
- To verbally count to at least 10

Materials

*Shape Sets

Vocabulary

When you *match* things, you find things that are the same.

Math Throughout the Year

Review activity directions at the top of page 51, and complete each in class whenever appropriate.

Looking Ahead

For tomorrow, ask children to bring in items with circle faces, and you should provide a hula hoop or other large circle.

*provided in Manipulative Kit

Monday

1 Whole Group

Warm-Up: Count and Move in Patterns

- In patterns of two, have all children count from 1 to 10, or another appropriate even number. For example, say 1 softly, 2 loudly (pause), 3 softly, 4 loudly (pause), and so on.
- When children are able, incorporate marching into the pattern.

Match and Name Shapes

- Sit in a circle with children. Give each child a shape from the Shape Set.
- Choose another Shape Set shape to match a child's shape exactly. Ask children to name who has an exact match for your shape.
- After a correct response is given, follow up by asking how the child knows his or her shape is a match. The child might offer to fit his or her shape on top of your shape to "prove" the match.
- Have children show their shapes to others seated near them, naming the shape whenever they can. Observe and assist as needed.
- Repeat once or twice. Afterward, tell children they will be able to explore and match shapes later during Work Time.

2 Work Time

Computer Center Building Blocks

Introduce Mystery Pictures 1: Match Shapes from the ***Building Blocks*** software. In this activity, children match shapes to outlines in order to construct a mystery picture; each mystery picture is revealed once the final correct shape has been placed. The software states each shape's name throughout the activity, thus teaching and reinforcing shape identification and vocabulary. Children typically love to guess what each picture is as they go. Each child should have an opportunity to complete Mystery Pictures 1 this week.

Monitoring Student Progress

If . . . children struggle with the computer activity,	**Then . . .** place corresponding Shape Set shapes for children to manipulate while at the computer.

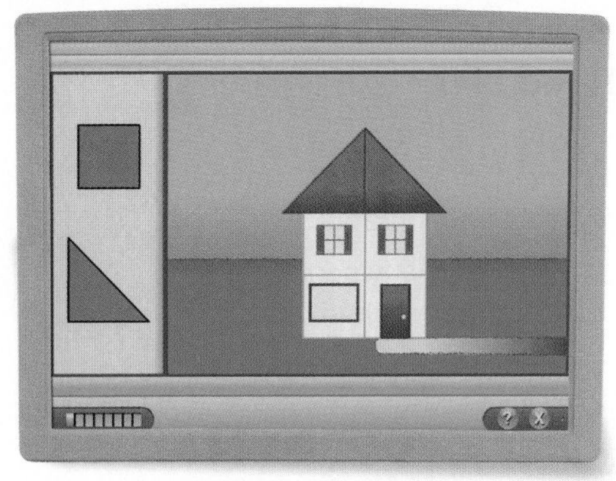

Mystery Pictures 1

Hands On Math Center

Explore Shape Sets

- Children investigate two complete Shape Sets. Observe what children do and say.
- If prompting is needed, encourage children to match shapes from one set to corresponding shapes from the other set to make pictures with the shapes (perhaps mimicking what they did on the computer), to feel shape contours and name them, and/or to place shapes in order by size (smallest to largest or largest to smallest).

RESEARCH IN ACTION

There is more than one way to determine that shapes are the same. Help children understand that one way to match shapes is when they are exactly the same size and same shape (congruent) and another way is to classify shapes into a category with the same attributes, such as triangles—a closed shape with three straight sides.

3 Reflect

5

Ask children:

- **How do you know whether two shapes are exactly the same size and shape?**

Children might say: By placing one shape on top of the other.

If children say "Look," follow up by asking them how they know for sure. If children match only one side, praise their thinking, but explain that sometimes one side matches while the others do not.

4 Assess

Use the Weekly Record Sheet from *Assessment* to record children's progress. Use their time at the centers as an opportunity to complete your observations.

Tuesday Planner

Objectives

- To name familiar two-dimensional shapes, such as circles and squares
- To describe why certain figures are or are not circles
- To match congruent shapes

Materials

- blocks
- hula hoop
- *Shape Sets

Math Throughout the Year

Review activity directions at the top of page 51, and complete each in class whenever appropriate.

Looking Ahead

For tomorrow, make sure you have clean cans in a variety of sizes, such as food cans (if empty, use tape to cover sharp edges). If necessary, ask children's families for donations. Preview ***Big Book Building Shapes*** as well.

*provided in Manipulative Kit

Tuesday

Whole Group

15 

Warm-Up: Match Blocks

- The goal is for children to match various block shapes to objects in the classroom. Have different block shapes in front of you with all the children in a circle around you.
- Show one face of a block, and ask children what things in the classroom are the same shape. Accept all reasonable answers, including those that name items which are not in the room itself but are indeed correct. Talk children through any incorrect responses, such as choosing something triangular but saying it has the shape of a quarter circle.
- Repeat with different-shaped blocks.

Circle Time!

- Have children sit in the best circle they can make. Show and name a large, flat circle, such as a hula hoop. As you trace the circle with your finger, discuss how it is perfectly round; it is a curved line that always curves the same.
- Ask children to talk about circles they know, such as those found in toys, buildings, books, tri- or bicycles, and clothing. Distribute a variety of circles for children's exploration—rolling, stacking, tracing, and so on.
- Have children make circles with their fingers, hands, arms, and mouths. Review a circle's attributes: round and curves the same without breaks.

Work Time

20 

Small Group

Match and Name Shapes

- Display a complete Shape Set.
- Choose a shape and show the entire small group. Ask them to find the exact match to your shape from the Shape Set on display. Follow up by asking why the shapes match. A child might offer to fit the shape on top of the other to prove its same size and shape.

- Now ask children to name the shape aloud; ask how they know what shape it is. To describe shapes, encourage children to feel a shape and its contours while their eyes are shut.
- Show a different shape. Ask children again to match and name the shape. After they do, ask how they know. Guide children in describing the shapes. Again, have them feel all around the shapes—discussing any curves, straight sides, and corners.

Monitoring Student Progress

If . . . children struggle during Match and Name Shapes,	**Then . . .** guide their hands around a shape to feel its full perimeter, and help children connect shapes to objects they know, such as circles to wheels and tires and rectangles to desks and doors.
If . . . children excel at Match and Name Shapes,	**Then . . .** name and discuss less familiar shapes, such as "skinny" triangles and rectangles, rhombuses (diamonds), and hexagons.

Computer Center

Continue to provide each child with a chance to complete Mystery Pictures 1.

Hands On Math Center

Continue to allow children to Explore Shape Sets from yesterday's Hands On Math Center.

3 Reflect

5

Ask children:

- **How do you know some shapes are circles and some are not?**

Children might say: Circles are all round, or otherwise explain that circles are perfectly round and continual.

4 Assess

During Small Group activities, use the Small Group Record Sheet from *Assessment* to observe and record children's progress.

Wednesday Planner

Objectives

- To match the face of a three-dimensional object to its congruent two-dimensional outline
- To describe why certain figures are or are not circles
- To match congruent shapes

Materials

- various food cans
- large paper
- *Shape Sets

Math Throughout the Year

Review activity directions at the top of page 51, and complete each in class whenever appropriate.

Looking Ahead

Using the food cans you gathered for today's class, trace their bases on large sheets of paper for Whole Group.

*provided in Manipulative Kit

Wednesday

1 Whole Group 20

Warm-Up: *Building Shapes*

Read ***Big Book Building Shapes*** to the class. Discuss familiar shapes, and introduce any new ones. Explain that a diamond can also be called a rhombus.

Circles and Cans

- Display several food cans, and discuss their shape (round) with children.
- Shift focus to the bottom and top, collectively the *bases,* of each can. Point out to children that these areas are circular; the edges are circles.
- Show the large sheets of paper on which you have traced the bases of a few cans that vary substantially in size. Trace one or two other cans to show children what you did, and then shuffle the papers and cans.
- Ask children to match the cans to the traced circles. For children who are unsure of their choice, have them place the can directly on the traced circle to check.
- Tell children they can all have a turn matching circles and cans at the Hands On Math Center, and store the activity's materials there.

2 Work Time 15

Computer Center Building Blocks

Continue to provide each child with a chance to complete Mystery Pictures 1.

Monitoring Student Progress

If . . .	Then . . .
If . . . a child has mastered Mystery Pictures 1,	**Then . . .** help him or her enter the activity's free explore level for additional matching and problem-posing practice.

 Hands On Math Center

Though both are new center activities, you may choose to focus on only one; both will continue the rest of this week.

Match Shape Sets

Using Shape Sets, children find exact matches (same size and shape).

Circles and Cans

Using the cans and papers from today's Whole Group, children match cans to traced circles. If an adult is available to help, children themselves can trace the cans, and then shuffle and match them.

Reflect

5 

Draw an oval, or show an oval from ***Big Book Building Shapes,*** and ask children:

- **Why is this shape *not* a circle?**

Children might say: Because it is like an egg; it is squashed, crushed, or the like; or it is not round all the way around.

Assess

Use the Weekly Record Sheet from ***Assessment*** to record children's progress. Use their time at the centers as an opportunity to complete your observations.

Thursday Planner

Objectives

- To name familiar two-dimensional shapes, such as circles and squares
- To match the face of a three-dimensional object to its congruent two-dimensional outline
- To match congruent shapes

Materials

- blocks
- *Shape Sets

Math Throughout the Year

Review activity directions at the top of page 51, and complete each in class whenever appropriate.

Looking Ahead

If you have not already, make copies of Family Letter Week 4 from the ***Teacher's Resource Guide*** for tomorrow.

*provided in Manipulative Kit

Thursday

1 Whole Group

10

Warm-Up: "Wake Up, Jack-in-the-Box"

Help children suit their actions to the finger play's words.

> Jack-in-the-Box, Jack-in-the-Box,
> Wake up, wake up, somebody knocks.
> One time, two times, three times, four,
> Jack pops out of his little round door.

Match Blocks

- Place different block shapes in front of you with all children in a circle around you.
- Show one block, and ask children what things in the classroom are the same shape. Remember to accept all reasonable answers, even those items which are not in the room but are correct. Talk children through incorrect responses, such as choosing a "square" that is actually rectangular and not square.
- Repeat with different-shaped blocks.

2 Work Time

25

Small Group

Match and Name Shapes

- Display a complete Shape Set.
- Choose a shape and show the entire small group. Ask them to find the exact match to your shape from the Shape Set on display. Follow up by asking why the shapes match. A child might offer to fit the shape on top of the other to prove its same size and shape.
- Now ask children to name the shape aloud; ask how they know what shape it is. To describe shapes, encourage children to feel a shape and its contours while their eyes are shut.
- Show a different shape. Ask children again to match and name the shape. After they do, ask how they know. Guide children in describing the shapes. Again, have them feel all around the shapes—discussing any curves, straight sides, and corners.

Monitoring Student Progress

If . . . children struggle during Match and Name Shapes,	**Then . . .** guide their hands around a shape to feel its full perimeter, and help children connect shapes to objects they know, such as circles to wheels and tires and rectangles to desks and doors.
If . . . children excel at Match and Name Shapes,	**Then . . .** name and discuss less familiar shapes, such as "skinny" triangles and rectangles, rhombuses (diamonds), and hexagons.

Computer Center Building Blocks

Continue to provide each child with a chance to complete Mystery Pictures 1.

Hands On Math Center

Based on what they benefit from more, continue to allow children to Match Shape Sets and/or complete the Circle and Cans activity from yesterday's Hands On Math Center.

3 Reflect

5

Ask children:

- **How did you match circles (or other shapes)?**

Children might say: I put the shape on other shapes until I found its match.

You may choose to follow up by asking:

- **If all shapes were circles, how could you match them?**

Children might say: Some are bigger (or smaller), and you have to find the circle that is the right size.

4 Assess

During Small Group activities, use the Small Group Record Sheet from *Assessment* to observe and record children's progress.

Friday Planner

Objectives

- To match the face of a three-dimensional object to its congruent two-dimensional outline
- To describe why certain figures are or are not circles
- To match congruent shapes

Materials

hula hoop

Math Throughout the Year

Review activity directions at the top of page 51, and complete each in class whenever appropriate.

Looking Ahead

For next week, familiarize yourself with Mystery Pictures 2: Name Shapes and Number Snapshots 1: 1–3 Dots from the ***Building Blocks*** software.

Friday

1 Whole Group

20

Warm-Up: "Circle"

Here are the finger play's words and actions:

> Here's a circle, *(Make a circle with index finger and thumb.)*
> And here's a circle, *(Make a circle by touching both index fingers and thumbs together.)*
> And a great big circle I see! *(Make a circle with both arms.)*
> Now let's count the circles... 1, 2, 3. *(Make each corresponding circle again as you count.)*

Circle Time!

- As children sit in the best circle they can make, show and name a large, flat circle, such as a hula hoop. Trace the circle with your finger, explaining that it is perfectly round and it is a curved line that always curves the same.
- Ask children to talk about circles they know, such as those found in toys, buildings, books, and so on. Distribute a variety of circles for children's exploration.
- Have children make circles with their fingers, hands, arms, and mouths. Review a circle's attributes: round and curves the same without breaks.

Circle or Not?

- Draw a true circle on a surface where the entire class can view it. Ask children to name it, and then tell why it is a circle.
- Draw an ellipse (an oval) on the same surface. Ask children what it looks like, and then ask them to tell why it is not a circle.
- Draw several other circles and shapes that are not circles but could be mistaken for them, and discuss their differences.
- Summarize by reviewing that a circle is perfectly round and consists of a curved line that always curves the same.

Monitoring Student Progress

If . . . children need help during Circle or Not?	**Then . . .** have children use their fingers to trace shapes and/or model desired responses, and make non-circles very obvious, such as long ovals and shapes that are not closed.

Eclipse Studios

Work Time

Computer Center Building Blocks

Continue to provide each child with a chance to complete Mystery Pictures 1.

Hands On Math Center

Based on what they benefit from more, continue to allow children to Match Shape Sets and/or complete the Circle and Cans activity from this week's previous Hands On Math Center.

Reflect

Briefly discuss with children what you have done in class this week, such as identifying circles, and ask:

- **How do you know for sure something is a circle?**

Children might say: It has to be just round, it has to be turning the same all around, or the like.

Assess

Use the Weekly Record Sheet from *Assessment* to record children's progress. Use their time at the centers as an opportunity to complete your observations.

Home Connection

From the ***Teacher's Resource Guide,*** distribute to children copies of Family Letter Week 4 to share with their family. Each letter has an area for children to show families an example of what they have been doing in class.

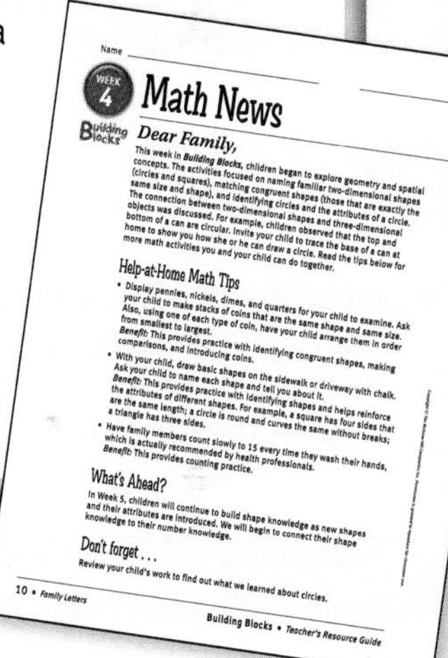

Name

WEEK 4 **Math News**

Building Blocks

Dear Family,

This week in ***Building Blocks***, children began to explore geometry and spatial concepts. The activities focused on naming familiar two-dimensional shapes (circles and squares), matching congruent shapes (those that are exactly the same size and shape), and identifying circles and the attributes of a circle. The connection between two-dimensional shapes and three-dimensional objects was discussed. For example, children observed that the top and bottom of a can are circular. Invite your child to trace the base of a can at home to show you how she or he can draw a circle. Read the tips below for more math activities you and your child can do together.

Help-at-Home Math Tips

- Display pennies, nickels, dimes, and quarters for your child to examine. Ask your child to make stacks of coins that are the same shape and same size. Also, using one of each type of coin, have your child arrange them in order from smallest to largest. *Benefit:* This provides practice with identifying congruent shapes, making comparisons, and introducing coins.
- With your child, draw basic shapes on the sidewalk or driveway with chalk. Ask your child to name each shape and tell you about it. *Benefit:* This provides practice with identifying shapes and helps reinforce the attributes of different shapes. For example, a square has four sides that are the same length; a circle is round and curves the same without breaks; a triangle has three sides.
- Have family members count slowly to 15 every time they wash their hands, which is actually recommended by health professionals. *Benefit:* This provides counting practice.

What's Ahead?

In Week 5, children will continue to build shape knowledge as new shapes and their attributes are introduced. We will begin to connect their shape knowledge to their number knowledge.

Don't forget . . .

Review your child's work to find out what we learned about circles.

10 • *Family Letters* Building Blocks • *Teacher's Resource Guide*

WEEK 4

Wrap-Up

Assess and Differentiate

A Gather Evidence

Review children's progress in mathematics by looking at the Weekly Record Sheets (Monday, Wednesday, Friday) and the Small Group Record Sheets (Tuesday, Thursday) from this past week.

B Summarize Findings

Using *Online Assessment,* summarize and analyze assessment data for each child based on your weekly observations and Record Sheets. Such information helps to determine where each child is on the math trajectory for geometry and counting. See *Assessment* for the print companion to each Learning Trajectory Record.

C Differentiate Instruction

Once you have seen a child exhibit specific levels of the trajectory, begin to encourage and work with that child toward the next level. Refer to Appendix A for individualized instruction opportunities, including Special Education concerns.

Shapes: Matching

If . . .	Then . . .
If . . . If the child can match basic shapes of the same size and orientation,	**Then . . .** *Shape Matcher, Identical* Matches familiar shapes (circle, square, typical triangle) of same size and orientation.
If . . . the child can match basic shapes of various sizes,	**Then . . .** *Shape Matcher, Sizes* Matches familiar shapes of different sizes.
If . . . the child can match basic shapes of different orientations,	**Then . . .** *Shape Matcher, Orientations* Matches familiar shapes of different orientations.

Shapes: Naming

If . . .	Then . . .
If . . . the child can identify basic shapes, some of different size and orientation,	**Then . . .** *Shape Recognizer, Typical* Recognizes and names typical circle, square, and, less often, triangle; some children correctly name different sizes, shapes, and orientations of rectangles but also call some shapes rectangles that look rectangular but are not rectangles.
If . . . the child can identify some less typical shapes,	**Then . . .** *Shape Recognizer, Circles, Squares, and Triangles+* Recognizes some less typical squares and triangles and may recognize some rectangles but usually not rhombuses (diamonds), often does not differentiate sides/corners.

Verbal Counting

If . . .	Then . . .
If . . . the child can accurately recite number words up to 10,	**Then . . .** *Reciter (10)* Verbally counts to ten with some correspondence to objects. For example, the child says: 1 (points to it), 2 (points to it), 3 (starts to point), 4 (finishes pointing but is pointing to third object), 5, and so on.

WEEK 5

Overview

Big Ideas

- recognizing two-dimensional shapes
- distinguishing among two-dimensional shapes
- subitizing

Teaching for Understanding

As explained last week, geometry and spatial concepts are important mathematical topics. They form the foundation of much mathematical learning, as well as other subjects. Spatial concepts are essential to daily living, such as understanding directions ("Please put that toy on top of the shelf under the windows.") and maps in social studies, visual arts, sciences, and so on.

Geometric Shapes

Building on children's matching and exploration of familiar shapes last week, this week's tasks require children to name such shapes, find them in the environment, and build them from parts. Children will also begin to discriminate among shapes, comparing those that are related but not true members of a class. For example, a triangular shape with curved sides is not actually a triangle.

Shapes in Life

One of the most important associations we help children make is that with their everyday life. Seeing shapes in everyday things helps children better understand geometry but also helps them see the "everyday things" in new ways. Similarly, when children discover how they can use two- and three-dimensional shapes to represent everyday objects, it helps them see the spatial structure of their world. For example, their teacher's desk is a rectangular shape.

Meaningful Connections

We link geometry to number in several ways. We count the sides of a shape to check that it has the correct attributes. We determine the shapes in a class, such as triangles, which involves problem solving, reasoning, and communication. We use appropriate vocabulary early on. Mathematics makes a special contribution to language because it is an area in which children can meaningfully and enthusiastically talk about the definitions and significance of words.

What's Ahead?

In the weeks to come, children will return to solving number problems, matching groups by number, counting to find how many in a group, and making groups of a given number. The focus of such skills and activities will be numbers up to 5.

Overview

How Children Learn about Shapes

Knowing where children are on the learning trajectory for geometry and what the next steps are helps facilitate their development. Children who easily surpass these trajectory levels might be challenged by less familiar shapes and encouraged to assist other children.

Shapes: Naming

What to Look For Does the child accurately recognize less typical examples of shapes?

Shape Recognizer, All Rectangles Recognizes more rectangle sizes, shapes, and orientations.

Shape Recognizer, More Shapes Recognizes most familiar shapes and typical examples of other shapes, such as hexagon, rhombus (diamond), and trapezoid.

Shapes: Representing

What to Look For Can the child make a goal shape using manipulatives that act as parts of a shape?

Constructor of Shapes from Parts, Looks Like Uses manipulatives representing parts of shapes, such as sides, to make a shape that "looks like" a goal shape.

Object Counting

What to Look For How many items can the child count accurately?

Counter (Small Numbers) Accurately counts objects in a line to 5, and answers "how many?" with the last number counted. When objects are visible, especially with small numbers, begins to understand cardinality.

Producer (Small Numbers) Counts out objects to 5.

Recognition of Number and Subitizing

What to Look For Does the child instantly recognize small groups?

Maker of Small Collections Nonverbally makes a small collection (no more than 4, usually up to 3) using the number of another collection, such as a mental or verbal model, not necessarily by matching.

Perceptual Subitizer to 4 Instantly recognizes collections up to 4 and verbally names the number of items.

English Learner

Building background accelerates English acquisition. Show English learners a Jack-in-the-Box, and discuss what it does. Review *uncover, angle,* and *opposite.*

Refer to the ***Teacher's Resource Guide.***

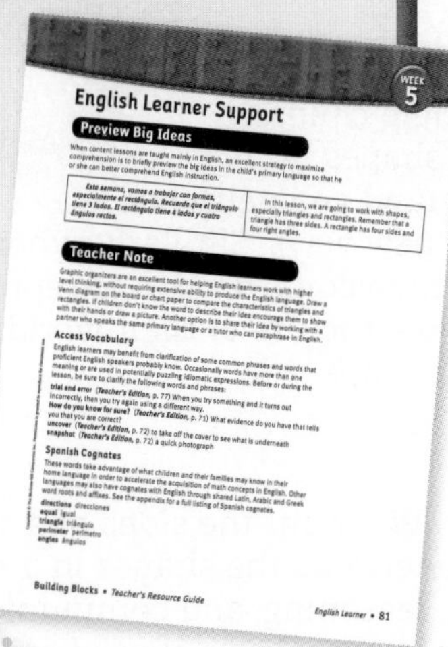

WEEK 5

English Learner Support

Preview Big Ideas

When content lessons are taught mainly in English, an excellent strategy to maximize comprehension is to briefly preview the big ideas in the child's primary language so that he or she can better comprehend English instruction.

Esta semana, vamos a trabajar con formas, especialmente el rectángulo. Recuerda que el triángulo tiene 3 lados. El rectángulo tiene 4 lados y cuatro ángulos rectos.	In this lesson, we are going to work with shapes, especially triangles and rectangles. Remember that a triangle has three sides. A rectangle has four sides and four right angles.

Teacher Note

Graphic organizers are an excellent tool for helping English learners work with higher level thinking, without requiring extensive ability to produce the English language. Draw a Venn diagram on the board or chart paper to compare the characteristics of triangles and rectangles. If children don't know the word to describe their idea encourage them to show with their hands or draw a picture. Another option is to share their idea by working with a partner who speaks the same primary language or a tutor who can paraphrase in English.

Access Vocabulary

English learners may benefit from clarification of some common phrases and words that proficient English speakers probably know. Occasionally words have more than one meaning or are used in potentially puzzling idiomatic expressions. Before or during the lesson, be sure to clarify the following words and phrases:

trial and error (*Teacher's Edition,* p. 77) When you try something and it turns out incorrectly, then you try again using a different way.
How do you know for sure? (*Teacher's Edition,* p. 71) What evidence do you have that tells you that you are correct?
uncover (*Teacher's Edition,* p. 72) to take off the cover to see what is underneath
snapshot (*Teacher's Edition,* p. 72) a quick photograph

Spanish Cognates

These words take advantage of what children and their families may know in their home language in order to accelerate the acquisition of math concepts in English. Other languages may also have cognates with English through shared Latin, Arabic and Greek word roots and affixes. See the appendix for a full listing of Spanish cognates.

directions direcciones
equal igual
triangle triángulo
perimeter perímetro
angles ángulos

Building Blocks • *Teacher's Resource Guide*

English Learner • 81

Technology Project

Children may get additional practice with shapes and making pictures with shapes using Mystery Pictures Free Explore. Encourage children to drag as many shapes as they would like to create their own pictures. If children click the play button, they should ask a friend to complete their picture. This is challenging because the friend must get the correct shape in the correct orientation. Ask children to name the shapes and why they chose them.

Math Throughout the Year

Math Throughout the Year activities are recommended to build on the mathematical skills highlighted in each week. Here are suggested activities for **Week 5.**

Name Faces of Blocks

During circle or free time, children name the faces (sides) of different building blocks. Tell or ask children which classroom items are the same shape.

Shape Walk

Go for a walk outside of the classroom to search for a specific shape. Consider taking with you some shapes from the Shape Sets to provide a quick and easy reference for children.

Center Preview

Computer Center Building Blocks

Get your classroom Computer Center ready for Mystery Pictures 2: Name Shapes and Number Snapshots 1: 1–3 Dots from the ***Building Blocks*** software.

After you introduce Mystery Pictures 2 and Number Snapshots 1, each child should complete the activities individually as you (or an assistant) monitor and guide him or her periodically. Ideally, each child will have at least ten minutes of computer time at least twice during the week. Use children's center time as an opportunity for assessment.

Hands On Math Center

This week's Hands On Math Center activities are Shape Hunt and Straw Shapes, each focusing on one of three target shapes per day. Supply the center with these materials: common objects, Shape Sets, colored tape, and plastic beverage stirrers cut to various lengths.

Literature Connections

These books help develop shape recognition.

- *What Is Square?* by Rebecca K. Dotlich
- *Village of Round and Square Houses* by Ann Grifalconi
- *Harold and the Purple Crayon* by Crockett Johnson
- *The Silly Story of Goldie Locks and the Three Squares* by Grace Maccarone
- *There's a Square: A Book about Shapes* by Mary

(t)Matt Meadows (b)Eclipse Studios

Weekly Planner

Learning Trajectories

Week 5 Objectives

- To locate, name, and build familiar two-dimensional shapes, including triangles, rectangles, and squares
- To distinguish between visually-similar non-examples of familiar two-dimensional shapes
- To name the number of objects in a group up to 3

	Developmental Path	Instructional Activities
Shapes: Naming	*Shape Recognizer, Circles, Squares, and Triangles+*	
	Shape Recognizer, All Rectangles	Shape Show (rectangles) Mystery Pictures 2
	Shape Recognizer, More Shapes	***Building Shapes*** Shape Show (all) Mystery Pictures 2 Shape Hunt Is It or Not?
	Shape Identifier	
Shapes: Representing	*Constructor of Shapes from Parts, Looks Like*	Straw Shapes
	Constructor of Shapes from Parts, Exact	
Object Counting	*Corresponder*	
	Counter (Small Numbers)	Shape Hunt
	Producer (Small Numbers)	"Wake Up, Jack-in-the-Box"
	Counter (10)	
Subitizing	*Maker of Small Collections*	"Wake Up, Jack-in-the-Box"
	Perceptual Subitizer to 4	Snapshots Number Snapshots 1
	Perceptual Subitizer to 5	

Use this chart to plan for your specific class schedule. If you have your prekindergarteners for only three days, complete Monday, Tuesday, and Thursday of the week.

Pacing

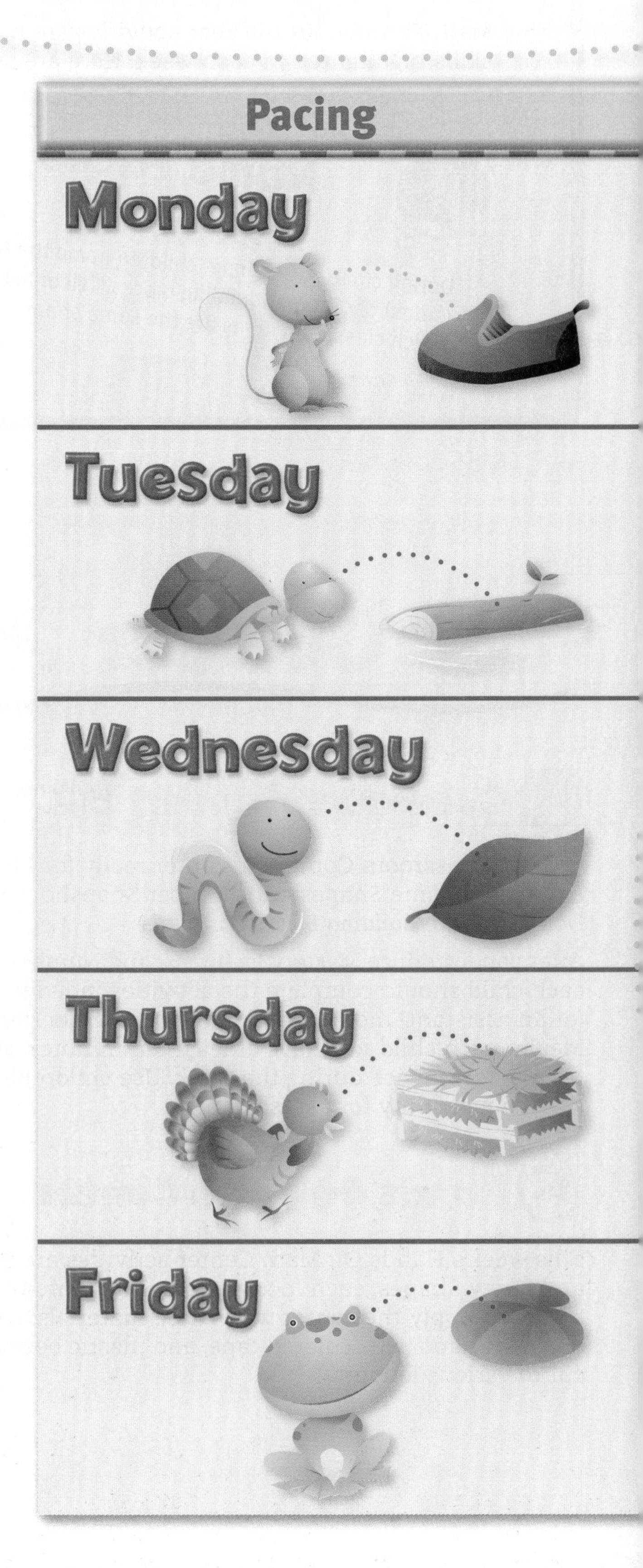

Work Time

Whole Group	Small Group	Computer	Hands On	Program Resources
Building Shapes **Shape Show** *Materials:* big, flat triangle triangular items colored tape		**Mystery Pictures 2**	**Triangles** **Shape Hunt** *Materials:* *Shape Set triangles **Straw Shapes** *Materials:* plastic stirrers	***Big Book Building Shapes*** Building Blocks ***Assessment*** Weekly Record Sheet
"Wake Up, Jack-in-the-Box" **Snapshots** *Materials:* *counters paper plate dark cloth	**Straw Shapes** *Materials:* plastic stirrers **Is It or Not?** *Materials:* plastic stirrers	**Number Snapshots 1**	**Triangles** **Shape Hunt** **Straw Shapes**	Building Blocks ***Teacher's Resource Guide*** **Straws** ***Assessment*** Small Group Record Sheet
Building Shapes **Shape Show** *Materials:* big, flat rectangle rectangular items		• **Mystery Pictures 2** • **Number Snapshots 1**	**Rectangles** **Shape Hunt** **Straw Shapes** *Materials:* plastic stirrers	***Big Book Building Shapes*** Building Blocks ***Assessment*** Weekly Record Sheet
"Wake Up, Jack-in-the-Box" **Snapshots** *Materials:* *counters paper plate dark cloth	**Straw Shapes** *Materials:* plastic stirrers **Is It or Not?** *Materials:* plastic stirrers	• **Mystery Pictures 2** • **Number Snapshots 1**	**Rectangles** **Shape Hunt** **Straw Shapes**	Building Blocks ***Assessment*** Small Group Record Sheet
Building Shapes **Shape Show** *Materials:* big, flat square square items		• **Mystery Pictures 2** • **Number Snapshots 1**	**Squares** **Shape Hunt** **Straw Shapes** *Materials:* plastic stirrers	***Big Book Building Shapes*** Building Blocks ***Teacher's Resource Guide*** Family Letter Week 5 ***Assessment*** Weekly Record Sheet

*provided in Manipulative Kit

Monday Planner

Objectives

- To locate, name, and build familiar two-dimensional shapes, including circles, squares, and rectangles
- To distinguish between visually-similar non-examples of familiar two-dimensional shapes

Materials

- big, flat triangle
- triangular items
- colored tape
- *Shape Set triangles
- plastic stirrers

Vocabulary

angle two lines that are connected to make a corner

triangle a closed shape that has three straight sides

Math Throughout the Year

Review activity directions at the top of page 67, and complete each in class whenever appropriate.

Looking Ahead

Cut plastic beverage stirrers to match the straws on ***Teacher's Resource Guide*** page 177 with at least two of each length and, if possible, enlist an adult helper for today's Hands On Math Center. For tomorrow, ask children to bring in items with triangle faces.

*provided in Manipulative Kit

Monday

1 Whole Group

15

Warm-Up: *Building Shapes*

- Read the page about triangles from ***Big Book Building Shapes*** to children.
- Together, look for and identify triangles in the classroom.

Shape Show: Triangles

- Show and name a large, flat triangle. Walk your fingers around its perimeter, describing and exaggerating your actions: straaiiight side...turn, straaiiight side...turn, straaiiight side...stop.
- Ask children how many sides the triangle has, and count the sides with them. Emphasize that a triangle's sides and angles can be different sizes; what matters is that its sides are straight and connected to make a closed shape (no openings or gaps).
- Ask children what things they have at home that are triangles. Show different examples of triangles.
- Have children draw triangles in the air. If available, have children walk around a large triangle, such as one marked with colored tape on the floor.

Monitoring Student Progress

If . . . children need help during Shape Show,	**Then . . .** guide their hands to feel around the full perimeter of the triangle, and count its sides again.
If . . . children need a challenge during Shape Show,	**Then . . .** name and discuss less typical shapes (you may use Shape Sets), including "skinny" or "fat" triangles and all triangles shown in different orientations (turned different ways).

2 Work Time

20

 Computer Center

Building Blocks

Introduce Mystery Pictures 2: Name Shapes from the ***Building Blocks*** software. In this activity, children click on the shape that is verbally named to construct more mystery pictures; thus they need to identify, not just match, shapes. Encourage children to guess what each picture is if they are not already. Each child should have an opportunity to complete Mystery Pictures 2 this week.

Mystery Pictures 2

 Hands On Math Center

These activities occur all week with a focus on a specific shape each day.

Shape Hunt: Triangles

- Tell children to find one or two items in the room with at least one triangle face. For variety, hide Shape Set triangles throughout the room beforehand.
- Encourage children to count the shape's sides and, if possible, show the triangle to an adult, discussing its shape. For example, triangles have three sides, but the sides are not always the same length. After discussion, have the child replace the triangle so other children can find it.
- You may choose to photograph the triangles for a class shape book.

Straw Shapes: Triangles

Children use plastic stirrers to make triangles and/or to create pictures and designs that include triangles.

RESEARCH IN ACTION

Triangles must have three straight lines and be closed. For example, a child sees a musical triangle, which typically has curved corners with one opening, and calls it a triangle; explain that, though it is triangular, it is not a true triangle based on the aforementioned attributes.

3 Reflect

5

Ask children:

■ **How do you know for sure a shape is a triangle?**

Children might say: It has to have three straight sides and no holes.

Show a triangle, and ask:

■ **Is this a triangle? Why or why not?**

Children might say: Yes, because it has three straight sides; or I see three sides, and they are all together, closed, or the like.

4 Assess

Use the Weekly Record Sheet from ***Assessment*** to record children's progress. Use their time at the centers as an opportunity to complete your observations.

Tuesday Planner

Objectives

- To locate, name, and build familiar two-dimensional shapes, including triangles, rectangles, and squares
- To distinguish between visually-similar non-examples of familiar two-dimensional shapes
- To name the number of objects in a group up to 3

Materials

- *counters
- paper plate
- dark cloth
- plastic stirrers

Math Throughout the Year

Review activity directions at the top of page 67, and complete each in class whenever appropriate.

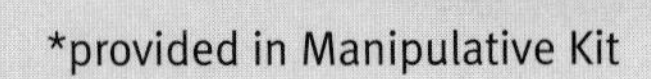
*provided in Manipulative Kit

Tuesday

Whole Group

Warm-Up: "Wake Up, Jack-in-the-Box"

Help children suit their actions to the following words:

Jack-in-the-Box, Jack-in-the-Box,
Wake up, wake up, somebody knocks.
One time, two times, three times, four,
Jack pops out of his little round door.

Snapshots

- Have prepared two counters on a paper plate (or construction paper), and cover the plate with a cloth you cannot see through.
- Ask children to tell about times when their family took photographs or "snapshots" with a camera. Tell children when they play Snapshots they will use their eyes and minds like a camera by "taking a snapshot" of what they see.
- Show your covered plate to children, and explain that counters are hidden under the cloth. Tell children to watch carefully, quietly, and with hands in their laps as you quickly uncover the plate to expose the counters. Uncover your plate for two seconds (count silently to yourself, "one thousand one, one thousand two"), and cover it again.
- Have children show with their fingers how many counters they saw, and, once you have seen all children's fingers, ask how many counters they saw. Repeat the uncovering if needed.
- Uncover the plate indefinitely and, to check their answers, ask children what they see. Allow children to say their answers (as opposed to showing a physical response), and recognize their contributions.
- Depending on children's ability, repeat the activity with up to four counters lined up.

Work Time

Small Group

Straw Shapes

- Children use the stirrers you cut based on ***Teacher's Resource Guide*** page 177 to make shapes. Ensure that they build shapes with correct attributes, such as all sides the same length and all right angles for squares. All stirrers should be "connected" (touching) at their endpoints. Discuss attributes as children build.

- If children need help, provide a model for them to copy or a drawing on which to place stirrers.

Is It or Not?

- Tell children you are going to try to fool them. Say *triangle,* and show a drawing or a straw shape of your own where all can see it. Ask children whether the shape is a triangle; then ask why or why not.
- Repeat with shapes that are true triangles and some that are not.
- Review that triangles have three straight sides that are closed.
- Repeat the activity using rectangles and shapes that are not rectangles. Review the attributes of rectangles.

Monitoring Student Progress

If . . . children struggle during Is It or Not?	**Then . . .** help them trace various triangles with their fingers, comparing triangle attributes to the attributes of shapes that are not triangles.
If . . . children excel during Is It or Not?	**Then . . .** challenge them to describe exactly why a triangle is a triangle, naming all its attributes. Show triangles that are slightly incorrect, and discuss what is wrong with each.

Computer Center Building Blocks

Introduce Number Snapshots 1: 1–3 Dots from the ***Building Blocks*** software, where children review groups up to three dots to match to another group. Each child should complete the activity this week.

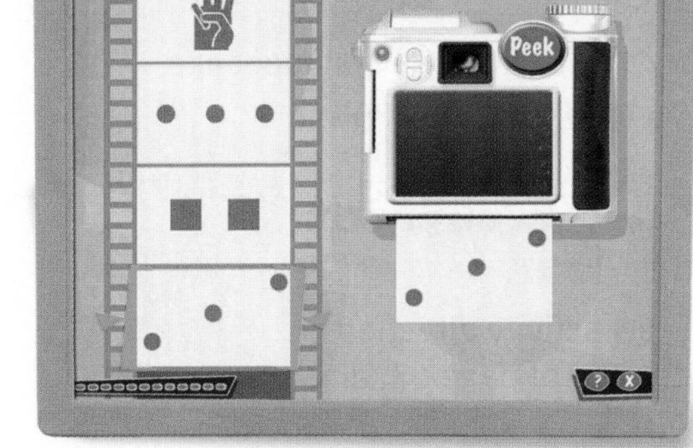

Number Snapshots 1

Hands On Math Center

Based on what they benefit from more, have children complete Shape Hunt or Straw Shapes, focusing on triangles, from yesterday's Hands On Math Center.

3 Reflect 5

Review with children, and ask:

■ **How do you know for sure a shape is a triangle?**

Children might say: It has to have three straight sides all closed.

4 Assess

During Small Group activities, use the Small Group Record Sheet from ***Assessment*** to observe and record children's progress.

Eclipse Studios

Wednesday Planner

Objectives

- To locate, name, and build familiar two-dimensional shapes, including triangles, rectangles, and squares
- To distinguish between visually-similar non-examples of familiar two-dimensional shapes
- To name the number of objects in a group up to 3

Materials

- big, flat rectangle
- rectangular items
- plastic stirrers

Vocabulary

rectangle a shape that has four sides and all right angles; opposite sides are the same length and parallel

right angle two lines that meet like the corner of a doorway

Math Throughout the Year

Review activity directions at the top of page 67, and complete each in class whenever appropriate.

Looking Ahead

If possible, for today's Hands On Math Center, enlist an adult helper. For tomorrow, ask children to bring in items with rectangle faces.

Wednesday

1 Whole Group

Warm-Up: *Building Shapes*

- Read the page about rectangles from ***Big Book Building Shapes*** to children.
- Together, look for and identify rectangles in the classroom.

Shape Show: Rectangles

- Show and name a large, flat rectangle. Walk your fingers around its perimeter, describing and exaggerating your actions: straaiiight side...turn, straaiiight side...turn, straaiiight side...turn, straaiiight side...stop.
- Ask children how many sides the rectangle has, and count the sides with them. Emphasize that opposite sides of a rectangle are the same lengths, and all "turns" are right angles. To model this, you may place a stirrer that is the same length as one pair of sides on top of each of those sides, and repeat for the other pair of opposite sides.
- To illustrate right angles, talk about the angle—like an uppercase *L*—in a doorway. Make uppercase *L*s with children using thumbs and index fingers. Fit your *L* on the angles of the rectangle.
- Ask children what things they have at home that are rectangles. Show different examples of rectangles.
- Have children walk around a large, flat rectangle, such as a rug. Once seated, have children draw rectangles in the air.

2 Work Time

Computer Center Building Blocks

Continue to provide each child with a chance to complete Mystery Pictures 2 and Number Snapshots 1.

 Hands On Math Center

Shape Hunt: Rectangles

- Tell children to find an item in the room with at least one rectangle face.
- Encourage children to count the shape's sides and, if possible, show the rectangle to an adult, discussing its shape. For example, rectangles have four sides and four right angles. After discussion, have the child replace the rectangle so other children can find it.
- You may choose to photograph the rectangles for a class shape book.

Straw Shapes: Rectangles

Children use plastic stirrers to make rectangles and/or to create pictures and designs that include rectangles.

Monitoring Student Progress

If . . . children need help during Straw Shapes,	**Then . . .** guide their hands to feel the sides of a rectangle and/or provide a rectangular model for them to copy or a rectangular drawing on which to place their stirrers.
If . . . children need a challenge during Straw Shapes,	**Then . . .** see if they can choose beforehand the correct amount and sizes of stirrers to make a shape.

RESEARCH IN ACTION

All squares are rectangles because rectangles can be defined as shapes with pairs of parallel sides that are equal in length with all right angles (not "two long sides and two short sides"). However, all rectangles are not squares—only those with all four sides of equal length are squares.

 Reflect

5 

Ask children:

- **How do you know for sure a shape is a rectangle?**

Children might say: It is like a square but longer, or it has corners like a square.

Show a rectangle, and ask:

- **Is this a rectangle? Why or why not?**

Children might say: Those two sides are the same and so are these two.

 Assess

Use the Weekly Record Sheet from ***Assessment*** to record children's progress. Use their time at the centers as an opportunity to complete your observations.

Thursday Planner

Objectives

- To locate, name, and build familiar two-dimensional shapes, including triangles, rectangles, and squares
- To distinguish between visually-similar non-examples of familiar two-dimensional shapes
- To name the number of objects in a group up to 3

Materials

- *counters
- paper plate
- dark cloth
- plastic stirrers

Math Throughout the Year

Review activity directions at the top of page 67, and complete each in class whenever appropriate.

Looking Ahead

For tomorrow, ask children to bring in items with square faces, and, if you have not already, make copies of Family Letter Week 5 from the ***Teacher's Resource Guide***.

*provided in Manipulative Kit

Thursday

Whole Group

Warm-Up: "Wake Up, Jack-in-the-Box"

Help children suit their actions to the following words:

Jack-in-the-Box, Jack-in-the-Box,
Wake up, wake up, somebody knocks.
One time, two times, three times, four,
Jack pops out of his little round door.

Snapshots

- Have prepared two counters on a paper plate, and cover the plate with a cloth you cannot see through.
- Review what "snapshots" are (pictures taken with a camera), and remind children that, while playing Snapshots, they will use their eyes and minds like a camera by "taking a snapshot" of what they see.
- Show your covered plate to children, and tell them counters are hidden under the cloth. Remind children to watch carefully, quietly, and with hands in their laps as you quickly uncover the plate to expose the counters. Uncover your plate for two seconds, and then cover it again.
- Have children show with their fingers how many counters they saw, and, once you have seen all children's fingers, ask how many counters they saw. Repeat the uncovering if needed.
- Uncover the plate indefinitely and, to check their answers, ask children what they see. Allow children to say their answers, recognizing their contributions.
- Depending on children's ability, repeat the activity with up to four counters lined up.

Small Group

Straw Shapes

- Children use the stirrers you cut based on ***Teacher's Resource Guide*** page 177 to make shapes. Ensure that they build shapes with correct attributes, such as three straight sides that touch without gaps for triangles. All stirrers should touch at their endpoints. Discuss attributes as children build.

- If children need help, give constructive feedback; guide their hands; and/or provide a model for them to copy or a drawing on which to place stirrers. Can they choose the correct amount and sizes of stirrers they will need to make a given shape? If children excel, challenge them to get a shape's corners and angles "just right." Can they place pieces immediately without much trial and error?

Is It or Not?

- Tell children you are going to try to fool them. Say rectangle, and show a drawing or a straw shape of your own where all can see it. Ask children whether the shape is a rectangle; then ask why or why not.
- Repeat with shapes that are true rectangles and some that are not.
- Repeat with shapes that are true triangles and some that are not.
- Summarize by reviewing a rectangle's attributes: opposite sides are the same lengths, all "turns" are right angles, and all sides are straight.

Computer Center Building Blocks

Continue to provide each child with a chance to complete Mystery Pictures 2 and Number Snapshots 1.

Monitoring Student Progress

If . . . a child has mastered Mystery Pictures 2,	**Then . . .** help him or her enter the activity's free explore level for additional practice.

Hands On Math Center

Based on what they benefit from more, have children complete Shape Hunt or Straw Shapes, focusing on rectangles, from yesterday's Hands On Math Center.

3 Reflect

5

Review with children, and ask:

■ **How do you know for sure a shape is a rectangle?**

Children might say: It is like a square but longer; it has corners like a square; or those two sides are the same.

4 Assess

During Small Group activities, use the Small Group Record Sheet from *Assessment* to observe and record children's progress.

Friday Planner

Objectives

- To locate, name, and build familiar two-dimensional shapes, including triangles, rectangles, and squares
- To distinguish between visually-similar non-examples of familiar two-dimensional shapes
- To name the number of objects in a group up to 3

Materials

- big, flat square
- square items
- plastic stirrers

Vocabulary

square a shape that has four equal sides and all right angles

Math Throughout the Year

Review activity directions at the top of page 67, and complete each in class whenever appropriate.

Looking Ahead

If possible, for today's Hands On Math Center, enlist an adult helper. For next week, familiarize yourself with Pizza Pizzazz 2: Make Matches and Road Race Counting Game from the ***Building Blocks*** software.

1 Whole Group

Warm-Up: *Building Shapes*

- Read the page about squares from ***Big Book Building Shapes*** to children.
- Together, look for and identify squares in the classroom.

Shape Show: Squares

- Show and name a large, flat square. Walk your fingers around its perimeter, describing and exaggerating your actions: straaiiight side...turn, straaiiight side...turn, straaiiight side...turn, straaiiight side...stop.
- Ask children how many sides the square has, and count the sides with them. Review that all sides of a square are the same length, and all "turns" are right angles. To model this, you may place stirrers that are the same length as each side on each side.
- Remind children about right angles (uppercase Ls or the corner of a doorway). Make uppercase Ls with children using thumbs and index fingers. Fit your *L* on the angles of the square.
- Ask children what things they have at home that are squares. Show different examples of squares.
- Have children walk around a large, flat square, such as a floor tile. Once seated, have children draw squares in the air.

Monitoring Student Progress

If . . . children have difficulty recognizing squares,	**Then . . .** help them make sure all side lengths are equal and all corners are right angles.

2 Work Time

20

Building Blocks

Continue to provide each child with a chance to complete Mystery Pictures 2 and Number Snapshots 1.

 Hands On Math Center

Shape Hunt: Squares

- Tell children to find an item in the room with at least one square face.
- Encourage children to count the shape's sides and, if possible, show the square to an adult, discussing its shape. For example, squares have four sides of equal length and four right angles. After discussion, have the child replace the square so other children can find it.
- You may choose to photograph the squares for a class shape book.

Straw Shapes: Squares

Children use plastic stirrers to make squares and/or to create pictures and designs that include squares.

 Reflect

Briefly discuss with children what you have done in class this week, such as identifying triangles, rectangles, and squares, and ask:

■ **How do you know for sure a shape is a square?**

Children might say: Squares have sides all the same, or squares have right corners (angles).

Show a square, and ask:

■ **Is this a square? Why or why not?**

Children might say: It is because the sides and corners are all alike.

 Assess

Use the Weekly Record Sheet from ***Assessment*** to record children's progress. Use their time at the centers as an opportunity to complete your observations.

RESEARCH IN ACTION

All squares are rectangles because rectangles can be defined as shapes with pairs of parallel sides that are equal in length with all right angles (not "two long sides and two short sides"). However, all rectangles are not squares—only those with all four sides of equal length are squares.

Home Connection

From the ***Teacher's Resource Guide,*** distribute to children copies of Family Letter Week 5 to share with their family, which includes directions for a 2D Shape Hunt at home. Each letter has an area for children to show families an example of what they have been doing in class.

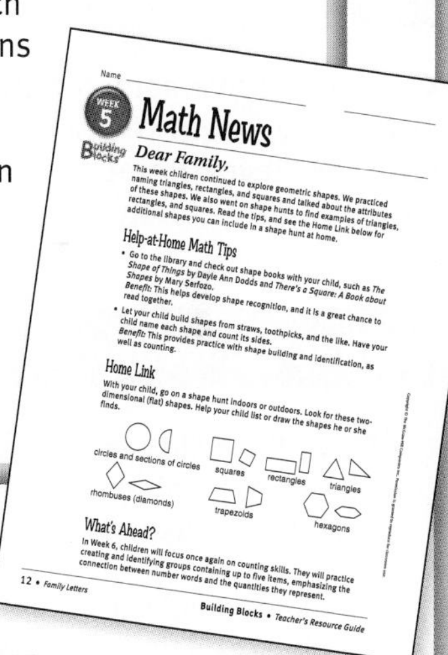

Name

WEEK 5 **Math News**

Building Blocks

Dear Family,

This week children continued to explore geometric shapes. We practiced naming triangles, rectangles, and squares and talked about the attributes of these shapes. We also went on shape hunts to find examples of triangles, rectangles, and squares. Read the tips, and see the Home Link below for additional shapes you can include in a shape hunt at home.

Help-at-Home Math Tips

- Go to the library and check out shape books with your child, such as *The Shape of Things* by Dayle Ann Dodds and *There's a Square: A Book about Shapes* by Mary Serfozo.
Benefit: This helps develop shape recognition, and it is a great chance to read together.
- Let your child build shapes from straws, toothpicks, and the like. Have your child name each shape and count its sides.
Benefit: This provides practice with shape building and identification, as well as counting.

Home Link

With your child, go on a shape hunt indoors or outdoors. Look for these two-dimensional (flat) shapes. Help your child list or draw the shapes he or she finds.

circles and sections of circles
squares
rectangles
triangles
rhombuses (diamonds)
trapezoids
hexagons

What's Ahead?

In Week 6, children will focus once again on counting skills. They will practice creating and identifying groups containing up to five items, emphasizing the connection between number words and the quantities they represent.

12 • *Family Letters*

Building Blocks • *Teacher's Resource Guide*

WEEK 5 Wrap-Up

Assess and Differentiate

A Gather Evidence

Review children's progress in mathematics by looking at the Weekly Record Sheets (Monday, Wednesday, Friday) and the Small Group Record Sheets (Tuesday, Thursday) from this past week.

B Summarize Findings

Using ***Online Assessment***, summarize and analyze assessment data for each child based on your weekly observations and Record Sheets. Such information helps to determine where each child is on the math trajectory for geometry. See ***Assessment*** for the print companion to each Learning Trajectory Record.

C Differentiate Instruction

Once you have seen a child exhibit specific levels of the trajectory, begin to encourage and work with that child toward the next level. Refer to Appendix A for individualized instruction opportunities, including Special Education concerns.

Shapes: Naming

If . . . the child can identify less typical examples of rectangles, **Then . . .** *Shape Recognizer, All Rectangles* Recognizes more rectangle sizes, shapes, and orientations.

If . . . the child can identify most familiar shapes in varied orientations, **Then . . .** *Shape Recognizer, More Shapes* Recognizes most familiar shapes and typical examples of other shapes, such as hexagon, rhombus (diamond), and

Shapes: Representing

If . . . the child can construct shapes into goal shapes, **Then . . .** *Constructor of Shapes from Parts, Looks Like* Uses manipulatives representing parts of shapes, such as sides, to make a shape that "looks like" a goal shape.

Object Counting

If . . . the child can count five items and tell how many with the last number counted, **Then . . .** *Counter (Small Numbers)* Accurately counts objects in a line to 5, and answers "how many?" with the last number counted. When objects are visible, especially with small numbers, begins to understand cardinality.

If . . . the child can produce (count out) up to five items, **Then . . .** *Producer (Small Numbers)* Counts out objects to 5.

Subitizing

If . . . the child can make a collection up to 4 based on a model of such a grouping, **Then . . .** *Maker of Small Collections* Nonverbally makes a small collection (no more than 4, usually up to 3) using the number of another collection, such as a mental or verbal model, not necessarily by matching.

If . . . the child can instantly recognize and name small collections up to 4, **Then . . .** *Perceptual Subitizer to 4* Instantly recognizes collections up to 4 and verbally names the number of items.

WEEK 6

Overview

Big Ideas

- counting small groups of objects
- producing groups of a specific amount
- comparing and ordering small groups
- subitizing

Teaching for Understanding

Week 6 builds on counting skills from the early weeks, maintaining the emphasis on four counting components: verbal counting, the counting of small collections, counting out (producing) small collections, and comparing small quantities. When children build these skills meaningfully, they understand that counting helps tell how many and describes order.

Games

Research shows that games are particularly effective for teaching and learning number skills. Games provide practice and experience in a motivational setting; they serve as a model for a meaningful reason to use number skills—that is, number is inherent in math games. Games can also connect math to other areas of development, such as social skills by taking turns and technological habits by using a computer and its tools. Additionally, different game settings link counting to a variety of familiar contexts, such as counting pizza toppings and moving along a path to finish a race.

Language

Several counting activities incorporate language skills. When children place manipulatives or small toys on a Places Scene, they tell stories about them which include numbers; this connects numbers to language. More profoundly, children begin to link scientific, quantitative knowledge to humanistic, narrative knowledge.

Object Counting

Most activities this week continue to develop a child's ability to understand that the last number word in a counting sequence tells how many, and the games require "counting out," or producing, a certain number of items or moves. Producing objects is more difficult than merely counting the objects in a group because children have to continually recall how many they were supposed to produce and compare that to each number word they say in order to stop at just the right number.

What's Ahead?

In the weeks to come, we will extend children's range of strategies for comparing numbers. This consists of counting, using one-to-one correspondence to compare separate groups of objects, and learning about numerals (actual written numbers such as 5).

Eclipse Studios

Overview

How Children Learn to Count

Knowing where children are on the learning trajectory for counting and what the next steps are helps facilitate their development. Children who easily surpass these trajectory levels might be challenged by larger quantities and encouraged to assist other children.

Verbal Counting

What to Look For Can the child count with some one-to-one correspondence?

Reciter (10) Verbally counts to ten with some correspondence to objects. For example, the child says: 1 (points to it), 2 (points to it), 3 (starts to point), 4 (finishes pointing but is pointing to third object), 5, and so on.

Object Counting

What to Look For How many items can the child count accurately?

Counter (Small Numbers) Accurately counts objects in a line to 5, and answers "how many?" with the last number counted. When objects are visible, especially with small numbers, begins to understand cardinality.

Producer (Small Numbers) Counts out objects to 5.

Subitizing

What to Look For Does the child instantly recognize groups of at least 5?

Perceptual Subitizer to 5 Instantly recognizes collections of up to 5 when shown briefly and verbally names the number of items.

Conceptual Subitizer to 5+ Verbally labels all arrangements to about 5 when shown briefly using groups: "I saw 2 and 2 so I said 4" or "I made 2 groups of 3 plus 1 in my mind so that is 7."

English Learner

For English learners, use real-life situations to provide meaningful practice. Review the following: *baker* and *blank cube*. Refer to the ***Teacher's Resource Guide.***

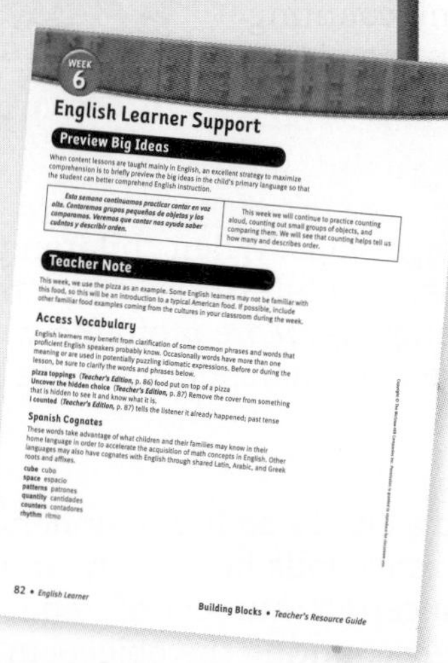

WEEK 6

English Learner Support

Preview Big Ideas

When content lessons are taught mainly in English, an excellent strategy to maximize comprehension is to briefly preview the big ideas in the child's primary language so that the student can better comprehend English instruction.

Esta semana continuamos practicar contar en voz alta. Contaremos grupos pequeños de objetos y los comparamos. Veremos que contar nos ayuda saber cuántos y describir orden.	This week we will continue to practice counting aloud, counting out small groups of objects, and comparing them. We will see that counting helps tell us how many and describes order.

Teacher Note

This week, we use the pizza as an example. Some English learners may not be familiar with this food, so this will be an introduction to a typical American food. If possible, include other familiar food examples coming from the cultures in your classroom during the week.

Access Vocabulary

English learners may benefit from clarification of some common phrases and words that proficient English speakers probably know. Occasionally words have more than one meaning or are used in potentially puzzling idiomatic expressions. Before or during the lesson, be sure to clarify the words and phrases below.

pizza toppings (***Teacher's Edition,*** p. 86) food put on top of a pizza
Uncover the hidden choice (***Teacher's Edition,*** p. 87) Remove the cover from something that is hidden to see it and know what it is.
I counted (***Teacher's Edition,*** p. 87) tells the listener it already happened; past tense

Spanish Cognates

These words take advantage of what children and their families may know in their home language in order to accelerate the acquisition of math concepts in English. Other languages may also have cognates with English through shared Latin, Arabic, and Greek roots and affixes.

cube cubo
space espacio
patterns patrones
quantity cantidades
counters contadores
rhythm ritmo

82 • *English Learner* Building Blocks • *Teacher's Resource Guide*

Technology Project

Children may get additional counting practice using Pizza Pizzazz Free Explore. Children can count with the program as they place toppings. Or, working cooperatively, one child names a number and the other puts that many toppings on a pizza. Challenge children to think of other ways to use the program. As you visit children at the computer, ask about the connection between the numerals and the number of toppings.

Math Throughout the Year

Math Throughout the Year activities are recommended to build on the mathematical skills highlighted in each week. Here are suggested activities for **Week 6.**

Simon Says Numbers

Play traditional Simon Says using only number commands, such as "Jump two times" and "Pat your head six times."

Snack Time

Tell children to take a specified amount of snack (and anything they might need to eat the snack), such as three crackers or five pretzels. Demonstrate what you mean, counting out the items and saying at the end how many there are.

Center Preview

Computer Center Building Blocks

Get your classroom Computer Center ready for Pizza Pizzazz 2: Make Matches and Road Race, a counting game, from the *Building Blocks* software.

After you introduce Pizza Pizzazz 2 and Road Race, each child should complete the activities individually as you (or an assistant) monitor and guide him or her periodically. Ideally, each child will have at least ten minutes of computer time at least twice during the week. Use children's center time as an opportunity for assessment.

Hands On Math Center

This week's Hands On Math Center activities are Make Number Pizzas, Find the Number, Draw Numbers, and Places Scenes. Supply the center with these materials: various counters (including round), paper plates, several opaque containers, Counting Cards, drawing paper, crayons, and Places Scenes.

Literature Connections

These books help develop counting.

- *The Very Hungry Caterpillar* by Eric Carle
- *Five Little Monkeys Jumping on the Bed* by Eileen Christelow
- *Feast for 10* by Cathryn Falwell
- *Five Little Ducks: An Old Rhyme* by Pamela Paparone
- *One Was Johnny: A Counting Book* by Maurice Sendak

(t)Matt Meadows (b)Eclipse Studios

WEEK 6

Weekly Planner

Use this chart to plan for your specific class schedule. If you have your prekindergarteners for only three days, complete Monday, Tuesday, and Thursday of the week.

Learning Trajectories

Week 6 Objectives

- To participate in rhythmic patterns
- To connect number words to the quantities they represent
- To make groups of up to five items
- To count verbally to 10 with understanding
- To name the number of objects in a group up to 5

	Developmental Path	Instructional Activities
Verbal Counting	*Reciter*	
	Reciter (10)	Count and Move in Patterns "Five Dancing Dolphins" *Where's One?*
	Counter Backward from 10	
Object Counting	*Corresponder*	
	Counter (Small Numbers)	*Where's One?* Road Race Find the Number Pizza Game 1 Number Me
	Producer (Small Numbers)	"Baker's Truck" Make Number Pizzas Pizza Pizzazz 2 Road Race Draw Numbers Places Scenes
	Counter (10)	
Subitizing	*Perceptual Subitizer to 4*	
	Perceptual Subitizer to 5	Snapshots Pizza Game 1
	Conceptual Subitizer to 5+	Snapshots
	Conceptual Subitizer to 10	

Work Time

Whole Group	Small Group	Computer	Hands On	Program Resources
"Baker's Truck" **Make Number Pizzas** *Materials:* *round counters paper plates		**Pizza Pizzazz 2**	**Make Number Pizzas** *Materials:* *round counters paper plates **Find the Number** *Materials:* opaque containers *round counters paper plates *Counting Cards	Building Blocks ***Assessment*** Weekly Record Sheet
"Baker's Truck" **Places Scenes** *Materials:* *counters *Counting Cards	**Pizza Game 1** *Materials:* *round counters *Number Cubes **Make Number Pizzas** *Materials:* *round counters paper plates	**Road Race**	**Draw Numbers** **Places Scenes** *Materials:* *counters *Counting Cards **Make Number Pizzas** **Find the Number**	***Teacher's Resource Guide*** • Places Scenes • Pizza Game 1 Building Blocks ***Assessment*** Small Group Record Sheet
Count and Move in Patterns ***Where's One?*** **Snapshots** *Materials:* paper plates *counters dark cloth		• **Pizza Pizzazz 2** • **Road Race**	**Places Scenes** **Make Number Pizzas** **Find the Number** **Draw Numbers**	***Big Book Where's One?*** Building Blocks ***Teacher's Resource Guide*** Places Scenes ***Assessment*** Weekly Record Sheet
Count and Move in Patterns **Number Me**	**Pizza Game 1** *Materials:* *round counters *Number Cubes **Make Number Pizzas** *Materials:* *round counters paper plates	• **Pizza Pizzazz 2** • **Road Race**	**Places Scenes** **Make Number Pizzas** **Find the Number** **Draw Numbers**	***Teacher's Resource Guide*** • Pizza Game 1 • Places Scenes Building Blocks ***Assessment*** Small Group Record Sheet
"Five Dancing Dolphins" **Snapshots** *Materials:* paper plates *round counters dark cloth		• **Pizza Pizzazz 2** • **Road Race**	**Places Scenes** **Make Number Pizzas** **Find the Number** **Draw Numbers**	Building Blocks ***Teacher's Resource Guide*** • Places Scenes • Family Letter Week 6 ***Assessment*** Weekly Record Sheet

*provided in Manipulative Kit

Monday Planner

Objectives

- To participate in rhythmic patterns
- To count verbally to 5 with understanding
- To make groups of up to five items
- To name the number of objects in a group up to 5
- To connect number words to the quantities they represent

Materials

- *round counters
- paper plates
- opaque containers
- *Counting Cards

Math Throughout the Year

Review activity directions at the top of page 83, and complete each in class whenever appropriate.

Looking Ahead

For today, enlist an adult helper for the Hands On Math Center if possible. For tomorrow, from the ***Teacher's Resource Guide,*** prepare Places Scenes for Whole Group, and make enough copies of the Pizza Game 1 activity sheet for Small Group (one for each child).

*provided in Manipulative Kit

Monday

1 Whole Group

10

Warm-Up: "Baker's Truck"

Here are the words and actions:

> The baker's truck drives down the street,
> Filled with everything good to eat.
> Two doors the baker opens wide. (*Outstretch arms.*)
> Let's look at the shelves inside. (*Cup hands around eyes to look.*)
> What do you see? What do you see?
> Five big pizzas for you and me! (*Show five fingers.*)

Make Number Pizzas

- Pretend a paper plate is a pizza and round counters are toppings. Tell children the goal is to put a certain number of "toppings" on their "pizzas." Explain that they will do this activity themselves during Work Time.
- While you prepare your pizza by placing five toppings on your paper plate, have children pretend to put that many toppings on their pizzas as everyone carefully counts together.
- Ask children how they know there are five toppings on the pizzas (because you counted to 5).
- Repeat the activity with different numbers up to 5 or more.

2 Work Time

25

Computer Center Building Blocks

Introduce Pizza Pizzazz 2: Make Matches from the **Building Blocks** software. In this activity, children put toppings on one of two pizzas to make the pizzas match; they make "twin" pizzas. Each child should have an opportunity to complete Pizza Pizzazz 2 this week.

Monitoring Student Progress

If . . .	Then . . .
children struggle during Pizza Pizzazz 2,	provide two paper plates and numerous counters for them to practice actually making "pizzas" before doing so on the computer.

Hands On Math Center

These activities and two others introduced tomorrow occur all week.

Make Number Pizzas

Follow the directions for the Whole Group version of this activity to introduce it to children at the center. You may display a Counting Card to remind children how many "toppings" to put on each "pizza."

Find the Number

- Conceal several pizzas (paper plates with counters), each under its own opaque container with a different number of toppings (counters).
- In plain view, display one pizza with three to five toppings, or use a Counting Card to represent the target number. The goal is for children to find the exposed pizza's hidden match by lifting each container and counting that pizza's toppings.
- Have children show their answers to you or another adult who assists your class. Initially, it is helpful to have an adult guide the activity closely. As a variation, if you have enough space, set up another table as already described but with a new target number.
- If needed, simplify the activity by reducing the number of containers children choose from, or leave all choices uncovered. For a challenge, have children work in pairs, determining their own amounts, and have each other, for example, "Find 10."

Pizza Pizzazz 2

3 Reflect

5

Ask children:

■ **How did you know the correct number of toppings during the pizza games?**
Children might say: Because I am smart; I counted the dots on the card and then got that many counters; or I just saw four.

4 Assess

Use the Weekly Record Sheet from *Assessment* to record children's progress. Use their time at the centers as an opportunity to complete your observations.

Tuesday Planner

Objectives

- To participate in rhythmic patterns
- To count verbally to 5 with understanding
- To make groups of up to five items
- To name the number of objects in a group up to 5
- To connect number words to the quantities they represent

Materials

- *counters (including round)
- *Number Cubes
- paper plates
- drawing paper
- crayons
- *Counting Cards

Math Throughout the Year

Review activity directions at the top of page 83, and complete each in class whenever appropriate.

Looking Ahead

Preview *Big Book Where's One?* for tomorrow.

*provided in Manipulative Kit

Tuesday

1 Whole Group

10

Warm-Up: "Baker's Truck"

Refer to Monday's Warm-Up for this finger play's words and actions.

Place Scenes

- Model placing five appropriate counters on the Places Scenes playground background from the *Teacher's Resource Guide* (or use another scene of your choice). Make up a story about your scene that includes number words, and share it with the children.
- Have children help you tell another story using a new scene.
- Tell students they will place five counters on a scene and tell their own story to a classmate or an adult at the Hands On Math Center.

2 Work Time

25

Small Group

Pizza Game 1

- Children play in pairs. Each player has a Pizza Game 1 activity sheet from the *Teacher's Resource Guide.*
- Player One rolls a Number Cube and puts that many counters on his or her plate. Player One asks Player Two, "Am I correct?" Player Two must agree that Player One is correct. Once correct, Player One moves the counters to the topping spaces on his or her pizza activity sheet.
- Players take turns until all the spaces on their pizzas have toppings. If the activity needs to be simplified, use a blank wooden cube to make a Number Cube with only one to three dots on each side. For a challenge, draw five to ten dots on each side of a blank cube.

Make Number Pizzas

- Put five counters on a paper plate, and then have children do the same.
- Ask children how they know there are five "toppings" on their "pizzas." Help them discuss their different arrangements of five. If children are able, repeat the activity with numbers greater than 5.

Monitoring Student Progress

If . . . children struggle during Make Number Pizzas,	**Then . . .** draw the appropriate number of circles, each on its own paper, and have children put a counter in each circle until they can reliably produce the target number. Follow up by verbally naming each group together, such as "This is two."
If . . . children excel during Make Number Pizzas,	**Then . . .** place them in pairs to produce their own groups of different amounts, asking each other, for example, "Where's the 10?"

Computer Center Building Blocks

Introduce Road Race, a counting game from the ***Building Blocks*** software, in which children identify amounts up to 5 and move forward that many spaces. Each child should have an opportunity to complete Road Race this week.

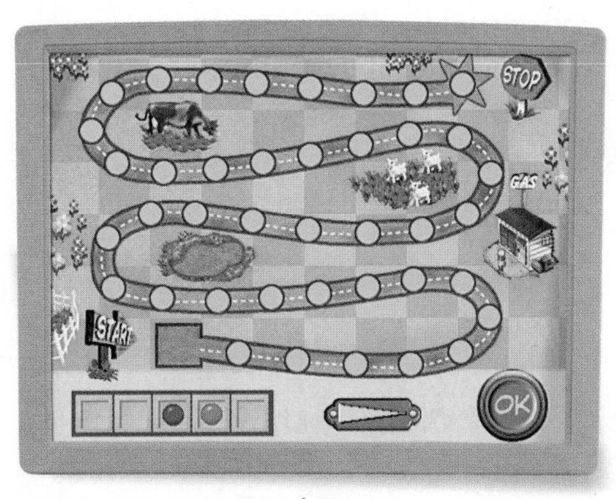

Road Race

Hands On Math Center

Introduce the activities below and/or continue yesterday's activities.

Draw Numbers

Building on a current class theme, have children draw five items on a sheet of paper. You may choose to model this on the board.

Places Scenes

Put a Counting Card on the table. Have children choose several Places Scenes backgrounds on which to place counters to match the card's amount. Children should also tell a story about one scene.

3 Reflect

Ask children:

- **How many toppings did you put on your pizza?**

Children might say: I put four; or while showing fingers, I put this many!

4 Assess

During Small Group activities, use the Small Group Record Sheet from ***Assessment*** to observe and record children's progress.

Wednesday Planner

Objectives

- To participate in rhythmic patterns
- To count verbally to 5 with understanding
- To make groups of up to five items
- To name the number of objects in a group up to 5
- To connect number words to the quantities they represent

Materials

- paper plates
- *round counters
- dark cloth

Math Throughout the Year

Review activity directions at the top of page 83, and complete each in class whenever appropriate.

*provided in Manipulative Kit

Wednesday

1 Whole Group

25

Warm-Up: Count and Move in Patterns

In patterns of 4, have all children count aloud from 1 to 16, 20, or more. For example, 1 (clap), 2 (clap), 3 (clap), 4 (jump then pause), 5 (clap), 6 (clap), 7 (clap), 8 (jump then pause), and so on.

Where's One?

- Read the first half of the story to children without interruption.
- Then return to several pages, asking children to find sets of objects that match numbers from 1 to 5.

Snapshots

- Put three to five round counters on your paper plate, making a "pizza," and cover it with a cloth you cannot see through. As an option, you may give children their own counters and paper plates.
- Show children your covered pizza, explaining that there are toppings on it. Tell children to watch carefully and quietly with hands in their laps while you quickly expose your pizza so they can take a "snapshot" of how many toppings they see. Cover the plate again.
- Have children show how many counters they saw with their fingers. Once you have seen everyone's fingers, ask children to *say* how many counters they saw. Repeat the reveal if needed. As an option, have children put that many counters on their plates.
- Uncover your pizza indefinitely, and ask children how many toppings there are.
- Based on children's ability, repeat with up to five counters in a row and/or up to four counters in a domino arrangement. If they are able, try scrambled arrangements.

Monitoring Student Progress

If . . . children struggle during Snapshots,	**Then . . .** reduce the number of counters; place counters in rows only; or reveal counters for a longer time.
If . . . children excel during Snapshots,	**Then . . .** increase the number of counters, or place counters in random arrangements only.

2 Work Time

10

Computer Center

Continue to provide each child with a chance to complete Pizza Pizzazz 2 and Road Race.

Hands On Math Center

Based on what children continue to learn and benefit from most, choose from this week's Hands On Math Center activities: Places Scenes, Make Number Pizzas, Find the Number, and Draw Numbers. Consult the Weekly Planner for corresponding materials and, if needed, previous days for activity directions.

3 Reflect

5

Ask children:

- **When you played Snapshots, how did you know how many counters there were?**

Children might say: I made a picture in my head and counted; or I just know the number by looking.

4 Assess

Use the Weekly Record Sheet from *Assessment* to record children's progress. Use their time at the centers as an opportunity to complete your observations.

RESEARCH IN ACTION

Moving items within a computer program is more difficult for children than "pointing and clicking." However, moving items allows them to do more interesting tasks and, therefore, learn more. An adult or an "expert" child should assist children who have difficulty operating within such a program.

Thursday Planner

Objectives

- To participate in rhythmic patterns
- To count verbally to 5 with understanding
- To make groups of up to five items
- To name the number of objects in a group up to 5
- To connect number words to the quantities they represent

Materials

- *round counters
- *Number Cubes
- paper plates

Math Throughout the Year

Review activity directions at the top of page 83, and complete each in class whenever appropriate.

Looking Ahead

For tomorrow, make copies of Family Letter Week 6 from the *Teacher's Resource Guide.*

*provided in Manipulative Kit

Thursday

1 Whole Group

10

Warm-Up: Count and Move in Patterns

In patterns of 5, have all children count aloud from 1 to 15, 20, or more. For example, 1 (clap), 2 (clap), 3 (clap), 4 (clap), 5 (spin then pause), 6 (clap), 7 (clap), 8 (clap), 9 (clap), 10 (spin then pause), and so on.

Number Me (5)

- Ask children how old they are, and have them show fingers to tell their age.
- Then ask children how many arms they have, and have them wave their arms. Repeat with hands, fingers, legs, feet, toes, head, nose, eyes, and ears. Include motion when possible, such as "Wiggle your ten fingers." Have fun!
- Make silly statements mixed with true statements, for example, "You have four ears...two legs...three feet...one nose...five eyes." Have children say whether or not you are correct. When children correctly disagree, ask: How many do you have? Can you prove it? (They can prove it by counting aloud.)

2 Work Time

25

Small Group

Pizza Game 1

- Children play in pairs. Each player has a Pizza Game 1 activity sheet from the *Teacher's Resource Guide.*
- Player One rolls a Number Cube and puts that many counters on his or her plate. Player One asks Player Two, "Am I correct?" Player Two must agree that Player One is correct. Once correct, Player One moves the counters to the topping spaces on his or her pizza activity sheet.
- Players take turns until all the spaces on their pizzas have toppings. If the activity needs to be simplified, use a blank wooden cube to make a Number Cube with only one to three dots on each side. For a challenge, draw five to ten dots on each side of a blank cube.

Make Number Pizzas

- Put five counters on a paper plate, and then have children do the same.
- Ask children how they know there are five "toppings" on their "pizzas." Help them discuss their different arrangements of five. If children are able, repeat the activity with numbers greater than 5.

 Computer Center Building Blocks

Continue to provide each child with a chance to complete Pizza Pizzazz 2 and Road Race.

 Hands On Math Center

Based on what children continue to learn and benefit from most, choose from this week's Hands On Math Center activities: Places Scenes, Make Number Pizzas, Find the Number, and Draw Numbers. Consult the Weekly Planner for corresponding materials and, if needed, previous days for activity directions.

Monitoring Student Progress

If . . . children struggle during Draw Numbers,	**Then . . .** reduce the number for them to represent, or have them paste cutouts if drawing is too difficult.
If . . . children excel during Draw Numbers,	**Then . . .** help them make pages for a number book with numerals to as high as they can count.

3 Reflect

5

Ask children:

- **How do you draw 5?**

Children might say: I make circles; or I keep drawing 1 more and count them until I get to 5.

4 Assess

During Small Group activities, use the Small Group Record Sheet from *Assessment* to observe and record children's progress.

Friday Planner

Objectives

- To participate in rhythmic patterns
- To count verbally to 5 with understanding
- To make groups of up to five items
- To name the number of objects in a group up to 5
- To connect number words to the quantities they represent

Materials

- paper plates
- *round counters
- dark cloth

Math Throughout the Year

Review activity directions at the top of page 83, and complete each in class whenever appropriate.

Looking Ahead

For next week, familiarize yourself with Party Time 1: Set the Table and Road Race: Shape Counting from the ***Building Blocks*** software.

*provided in Manipulative Kit

Friday

1 Whole Group

15

Warm-Up: "Five Dancing Dolphins"

For each dolphin that is added, show another "dancing" finger. Here are the words:

One dancing dolphin on a sea of blue,
She called her sister;
Then there were two.
Two dancing dolphins swimming in the sea,
They called for mother;
Then there were three.
Three dancing dolphins swimming close to shore,
They called for daddy;
Then there were four.
Four dancing dolphins in a graceful dive,
They called for baby;
Then there were five... five dancing dolphins on a sea of blue.

Snapshots

- Put three to five round counters on your paper plate, making a "pizza," and cover it with a cloth you cannot see through. As an option, you may give children their own counters and paper plates.
- Show children your covered pizza, explaining that there are toppings on it. Remind children to watch carefully and quietly with hands in their laps while you quickly expose your pizza so they can take a "snapshot" of how many toppings they see. Cover the plate again.
- Have children show how many counters they saw with their fingers. Once you have seen everyone's fingers, ask children to *say* how many counters they saw. Repeat the reveal if needed. As an option, have children put that many counters on their plates.
- Uncover your pizza indefinitely, and ask children how many toppings there are.
- Based on children's ability, repeat with up to five counters in a row and/or up to four counters in a domino arrangement. If they are able, try scrambled arrangements.

Eclipse Studios

2 Work Time

20

Computer Center Building Blocks

Continue to provide each child with a chance to complete Pizza Pizzazz 2 and Road Race.

Monitoring Student Progress

If . . . children excel at Pizza Pizzazz 2,	**Then . . .** allow them to play its Free Explore level as mentioned in the Technology Project on page 82.

Hands On Math Center

Based on what children continue to learn and benefit from most, choose from this week's Hands On Math Center activities: Places Scenes, Make Number Pizzas, Find the Number, and Draw Numbers. Consult the Weekly Planner for corresponding materials and, if needed, previous days for activity directions.

3 Reflect

5

Briefly discuss with children what you have done in class this week, such as learning about the numeral 5. Have children show you five fingers, and ask:

- **What if someone said that is 4, how could you prove you are correct?**

Children might say: You can see it is 5; everyone has five fingers; or I'd count.

4 Assess

Use the Weekly Record Sheet from ***Assessment*** to record children's progress. Use their time at the centers as an opportunity to complete your observations.

RESEARCH IN ACTION

Children who can quickly recognize numbers by taking a mental snapshot of them are more able to determine the number of items in a collection, represent that number, and eventually add such numbers.

Home Connection

From the ***Teacher's Resource Guide,*** distribute to children copies of Family Letter Week 6 to share with their family. Each letter has an area for children to show families an example of what they have been doing in class.

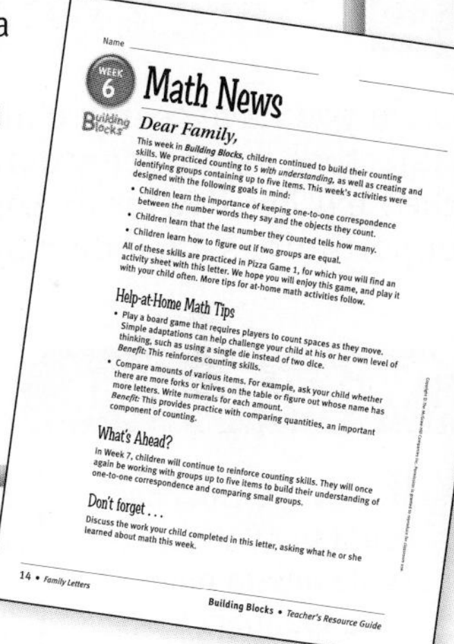

Name

WEEK 6 **Math News**

Building Blocks

Dear Family,

This week in *Building Blocks*, children continued to build their counting skills. We practiced counting to 5 *with understanding*, as well as creating and identifying groups containing up to five items. This week's activities were designed with the following goals in mind:

- Children learn the importance of keeping one-to-one correspondence between the number words they say and the objects they count.
- Children learn that the last number they counted tells how many.
- Children learn how to figure out if two groups are equal.

All of these skills are practiced in Pizza Game 1, for which you will find an activity sheet with this letter. We hope you will enjoy this game, and play it with your child often. More tips for at-home math activities follow.

Help-at-Home Math Tips

- Play a board game that requires players to count spaces as they move. Simple adaptations can help challenge your child at his or her own level of thinking, such as using a single die instead of two dice. *Benefit:* This reinforces counting skills.
- Compare amounts of various items. For example, ask your child whether there are more forks or knives on the table or figure out whose name has more letters. Write numerals for each amount. *Benefit:* This provides practice with comparing quantities, an important component of counting.

What's Ahead?

In Week 7, children will continue to reinforce counting skills. They will once again be working with groups up to five items to build their understanding of one-to-one correspondence and comparing small groups.

Don't forget . . .

Discuss the work your child completed in this letter, asking what he or she learned about math this week.

14 • *Family Letters* **Building Blocks** • *Teacher's Resource Guide*

Assess and Differentiate

A Gather Evidence

Review children's progress in mathematics by looking at the Weekly Record Sheets (Monday, Wednesday, Friday) and the Small Group Record Sheets (Tuesday, Thursday) from this past week.

B Summarize Findings

Using ***Online Assessment,*** summarize and analyze assessment data for each child based on your weekly observations and Record Sheets. Such information helps determine where each child is on the math trajectory for counting. See ***Assessment*** for the print companion to each Learning Trajectory Record.

C Differentiate Instruction

Once you have seen a child exhibit specific levels of the trajectory, begin to encourage and work with that child toward the next level. Refer to Appendix A for individualized instruction opportunities, including Special Education concerns.

Verbal Counting

If . . .	Then . . .
If . . . the child can count to 10 with some one-to-one correspondence,	**Then . . .** *Reciter (10)* Verbally counts to 10 with some correspondence to objects. For example, the child says: 1 (points to it), 2 (points to it), 3 (starts to point), 4 (finishes pointing but is pointing to third object), 5, and so on.

Object Counting

If . . .	Then . . .
If . . . the child can count five items and tell how many with the last number counted,	**Then . . .** *Counter (Small Numbers)* Accurately counts objects in a line to five, and answers "how many?" with the last number counted. When objects are visible, especially with small numbers, begins to understand cardinality.
If . . . the child can produce up to five items,	**Then . . .** *Producer (Small Numbers)* Counts out objects to 5.

Subitizing

If . . .	Then . . .
If . . . the child can instantly identify groups up to 5,	**Then . . .** *Perceptual Subitizer to 5* Instantly recognizes collections of up to 5 when shown briefly and verbally names the number of items.
If . . . the child can label all arrangements up to 10,	**Then . . .** *Conceptual Subitizer to 5+* Verbally labels all arrangements to about 5 when shown briefly using groups: "I saw 2 and 2 so I said 4" or "I made 2 groups of 3 plus 1 in my mind so that is 7."

Overview

Big Ideas

- counting to find out "how many?"
- comparing using one-to-one correspondence
- subitizing

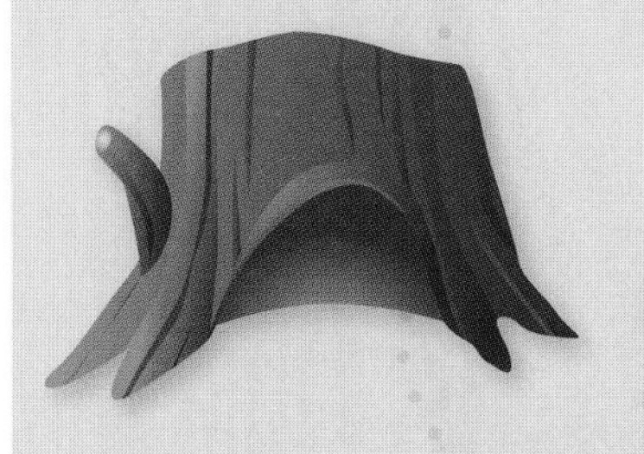

Teaching for Understanding

To a large extent, "understanding" an idea involves relating it to other ideas. When children realize they can compare the number of items in groups by matching the items one-to-one and they can also compare by counting and then comparing both groups, they relate the two strategies. Such relationships form a solid foundation for understanding number and quantitative reasoning.

Numerals

Is learning numerals (written symbols such as 4) important to young children's mathematical understanding? On one hand, just recognizing such a numeral is not as central to mathematics as it is to language arts and reading. On the other hand, numerals are another everyday experience young children have that relates to mathematics. As symbols, numerals can help children abstract the meaning of numbers, as long as they are connected to mathematical ideas and reasoning. Like letters, numerals can be introduced without pressure; unlike letters, which should be combined to signify true meaning, numerals can carry meaning independently.

Daily Environments

With all people, but especially with very young children, one of the main ways to develop understanding is to connect mathematics to everyday situations. For example, using a favorite or familiar story like *Goldilocks and the Three Bears*, in which the common task of setting the table is a focus, builds children's quantitative concepts. Children can use one-to-one correspondence between two groups of objects, such as placemats to dishes, to make groups that are equal in number.

Meaningful Connections

Children can also count the groups they compare to check their equality. Significant connections build strong mathematical concepts in a child's mind. Another example is when children play a game in which the main task is to count the number of sides a geometric shape has, thus linking number knowledge to geometry.

What's Ahead?

In the weeks to come, children will extend their abilities to compare numbers, hopefully beginning to differentiate between the idea of number as how many in a group and number as how to count and produce a certain amount. One-to-one correspondence will continue to be a focus skill. Children will play several number games to further develop their counting and reasoning abilities.

Eclipse Studios

WEEK 7 Overview

How Children Learn about Numbers

Knowing where children are on the learning trajectory for counting, comparing, and subitizing and what the next steps are helps facilitate their development. Children who easily surpass these trajectory levels might be challenged by larger numbers and encouraged to assist other children.

Object Counting

What to Look For What size groups can the child label and/or produce?

Counter (Small Numbers) Accurately counts objects in a line to 5, and answers "how many?" with the last number counted. When objects are visible, especially with small numbers, begins to understand cardinality.

Producer (Small Numbers) Counts out objects to 5.

Comparing and Ordering

What to Look For How does the child identify small groups?

Matching Comparer Compares groups up to 6 by matching.

Counting Comparer (Same Size) Accurate comparison via counting but only when objects are similar in size and in groups up to 5.

Counting Comparer (5) Compares with counting even when larger group's objects are smaller; figures out how many more or less.

Subitizing

What to Look For Does the child instantly recognize groups of at least 5?

Perceptual Subitizer to 5 Instantly recognizes collections of up to 5 when shown briefly and verbally names the number of items.

Conceptual Subitizer to 5+ Verbally labels all arrangements to about 5 when shown briefly using groups: "I saw 2 and 2 so I said 4," or "I made 2 groups of 3 plus 1 in my mind so that is 7."

English Learner

For English learners, make language comprehensible through role-playing. Review the following: *setting the table, form a numeral,* and *facedown.* Refer to the ***Teacher's Resource Guide.***

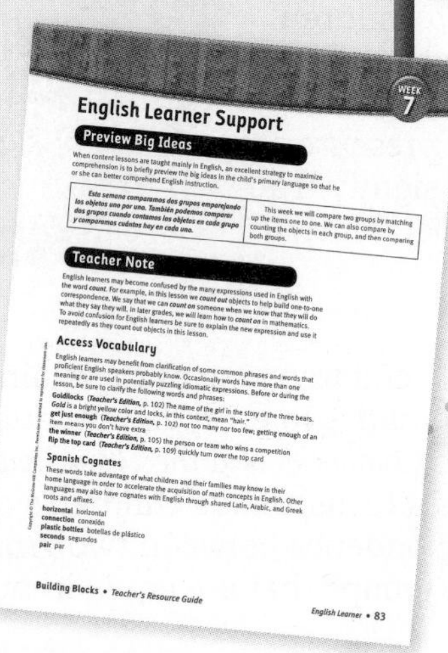

WEEK 7

English Learner Support

Preview Big Ideas

When content lessons are taught mainly in English, an excellent strategy to maximize comprehension is to briefly preview the big ideas in the child's primary language so that he or she can better comprehend English instruction.

Esta semana comparamos dos grupos emparejando los objetos uno por uno. También podemos comparar dos grupos cuando contamos los objetos en cada grupo y comparamos cuántos hay en cada uno.	This week we will compare two groups by matching up the items one to one. We can also compare by counting the objects in each group, and then comparing both groups.

Teacher Note

English learners may become confused by the many expressions used in English with the word *count.* For example, in this lesson we ***count out*** objects to help build one-to-one correspondence. We say that we can ***count on*** someone when we know that they will do what they say they will. In later grades, we will learn how to ***count on*** in mathematics. To avoid confusion for English learners be sure to explain the new expression and use it repeatedly as they count out objects in this lesson.

Access Vocabulary

English learners may benefit from clarification of some common phrases and words that proficient English speakers probably know. Occasionally words have more than one meaning or are used in potentially puzzling idiomatic expressions. Before or during the lesson, be sure to clarify the following words and phrases:

Goldilocks (***Teacher's Edition,*** p. 102) The name of the girl in the story of the three bears. *Gold* is a bright yellow color and locks, in this context, mean "hair."

get just enough (***Teacher's Edition,*** p. 102) not too many nor too few; getting enough of an item means you don't have extra

the winner (***Teacher's Edition,*** p. 105) the person or team who wins a competition

flip the top card (***Teacher's Edition,*** p. 109) quickly turn over the top card

Spanish Cognates

These words take advantage of what children and their families may know in their home language in order to accelerate the acquisition of math concepts in English. Other languages may also have cognates with English through shared Latin, Arabic, and Greek roots and affixes.

horizontal horizontal
connection conexión
plastic bottles botellas de plástico
seconds segundos
pair par

Building Blocks • *Teacher's Resource Guide*

English Learner • 83

Technology Project

Children may get additional experience counting, matching, and problem-solving using Party Time Free Explore. Here children create their own parties, as well as pose problems to each other. For example, a child turns off the program's sound, sets ten placemats, and challenges a partner to tell how many cups are needed (to match the number of placemats). But, what if each guest gets two beverages? The children can work together to figure out the answer is 20.

Math Throughout the Year

Math Throughout the Year activities are recommended to build on the mathematical skills highlighted in each week. Here are suggested activities for **Week 7**.

Numerals Every Day

Numerals are all around us. Help children notice and read numerals on common items throughout the day, such as clocks, food containers, street signs, room numbers, newspapers, and so on.

Set the Table

Children set a table for dolls/toy animals, possibly in the dramatic play area, using a real or pretend table. Children should set out just enough paper (or toy) plates, cloth napkins, and plastic (or toy) silverware for the dolls/toy animals. Work with children to establish the idea that one-to-one matching creates equal groups: when you know the number in one of the groups, then you know the number in the other.

Center Preview

Computer Center

Get your classroom Computer Center ready for Party Time 1: Set the Table and Road Race: Shape Counting from the ***Building Blocks*** software.

After you introduce Party Time 1 and Road Race, each child should complete the activities individually as you (or an assistant) monitor and guide him or her periodically. Ideally, each child will have at least ten minutes of computer time at least twice during the week. Use children's center time as an opportunity for assessment.

Hands On Math Center

This week's Hands On Math Center activities are *Goldilocks and the Three Bears,* Set the Table, Pizza Game 1, and Places Scenes. Supply the center with these materials: *Goldilocks and the Three Bears* flannel board characters and props, paper plates, cloth napkins, toy silverware (all used for Set the Table as well), dolls, toy animals, Pizza Game 1 activity sheet, round counters, Number Cubes, Places Scenes, various counters, and Counting Cards.

Literature Connections

These books help develop counting.

- *Feast for 10* by Cathryn Falwell
- *Fiesta* by Ginger Foglesong
- *Miss Spider's Tea Party* by David Kirk
- *Uno, Dos, Tres: One, Two, Three* by Pat Mora
- *One Hungry Monster: A Counting Book in Rhyme* by Susan Heyboer O'Keefe

(t)Matt Meadows (b)Eclipse Studios

WEEK 7

Weekly Planner

Learning Trajectories

Week 7 Objectives

- To produce a group of one to five objects
- To make a group equal in number to another group using one-to-one correspondence
- To count objects organized in a line up to 5
- To compare two groups to determine whether or not they have the same small number of objects

	Developmental Path	Instructional Activities
Object Counting	*Corresponder*	
	Counter (Small Numbers)	Road Race: Shape Counting Pizza Game 1
	Producer (Small Numbers)	Number Jump Places Scenes Pizza Game 1
	Counter and Producer (10+)	
Comparing and Ordering	*Nonverbal Comparer*	
	Matching Comparer	*Goldilocks and the Three Bears* Party Time 1
	Counting Comparer (Same Size)	Compare Snapshots Compare Game
	Counting Comparer (5)	Get Just Enough Compare Game
	Counting Comparer (10)	
Subitizing	*Perceptual Subitizer to 4*	
	Perceptual Subitizer to 5	Compare Snapshots
	Conceptual Subitizer to 5+	Compare Snapshots
	Conceptual Subitizer to 10	

Use this chart to plan for your specific class schedule. If you have your prekindergarteners for only three days, complete Monday, Tuesday, and Thursday of the week.

Work Time

Whole Group	Small Group	Computer	Hands On	Program Resources
Numeral 1 ***Goldilocks and the Three Bears*** *Materials:* flannel board with characters and props paper plates cloth napkins toy silverware		Party Time 1	***Goldilocks and the Three Bears*** *Materials*: See Whole Group. **Pizza Game 1** *Materials*: *round counters *Number Cubes paper plates	***Big Book Where's One?*** Building Blocks ***Teacher's Resource Guide*** Pizza Game 1 ***Assessment*** Weekly Record Sheet
Numeral 2 **Compare Snapshots** *Materials*: paper plates *counters dark cloth	**Get Just Enough** *Materials*: paintbrushes and cups nontoxic markers bowls and toy dogs boxes and toy cars dolls and hats toy cups and saucers **Compare Game**	**Road Race: Shape Counting**	**Places Scenes** *Materials*: *counters *Counting Cards ***Goldilocks and the Three Bears*** **Pizza Game 1**	***Big Book Where's One?*** Building Blocks ***Teacher's Resource Guide*** • Places Scenes • Pizza Game 1 ***Assessment*** Small Group Record Sheet
Numeral 3 **Number Jump**		• **Party Time 1** • **Road Race: Shape Counting**	**Places Scenes** ***Goldilocks and the Three Bears*** **Pizza Game 1**	***Big Book Where's One?*** Building Blocks ***Teacher's Resource Guide*** • Places Scenes • Pizza Game 1 ***Assessment*** Weekly Record Sheet
Numeral 4 **Compare Snapshots** *Materials*: paper plates *counters dark cloth	**Get Just Enough** *Materials*: See Tuesday. **Compare Game** *Materials*: *Counting Cards	• **Party Time 1** • **Road Race: Shape Counting**	**Places Scenes** ***Goldilocks and the Three Bears*** **Pizza Game 1**	***Big Book Where's One?*** Building Blocks ***Teacher's Resource Guide*** • Places Scenes • Pizza Game 1 ***Assessment*** Small Group Record Sheet
Numeral 5 **Number Jump**		• **Party Time 1** • **Road Race: Shape Counting**	**Places Scenes** ***Goldilocks and the Three Bears*** **Pizza Game 1**	***Big Book Where's One?*** Building Blocks ***Teacher's Resource Guide*** • Places Scenes • Pizza Game 1 • Family Letter Week 7 ***Assessment*** Weekly Record Sheet

*provided in Manipulative Kit

Monday Planner

Objectives

- To produce a group of one to five objects
- To make a group equal in number to another group using one-to-one correspondence
- To count objects organized in a line up to 5
- To compare two groups to determine whether or not they have the same small number of objects

Materials

- flannel board with characters and props
- paper plates
- cloth napkins
- toy silverware
- *round counters
- *Number Cubes
- *Counting Cards

Vocabulary

Vertical means "up and down."

Math Throughout the Year

Review activity directions at the top of page 99, and complete each in class whenever appropriate.

Looking Ahead

For today, provide ***Big Book Where's One?*** for Warm-Up and copies of Pizza Game 1 for the Hands On Math Center. For tomorrow, gather items for Get Just Enough.

*provided in Manipulative Kit

Monday

1 Whole Group

20

Warm-Up: Numeral 1

- Read ***Big Book Where's One?***. Return to the page with the numeral 1; show and point to the 1, and have children say 1.
- Identify together what there is only one of on the page. Explain that the number of things matches the numeral.
- Model and explain how the numeral 1 is formed; it is a straight, up-to-down (vertical) line. Children should practice forming it in the air with their fingers: "Start at the top, and then go straight down."

Goldilocks and the Three Bears

- Tell *Goldilocks and the Three Bears* as a flannel-board story.
- After completing the story, discuss and show the one-to-one correspondence of bears to other items in the story. Ask children: How many bowls are in the story? How many chairs? How do you know? Then ask: Were there just enough beds for the bears? How do you know?
- Summarize that one-to-one matching creates equal groups. For example, when you know the number of bears in one group, then you know the number of beds in the other group.
- Tell children they will retell the story and match props later at the Hands On Math Center.

Monitoring Student Progress

If . . . children struggle with one-to-one correspondence during *Goldilocks and the Three Bears*,	**Then . . .** use the children and their classroom chairs to model and reinforce the concept.

2 Work Time

15

Computer Center

Party Time 1

Introduce Party Time 1: Set the Table from the ***Building Blocks*** software, in which children use one-to-one correspondence to set a table. Talk to children about getting one item for each placemat and how that will give them the same number of items. Each child should complete Party Time 1 this week.

Hands On Math Center

Both activities may occur at the center all week.

Goldilocks and the Three Bears

Children use flannel-board characters and props, paper plates, cloth napkins, and toy silverware to reenact the story, focusing on one-to-one matching.

Pizza Game 1

- Children play in pairs. Each player has a Pizza Game 1 activity sheet from the ***Teacher's Resource Guide***.
- Player One rolls a Number Cube, and puts that many counters on his or her plate. Player One asks Player Two, "Am I correct?" Player Two must agree that Player One is correct. Once correct, Player One moves the counters to the topping spaces on his or her pizza activity sheet.
- Players take turns until all their pizza spaces have toppings. To simplify the activity, make a Number Cube with one to three dots on each side; for a challenge, draw five to ten dots.

RESEARCH IN ACTION

Using numerals to label numbers, as well as representing collections with written symbols, are key steps toward mathematical abstraction.

3 Reflect

5

Ask children:

- **When you set a table, what can you say about the number of plates and the number of people? Why?**

Children might say: If there are two, the other should have two, or every person gets a plate.

4 Assess

Use the Weekly Record Sheet from ***Assessment*** to record children's progress. Use their time at the centers as an opportunity to complete your observations.

Tuesday Planner

Objectives

- To produce a group of one to five objects
- To make a group equal in number to another group using one-to-one correspondence
- To count objects organized in a line up to 5
- To compare two groups to determine whether or not they have the same small number of objects

Materials

- paper plates
- *counters (including round)
- dark cloth
- items for Get Just Enough
- *Counting Cards

Vocabulary

Horizontal means "side to side."

Math Throughout the Year

Review activity directions at the top of page 99, and complete each in class whenever appropriate.

Looking Ahead

From the ***Teacher's Resource Guide,*** provide Places Scenes for today's Hands On Math Center.

*provided in Manipulative Kit

Tuesday

1 Whole Group

10 

Warm-Up: Numeral 2

- Review ***Big Book Where's One?***. Return to the page with the numeral 2; show and point to the 2, and have children say 2.
- Identify together how many things there are two of on the page. Explain that the number of things matches the numeral.
- Model and explain how the numeral 2 is formed; it curves from the left up and then curves back to the bottom, connecting to a straight, left-to-right horizontal line. Children should form it in the air with their fingers.

Compare Snapshots

- Have prepared three counters on one plate and five on another. Cover the plate with five counters using a dark cloth. Show children both plates. Tell them to watch carefully and quietly, as you quickly reveal the covered plate so they can compare it to the other plate.
- Uncover the plate for two seconds, and cover it again. Ask children: Do the plates have the same number of counters? Because the answer is "no," ask: Which plate has more? Have children point to or say the number on the plate. Then ask: Which plate has fewer counters?
- Uncover the plate indefinitely, and ask children how many counters are on each plate. Confirm that 5 is more than 3 because 5 comes after 3.
- As children's abilities allow, repeat with groupings such as one to five counters in a line, one to four counters placed randomly, and finally, up to six or more counters in your own arrangement. If needed, reduce the number of counters; place counters in lines only; or, show counters a longer time.

2 Work Time

25 

Small Group

Get Just Enough

- Children work in pairs to "get just enough" of one group of items to match another group. Start with obvious pairings, such as plastic bottles and their tops (other suggestions are in the Weekly Planner).
- Ask children about the number of each type of item. If children only use one-to-one correspondence, challenge them to get just enough objects on one side of the table to pair with another group on the other side.
- If children struggle, use fewer items. If children excel, use less obvious pairings, such as red and blue blocks, or use more items to match.

Compare Game

- For each pair of children playing, two sets of Counting Cards (start with one to five dots) are needed. Deal the cards evenly, facedown to both players.
- Players simultaneously flip their top cards and compare them to find out which is greater (or has more dots). The player with the greater amount says "I have more!" and takes the opponent's card. If card amounts are equal, however, players each flip another card to determine a result. The game is over when all cards have been played, and the "winner" has more cards.

Monitoring Student Progress

If . . . children struggle during Compare Game,	**Then . . .** use only one to three or four cards, and review the concepts of *more* and *less*.
If . . . children excel during Compare Game,	**Then . . .** use cards up to 10; use Numeral Cards; or make cards with unusual dot arrangements.

Computer Center Building Blocks

Introduce Road Race: Shape Counting from the ***Building Blocks*** software. Children race by counting the sides of geometric shapes. Each child should play this week.

Road Race: Shape Counting

Hands On Math Center

Children may continue Monday's activities and/or this one.

Places Scenes

- Put a Counting Card on the table. Children choose Places Scenes to place counters on to match the card's amount and then tell a story about it.
- To simplify, use a lesser Counting Card, or have children put counters on the card's dots before moving them to a scene. For a challenge, use a greater card.

3 Reflect

5

Ask children:

■ **How did you figure out how to get just enough of what you were matching?**

Children might say: I counted and that told me how many to get.

4 Assess

During Small Group activities, use the Small Group Record Sheet from *Assessment* to observe and record children's progress.

Wednesday Planner

Objectives

- To produce a group of one to five objects
- To make a group equal in number to another group using one-to-one correspondence
- To count objects organized in a line up to 5
- To compare two groups to determine whether or not they have the same small number of objects

Materials

no new materials

Math Throughout the Year

Read a book to children, such as *Miss Spider's Tea Party* by David Kirk, that addresses one-to-one correspondence. Have children tell you, for example, how many of an item should be set before reading the answer.

Wednesday

1 Whole Group

Warm-Up: Numeral 3

- Review ***Big Book Where's One?***. Return to the page with the numeral 3; show and point to the 3, and have children say 3.
- Identify together how many things there are three of on the page. Explain that the number of things matches the numeral.
- Model and explain how the numeral 3 is formed; it is two curved parts on top of each other. You might use this rhyme: "Around the tree and around the tree—3, 3, 3."
- Children should practice forming the numeral 3 in the air with their fingers.

Number Jump

- Show a number of fingers, and write that numeral for children to see. Tell children to jump safely that many times. Count the jumps in unison. Repeat with another appropriate numeral.
- As a variation, here is the subitizing version: hide your hands behind your back, tell children to jump *only* if you hold up three fingers, and show your fingers for just two seconds.

Monitoring Student Progress

If . . . children need help during Number Jump,	**Then . . .** use smaller numerals, and say each numeral as you show your fingers.
If . . . children need a challenge during Number Jump,	**Then . . .** use larger numerals, and show finger combinations using both hands.

Eclipse Studios

Work Time

Computer Center Building Blocks

Continue to provide each child with a chance to complete Party Time 1 and Road Race: Shape Counting.

Hands On Math Center

Based on what children continue to learn and benefit from most, choose from this week's Hands On Math Center activities: *Goldilocks and the Three Bears*, Pizza Game 1, and Places Scenes. Consult the Weekly Planner for corresponding materials and, if needed, previous days for activity directions.

Reflect

Ask children:

- **How did you figure out how to get just enough of what you were matching?**

Children might say: You can see it. With encouragement, children might be able to explain that "There is a bowl for every bear" or the like.

Assess

Use the Weekly Record Sheet from ***Assessment*** to record children's progress. Use their time at the centers as an opportunity to complete your observations.

RESEARCH IN ACTION

To recognize and write a numeral, children need to know its parts, how the parts fit together, and left and right. For example, 3 has two curves, one on top of the other, and the curves start on the left. Most children this age do not know left and right so the concepts have to be communicated in child-friendly, age-appropriate ways, such as temporarily referring to the "window side" of the board as the right or left, whichever is the case in your classroom.

Thursday Planner

Objectives

- To produce a group of one to five objects
- To make a group equal in number to another group using one-to-one correspondence
- To count objects organized in a line up to 5
- To compare two groups to determine whether or not they have the same small number of objects

Materials

- paper plates
- *counters
- dark cloth
- items for Get Just Enough
- *Counting Cards

Math Throughout the Year

Review activity directions at the top of page 99, and complete each in class whenever appropriate.

Looking Ahead

For tomorrow, make copies of Family Letter Week 7 from the *Teacher's Resource Guide.*

*provided in Manipulative Kit

Thursday

1 Whole Group 10

Warm-Up: Numeral 4

- Review ***Big Book Where's One?***. Return to the page with the numeral 4; show and point to the 4, and have children say 4.
- Identify together how many things there are four of on the page. Explain that the number of things matches the numeral.
- Model and explain how the numeral 4 is formed; it has three straight lines, two shorter—one up-to-down (vertical), another left-to-right (horizontal). You might use this rhyme: "Down, stop! Over and down some more—4, 4, 4."
- Children should practice forming the numeral 4 in the air with their fingers.

Compare Snapshots

- Have prepared three counters on one plate and five on another. Cover the plate with five counters using a dark cloth. Show children both plates. Remind children to watch carefully and quietly as you quickly reveal the covered plate so they can compare it to the other plate.
- Uncover the plate for two seconds, and cover it again. Ask children: Which plate has more? Have children point to the plate. Then ask: Which plate has fewer counters?
- Uncover the plate indefinitely, and ask children how many counters are on each plate. As children's abilities allow, repeat with groupings such as one to five counters in a line, one to four counters placed randomly, and finally, up to six or more counters in an arrangement of your choice. If children struggle, reduce the number of counters; place counters in lines only; or show counters for a longer time.

2 Work Time 25

Small Group

Get Just Enough

- Children work in pairs to "get just enough" of one group of items to match another group. Start with obvious pairings, such as markers and their caps.
- Ask children about the number of each type of item. Whenever children compare items, encourage them to count.

Monitoring Student Progress

If . . . children struggle during Get Just Enough,	**Then . . .** use fewer items.
If . . . children excel during Get Just Enough,	**Then . . .** use less obvious pairings, or use more items to match.

Compare Game

- For each pair of children playing the game, two or more sets of Counting Cards are needed. Deal the cards evenly, facedown to both players.
- Players simultaneously flip their top cards and compare them to find out which is greater. The player with the greater amount says "I have more!" and takes the opponent's card. If card amounts are equal, players each flip another card to determine a result. The game is over when all cards have been played, and the "winner" is the player with more cards.
- To simplify the game, use only one to three or four cards. For a challenge, ask children to count the pairs they have, not the cards.

Computer Center

Continue to provide each child with a chance to complete Party Time 1 and Road Race: Shape Counting.

Hands On Math Center

Based on what children continue to learn and benefit from most, choose from this week's Hands On Math Center activities: *Goldilocks and the Three Bears*, Pizza Game 1, and Places Scenes. Consult the Weekly Planner for corresponding materials and, if needed, Monday and Tuesday for activity directions.

3 Reflect 5

Ask children:

■ **When you played Compare Game, how did you figure out who had more?**

Children might say: I just saw it. With guidance, they might be able to explain that "I counted 3 for me and 5 for her. 5 comes after 3 when counting so she has more."

4 Assess

During Small Group activities, use the Small Group Record Sheet from ***Assessment*** to observe and record children's progress.

RESEARCH IN ACTION

The word *numerals* means "written numbers," such as 1 and 4. These symbols can also be written as number words (*one* and *four*). An actual number is the idea of, for example, four things. We might use the word *numerals* when reading or writing them, but we do not insist that children do so.

Friday Planner

Objectives

- To produce a group of one to five objects
- To make a group equal in number to another group using one-to-one correspondence
- To count objects organized in a line up to 5
- To compare two groups to determine whether or not they have the same small number of objects

Materials

no new materials

Math Throughout the Year

Review activity directions at the top of page 99, and complete each in class whenever appropriate.

Looking Ahead

For next week, familiarize yourself with Pizza Pizzazz 3: Make Number Pizzas (1–5) and Numeral Train Game from the ***Building Blocks*** software.

Friday

1 Whole Group

10 

Warm-Up: Numeral 5

- Review ***Big Book Where's One?***. Return to the page with the numeral 5; show and point to the 5, and have children say 5.
- Identify together how many things there are five of on the page. Explain that the number of things matches the numeral.
- Model and explain how the numeral 5 is formed. Starting at the top, it drops straight down part way, curves out like a sideways letter *u*, and ends with a side-to-side line back at the top. You might use this rhyme: "A short little line and a big round tummy—don't forget its hat. 5 is funny!"
- Children should practice forming the numeral 5 in the air with their fingers.

Monitoring Student Progress

If . . . children hesitate with the numeral 5 (perhaps not enough practice with 1 through 4),	**Then . . .** proceed at a slower pace, or spend more time on 5 specifically.
If . . . children show confidence with the early numerals,	**Then . . .** write and read very large numerals, such as 100, and have fun saying them together!

Number Jump

- Show a number of fingers, and write that numeral for children to see. Tell children to jump safely that many times. Count the jumps in unison. Repeat with another appropriate numeral.
- As a variation, here is the subitizing version: hide your hands behind your back, tell children to jump *only* if you hold up three fingers, and show your fingers for just two seconds.

Work Time

25

Computer Center Building Blocks

Continue to provide each child with a chance to complete Party Time 1 and Road Race: Shape Counting.

Hands On Math Center

Based on what children continue to learn and benefit from most, choose from this week's Hands On Math Center activities: *Goldilocks and the Three Bears*, Pizza Game 1, and Places Scenes. Consult the Weekly Planner for corresponding materials and, if needed, Monday and Tuesday for activity directions.

Home Connection

From the ***Teacher's Resource Guide,*** distribute to children copies of Family Letter Week 7 to share with their family. Each letter has an area for children to show families an example of what they have been doing in class.

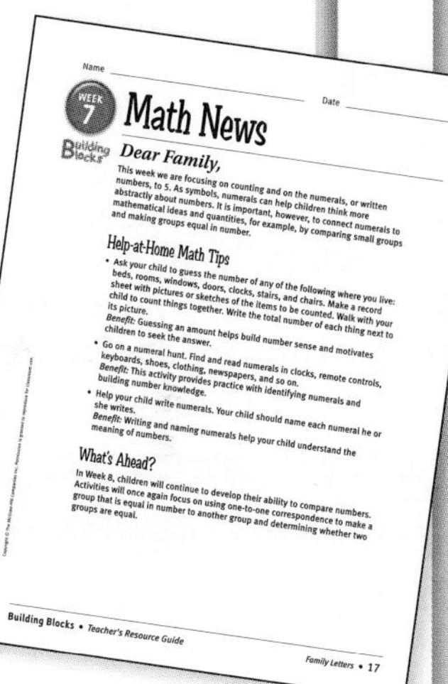

Name ______ Date ______

WEEK 7

Math News

Building Blocks

Dear Family,

This week we are focusing on counting and on the numerals, or written numbers, to 5. As symbols, numerals can help children think more abstractly about numbers. It is important, however, to connect numerals to mathematical ideas and quantities, for example, by comparing small groups and making groups equal in number.

Help-at-Home Math Tips

- Ask your child to guess the number of any of the following where you live: beds, rooms, windows, doors, clocks, stairs, and chairs. Make a record sheet with pictures or sketches of the items to be counted. Walk with your child to count things together. Write the total number of each thing next to its picture.
  Benefit: Guessing an amount helps build number sense and motivates children to seek the answer.
- Go on a numeral hunt. Find and read numerals in clocks, remote controls, keyboards, shoes, clothing, newspapers, and so on.
  Benefit: This activity provides practice with identifying numerals and building number knowledge.
- Help your child write numerals. Your child should name each numeral he or she writes.
  Benefit: Writing and naming numerals help your child understand the meaning of numbers.

What's Ahead?

In Week 8, children will continue to develop their ability to compare numbers. Activities will once again focus on using one-to-one correspondence to make a group that is equal in number to another group and determining whether two groups are equal.

Building Blocks • *Teacher's Resource Guide*   *Family Letters* • 17

Reflect

5

Briefly discuss with children what you have done in class this week, such as learning about one-to-one correspondence, and ask:

- **What numerals do you see in the classroom? Where are they?**

Children might say: I see 1, 2, 3, 4, and 5 on the clock, I see them in our room number, and the like.

Assess

Use the Weekly Record Sheet from ***Assessment*** to record children's progress. Use their time at the centers as an opportunity to complete your observations.

Assess and Differentiate

A Gather Evidence

Review children's progress in mathematics by looking at the Weekly Record Sheets (Monday, Wednesday, Friday) and the Small Group Record Sheets (Tuesday, Thursday) from this past week.

B Summarize Findings

Using *Online Assessment*, summarize and analyze assessment data for each child based on your weekly observations and Record Sheets. Such information helps to determine where each child is on the math trajectory for counting, comparing, and subitizing. See *Assessment* for the print companion to each Learning Trajectory Record.

C Differentiate Instruction

Once you have seen a child exhibit specific levels of the trajectory, begin to encourage and work with that child toward the next level. Refer to Appendix A for individualized instruction opportunities, including Special Education concerns.

Object Counting

If . . .	Then . . .
If . . . the child can count five items and tell how many with the last number counted,	**Then . . .** *Counter (Small Numbers)* Accurately counts objects in a line to 5, and answers "how many?" with the last number counted. When objects are visible, especially with small numbers, begins to understand cardinality.
If . . . the child can produce up to five items,	**Then . . .** *Producer (Small Numbers)* Counts out objects to 5.

Comparing and Ordering

If . . .	Then . . .
If . . . the child can compare small groups by matching,	**Then . . .** *Matching Comparer* Compares groups up to 6 by matching.
If . . . the child can compare only similarly-sized items in small groups,	**Then . . .** *Counting Comparer (Same Size)* Accurate comparison via counting but only when objects are similar in size and in groups up to 5.
If . . . the child can compare various small groups, sometimes labeling one as having more or less,	**Then . . .** *Counting Comparer (5)* Compares with counting even when larger group's objects are smaller; figures out how many more or less.

Subitizing

If . . .	Then . . .
If . . . the child can instantly identify groups up to 5,	**Then . . .** *Perceptual Subitizer to 5* Instantly recognizes collections of up to 5 when shown briefly and verbally names the number of items.
If . . . the child can label some arrangements up to 10,	**Then . . .** *Conceptual Subitizer to 5+* Verbally labels all arrangements to about 5 when shown briefly using groups: "I saw 2 and 2 so I said 4" or "I made 2 groups of 3 plus 1 in my mind so that is 7."

WEEK 8

Overview

Big Ideas

- counting
- one-to-one correspondence
- comparing number
- subitizing

Teaching for Understanding

Understanding number involves subitizing (the instant recognition of small groups) to determine how many in a group; counting to determine how far along a path; counting or subitizing to recognize how many actions in a sequence; and counting, subitizing, or one-to-one corresponding to compare amounts greater or less than. The first three might seem the same; however, children may understand number as how many in a group but not understand how to count how many steps along a path. When children connect the meanings, they form a strong foundation for all future mathematical learning.

Counting Games

Games can provide good experiences for children. Aside from providing a highly motivational setting for counting, games give meaning to counting. Children know if they roll a 3, they move 3 spaces. In doing so, they translate one representation of 3 (three dots) to a very different one (three moves on a game board). Thus learning the greater the number they roll, the farther they move. Games also help children develop social and language skills by taking turns and playing fairly.

Rhythmic Patterns and Number

A new type of activity this week involves observing a sequence of actions and reproducing them, and it introduces a new kind of subitizing—temporal subitizing, which is hearing a rhythmic pattern, learning it, and reproducing it. This is another and important way to understand number. Later, when children count up 3 from 6, they will be able to "feel three beats" and count "6...7, 8, 9."

Meaningful Connections

Children should start to literally see that numbers and the numerals that represent them carry meaning. As children identify amounts in small groups, they begin to recognize the parts of a whole and may begin to apply that knowledge to activities, games, and other daily routines.

What's Ahead?

In the next few weeks, children continue to build their understanding of number, yet the focus will shift to geometry. They will learn more about two-dimensional (flat) and three-dimensional shapes (solids, prisms). Children will build images of these shapes and will distinguish and describe why a certain shape is or is not a member of a class of shapes.

English Learner

For English learners, when using food examples, include culturally familiar items. Review the following: ***Number Cubes,*** *upside-down,* and *marbles.* Refer to the ***Teacher's Resource Guide.***

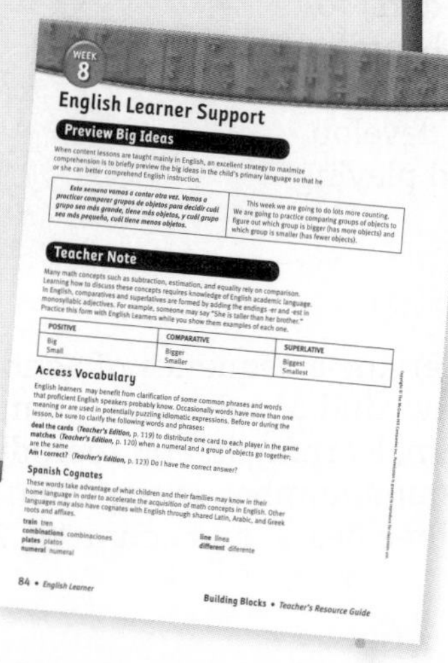

WEEK 8

English Learner Support

Preview Big Ideas

When content lessons are taught mainly in English, an excellent strategy to maximize comprehension is to briefly preview the big ideas in the child's primary language so that he or she can better comprehend English instruction.

Esta semana vamos a contar otra vez. Vamos a practicar comparar grupos de objetos para decidir cuál grupo sea más grande, tiene más objetos, y cuál grupo sea más pequeño, cuál tiene menos objetos.	This week we are going to do lots more counting. We are going to practice comparing groups of objects to figure out which group is bigger (has more objects) and which group is smaller (has fewer objects).

Teacher Note

Many math concepts such as subtraction, estimation, and equality rely on comparison. Learning how to discuss these concepts requires knowledge of English academic language. In English, comparatives and superlatives are formed by adding the endings -er and -est in monosyllabic adjectives. For example, someone may say "She is taller than her brother." Practice this form with English Learners while you show them examples of each one.

POSITIVE	COMPARATIVE	SUPERLATIVE
Big Small	Bigger Smaller	Biggest Smallest

Access Vocabulary

English learners may benefit from clarification of some common phrases and words that proficient English speakers probably know. Occasionally words have more than one meaning or are used in potentially puzzling idiomatic expressions. Before or during the lesson, be sure to clarify the following words and phrases:

deal the cards (*Teacher's Edition,* p. 119) to distribute one card to each player in the game
matches (*Teacher's Edition,* p. 120) when a numeral and a group of objects go together; are the same
Am I correct? (*Teacher's Edition,* p. 123) Do I have the correct answer?

Spanish Cognates

These words take advantage of what children and their families may know in their home language in order to accelerate the acquisition of math concepts in English. Other languages may also have cognates with English through shared Latin, Arabic, and Greek roots and affixes.

train tren
combinations combinaciones
plates platos
numeral numeral
line línea
different diferente

84 • *English Learner* Building Blocks • *Teacher's Resource Guide*

Technology Project

Children may get additional problem-posing and -solving experience using Pizza Pizzazz Free Explore. Children can count along with the program as they place toppings on either pizza, or, working cooperatively, one child can name a number for the other to put that many toppings on a pizza. Challenge children to think of other ways to explore the program. As you visit them, ask children about the connection between numerals and the number of toppings.

How Children Learn to Count and Compare

Knowing where children are on the learning trajectory for counting, comparing, and subitizing and what the next steps are helps facilitate their development. Children who easily surpass these trajectory levels might be challenged by larger quantities and encouraged to assist other children.

Object Counting

What to Look For What size groups can the child label and/or produce?

Counter (Small Numbers) Accurately counts objects in a line to 5, and answers "how many?" with the last number counted. When objects are visible, especially with small numbers, begins to understand cardinality.

Producer (Small Numbers) Counts out objects to 5.

Counter (10) Counts structured arrangements of objects to 10; may be able to write or draw to represent 10.

Comparing and Ordering

What to Look For What types of groups does the child identify?

Counting Comparer (Same Size) Accurate comparison via counting but only when objects are similar in size and in groups up to 5.

Counting Comparer (5) Compares with counting even when larger group's objects are smaller; figures out how many more or less.

Subitizing

What to Look For Does the child instantly recognize groups of at least 5?

Perceptual Subitizer to 5 Instantly recognizes collections of up to 5 when shown briefly and verbally names the number of items.

Conceptual Subitizer to 5+ Verbally labels all arrangements to about 5 when shown briefly using groups: "I saw 2 and 2 so I said 4," or "I made 2 groups of 3 plus 1 in my mind so that is 7."

Numerals

What to Look For Can the child read numerals?

Numeral Recognizer Reads single-digit numerals, recognizing them as symbolizing number words.

Math Throughout the Year

Math Throughout the Year activities are recommended to build on the mathematical skills highlighted in each week. Here are suggested activities for **Week 8.**

Counting Jar

A counting jar holds a specified number of items for children to count. Use the same jar all year, changing its small amount of items weekly. Have children spill the items to count them.

Numerals Every Day

Numerals are all around us. Help children notice and read numerals on common items throughout the day, such as clocks, food containers, street signs, room numbers, newspapers, and so on.

Center Preview

Computer Center Building Blocks

Get your classroom Computer Center ready for Pizza Pizzazz 3: Make Number Pizzas (1–5) and Numeral Train Game from the *Building Blocks* software.

After you introduce Pizza Pizzazz 3 and Numeral Train Game, each child should complete the activities individually as you (or an assistant) monitor and guide him or her periodically. Ideally, each child will have at least ten minutes of computer time at least twice during the week. Use children's center time as an opportunity for assessment.

Hands On Math Center

This week's Hands On Math Center activities are Compare Game, Get Just Enough, Find the Number, Places Scenes, and Pizza Game 1. Supply the center with these materials: Counting Cards, dark containers, counters (including round), paper plates, Numeral Cards, Pizza Game 1 activity sheets, and Number Cubes. Please refer to the Weekly Planner for Get Just Enough's materials.

Literature Connections

These books help develop counting.

- *Feast for 10* by Cathryn Falwell
- *Fiesta* by Ginger Foglesong
- *Miss Spider's Tea Party* by David Kirk
- *Uno, Dos, Tres: One, Two, Three* by Pat Mora
- *One Hungry Monster: A Counting Book in Rhyme* by Susan Heyboer O'Keefe

(t)Matt Meadows (b)Eclipse Studios

WEEK 8 Weekly Planner

Learning Trajectories

Week 8 Objectives

- To produce a group of one to five objects
- To make a group equal in number to another group using one-to-one correspondence
- To count objects (or "steps" in a path) organized in a line up to 5
- To compare two groups to determine whether or not they have the same small number of objects
- To quickly recognize the number of objects in a small group when shown only briefly

	Developmental Path	Instructional Activities
Object Counting	*Corresponder*	
	Counter (Small Numbers)	Listen and Count Number Race Pizza Pizzazz 3 Listen and Copy Pizza Game 1
	Producer (Small Numbers)	Number Race Pizza Pizzazz 3 Pizza Game 1 Number Jump Places Scenes
	Counter (10)	Number Jump Pizza Game 1
	Counter and Producer (10+)	
Comparing and Ordering	*Matching Comparer*	
	Counting Comparer (Same Size)	Compare Game
	Counting Comparer (5)	Compare Game Get Just Enough Find the Number
	Counting Comparer (10)	
Subitizing	*Perceptual Subitizer to 4*	
	Perceptual Subitizer to 5	Listen and Copy
	Conceptual Subitizer to 5+	Where's My Number?
	Conceptual Subitizer to 10	

Use this chart to plan for your specific class schedule. If you have your prekindergarteners for only three days, complete Monday, Tuesday, and Thursday of the week.

Work Time

Whole Group	Small Group	Computer	Hands On	Program Resources
Numeral 6 Listen and Count *Materials:* counting book coffee can marbles		**Numeral Train Game**	**Compare Game** *Materials:* *Counting Cards **Get Just Enough** *Materials:* paintbrushes and cups plastic bottles and tops small boxes and toy cars toy cups and saucers	***Big Book Where's One?*** Building Blocks ***Assessment*** Weekly Record Sheet
Listen and Copy **Where's My Number?** *Materials:* *Numeral Cards *counters	**Number Jump** *Materials:* *Numeral Cards **Number Race** *Materials:* *game board *game pieces *Number Cube	**Pizza Pizzazz 3**	**Find the Number** *Materials:* dark containers *counters paper plates *Numeral Cards **Compare Game** Get Just Enough	Building Blocks ***Assessment*** Small Group Record Sheet
Number Jump Listen and Count *Materials:* counting book coffee can marbles		• **Numeral Train Game** • **Pizza Pizzazz 3**	**Places Scenes** *Materials:* *counters *Numeral Cards **Pizza Game 1** *Materials:* *round counters *Number Cube paper plates	Building Blocks ***Teacher's Resource Guide*** • Places Scenes • Pizza Game 1 ***Assessment*** Weekly Record Sheet
Listen and Copy **Where's My Number?** *Materials:* *Numeral Cards *counters	**Number Jump** *Materials:* *Numeral Cards **Number Race** *Materials:* *game board *game pieces *Number Cube	• **Numeral Train Game** • **Pizza Pizzazz 3**	**Places Scenes** **Pizza Game 1** **Compare Game** **Get Just Enough** **Find the Number**	Building Blocks ***Teacher's Resource Guide*** • Places Scenes • Pizza Game 1 ***Assessment*** Small Group Record Sheet
Number Jump Listen and Count *Materials:* counting book coffee can marbles		• **Numeral Train Game** • **Pizza Pizzazz 3**	**Places Scenes** **Pizza Game 1** **Compare Game** **Get Just Enough** **Find the Number**	Building Blocks ***Teacher's Resource Guide*** • Places Scenes • Pizza Game 1 • Family Letter Week 8 ***Assessment*** Weekly Record Sheet

*provided in Manipulative Kit

Monday Planner

Objectives

- To produce a group of one to five objects
- To make a group equal in number to another group using one-to-one correspondence
- To count objects (or "steps" in a path) organized in a line up to 5
- To compare two groups to determine whether or not they have the same small number of objects
- To quickly recognize the number of objects in a small group when shown only briefly

Materials

- counting book
- coffee can
- marbles
- *Counting Cards
- items for Get Just Enough

Math Throughout the Year

Review activity directions at the top of page 115, and complete each in class whenever appropriate.

Looking Ahead

For today, provide ***Big Book Where's One?*** for Warm-Up and a counting book for Listen and Count, and make sure items for Get Just Enough (continuing from last week) are at the Hands On Math Center. For tomorrow, gather materials for Small Group's Number Race.

*provided in Manipulative Kit

Monday

1 Whole Group

15

Warm-Up: Numeral 6

- Read ***Big Book Where's One?***. Return to the page with the numeral 6; show and point to the 6, and have children say 6.
- Count together how many things there are six of on the page. Explain that the number of things matches the numeral.
- Model and explain how the numeral 6 is formed; it slants down to the left and then continues up into a loop. You may wish to use this rhyme: "Curve down and then around 'til it sticks. That's how you write the numeral 6."
- Children should practice forming the numeral 6 in the air with their fingers.

Listen and Count

- Read a book, such as *Blueberries for Sal* by Robert McCloskey, in which something specific is being counted. And, for example, tell children you are going to drop items into a "bucket" like Sal did. Ask them to listen quietly as you slowly drop marbles (or counters) into a clean, empty coffee can.
- When you finish, have children hold up their fingers to show how many marbles they think are in the can. After you have observed their responses, ask children to say the number.
- Spill the items out of the can, and count them as a whole group to check.
- Repeat twice more with other small numbers.

2 Work Time

20

Computer Center Building Blocks

Introduce Numeral Train Game from the **Building Blocks** software. In this activity, children identify numerals 1 to 5 and move forward the corresponding number of spaces on the program's game board. Each child should have an opportunity to complete Numeral Train Game this week.

Hands On Math Center

Today's activities, which continue from Week 7, occur at the center all week.

Compare Game

- For each pair of children playing the game, two or more sets of Counting Cards are needed. Children mix the cards and then deal them evenly facedown.
- Players simultaneously flip their top cards and compare them to find out which is greater. The player with the greater amount says "I have more!" and takes the opponent's card. If card amounts are equal, however, players each flip another card to determine a result.
- The game is over when all cards have been played, and the "winner" is the player with more cards.

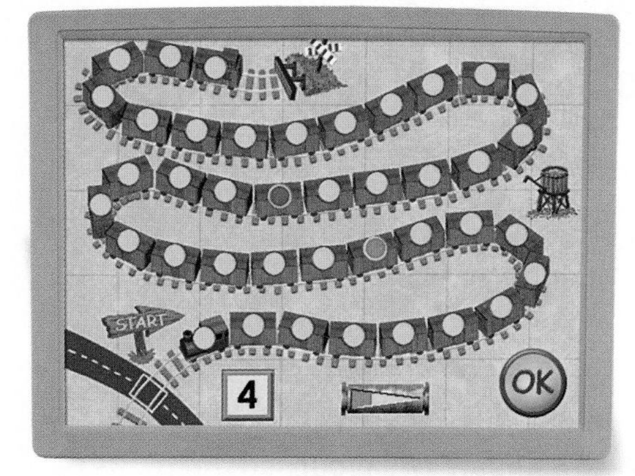

Numeral Train Game

Monitoring Student Progress

If . . .	Then . . .
If . . . children need help during Compare Game,	**Then . . .** use cards with fewer dots.
If . . . children need a challenge during Compare Game,	**Then . . .** use cards with more dots; use Numeral Cards; or have players count dot pairs instead of cards.

Get Just Enough

- Children "get just enough" of one group of items to match another group. Start with obvious pairings, such as nontoxic markers and their caps or toy dogs and plastic bowls (other suggestions are in the Weekly Planner). Whenever children compare items, encourage them to count.
- If children struggle, use fewer items to match. If children excel, use less obvious pairings, such as red and blue blocks, or use more items to match.

Reflect

Ask children:

■ **When you played Compare Game, how did you figure out who had more?**

Children might say: When one is big and one is little, I can just see; or she had 8 and I had 7 so I had to count.

Assess

Use the Weekly Record Sheet from *Assessment* to record children's progress. Use their time at the centers as an opportunity to complete your observations.

Tuesday Planner

Objectives

- To produce a group of one to five objects
- To make a group equal in number to another group using one-to-one correspondence
- To count objects (or "steps" in a path) organized in a line up to 5
- To compare two groups to determine whether or not they have the same small number of objects
- To quickly recognize the number of objects in a small group when shown only briefly

Materials

- *Numeral Cards
- *counters
- *game board
- *game pieces
- *Number Cube
- dark containers
- paper plates

Math Throughout the Year

Review activity directions at the top of page 115, and complete each in class whenever appropriate.

Looking Ahead

For tomorrow's Hands On Math Center, provide copies of the Places Scenes and Pizza Game 1 activity sheet from the ***Teacher's Resource Guide.***

*provided in Manipulative Kit

Tuesday

Whole Group

5

Warm-Up: Listen and Copy

Clap 1 to 5 times, and tell children to clap the same number of times in the same way. For example, clap quickly, slowly, or with pauses to create patterns, such as clap, clap, pause, clap. Repeat with different small numbers.

Where's My Number?

- Show a Numeral Card to children. Secretly put that many counters in one of your hands; put a different amount in your other hand. Hold out your closed hands, open them for two seconds, and then close them.
- Have children point to the hand with the number of counters that matches the Numeral Card. Repeat with other small numerals.

Work Time

30

Small Group

Number Jump

- Show a number of fingers, and write that numeral for children to see. Tell children to jump safely that many times. Count the jumps in unison. Repeat with another appropriate number.
- As a variation, here is the subitizing version: hide your hands behind your back, tell children to jump only if you show three fingers, and show your fingers for only two seconds.
- To simplify, use smaller numbers, and say each number as you show your fingers. For a challenge, show finger combinations on both hands.

Number Race

- Demonstrate the game by playing it with a child. The game has two players or two teams with two players. Each game is played with one game board, one Number Cube, and a game piece for each player.
- Player One rolls the Number Cube and announces the number that was rolled. Player Two checks whether Player One said the correct number. Player One moves that many spaces.
- Player Two takes his or her turn, following the same steps. The game continues until all players reach the end.

- Review directions as needed. When applicable, help players on each team take turns by guiding Player One on Team One, then Player One on Team Two, and so on until the teams can proceed without you.

Monitoring Student Progress

If . . . children struggle during Number Race,	**Then . . .** provide a Number Cube up to 3 only.
If . . . children excel during Number Race,	**Then . . .** provide a Number Cube 3 to 8, or ask players to tell you how many it would take to land on a particular space or to finish the game.

Computer Center *Building Blocks*

Introduce Pizza Pizzazz 3: Make Number Pizzas (1–5) from the ***Building Blocks*** software. Children are given a verbal pizza order, and they make a pizza with that many toppings. Each child should complete the activity this week.

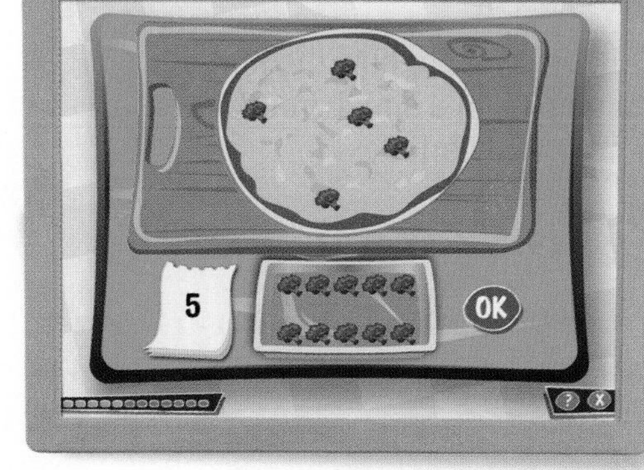

Pizza Pizzazz 3

Hands On Math Center

Children may continue Monday's activities and/or complete this one.

Find the Number

- Before children get to the center, conceal several pizzas (paper plates), each with a different number of toppings (round counters) under its own dark container.
- Place a Numeral Card in plain view. Children lift each container to count toppings until they find the pizza that matches the card's numeral. They can show their answer to a classmate or an adult.
- To simplify the activity, reduce the number of containers, or display the pizzas uncovered. For a challenge, have children work in pairs, making their own groups, and have one another, for example, "Find the 10."

3 Reflect 5

Show a familiar numeral, and ask children:

■ **What numeral is this? How do you know?**

Children might say: I know that is a (insert numeral) because of how it looks.

4 Assess

During Small Group activities, use the Small Group Record Sheet from *Assessment* to observe and record children's progress.

Wednesday Planner

Objectives

- To produce a group of one to five objects
- To make a group equal in number to another group using one-to-one correspondence
- To count objects (or "steps" in a path) organized in a line up to 5
- To compare two groups to determine whether or not they have the same small number of objects
- To quickly recognize the number of objects in a small group when shown only briefly

Materials

- counting book
- coffee can
- marbles
- *counters (including round)
- *Numeral Cards
- *Number Cube
- paper plates

Math Throughout the Year

Read a book to children, such as *Miss Spider's Tea Party* by David Kirk, that addresses one-to-one correspondence. Have children tell you, for example, how many of an item should be set before reading the answer.

*provided in Manipulative Kit

Wednesday

1 Whole Group

15 

Warm-Up: Number Jump

- Show a number of fingers, and write that numeral for children to see.
- Tell children to jump safely that many times, and count the jumps in unison. Repeat the activity with another appropriate numeral.

Listen and Count

- Read a book, such as *Blueberries for Sal* by Robert McCloskey, in which something specific is being counted. Remind children you are going to drop items into a can for them to count. Ask children to listen quietly as you slowly drop marbles (or counters) into a clean, empty coffee can.
- When you finish, have children hold up their fingers to show how many marbles they think are in the can. After you have observed their responses, ask children to say the number.
- Spill the items out of the can, and count them as a whole group to check.

2 Work Time

20 

Computer Center Building Blocks

Continue to provide each child with a chance to complete Numeral Train Game and Pizza Pizzazz 3.

Hands On Math Center

Based on what children learn and benefit from most, allow them to continue this week's previous activities, adding these recurring ones.

Places Scenes

Put a Numeral Card on the table, and have children choose several Places Scenes to place counters on each to match the card's amount. Children should tell a story about one scene.

Monitoring Student Progress

If . . . children struggle during Places Scenes,	**Then . . .** have them specifically use the space or beach scene where items up to 5 can be counted out on special spaces.
If . . . children excel during Places Scenes,	**Then . . .** use a greater Numeral Card.

RESEARCH IN ACTION

The computer can help children connect representations of ideas, such as the written numeral 4, the spoken number word *four*, and a collection of 4.

Pizza Game 1

- Each player has a Pizza Game 1 activity sheet from the ***Teacher's Resource Guide***.
- Player One rolls a Number Cube and puts that many counters on his or her plate. Player One asks Player Two, "Am I correct?" Player Two must agree that Player One is correct. Once correct, Player One moves the counters to the topping spaces on his or her pizza activity sheet. Players take turns until all the spaces on their pizzas have toppings.
- If needed, use a blank wooden cube to make a Number Cube that is easier for children, such as one to three dots on each side, or, for a challenge, draw up to ten dots on each side.

Reflect

Show children a list such as 4, 3, 5, 2, 1, and ask:

■ **Which is the numeral 2?**

Children either respond correctly or point to the 5. Respond with praise, or explain with encouragement that 5 looks almost like an upside-down 2.

Assess

Use the Weekly Record Sheet from ***Assessment*** to record children's progress. Use their time at the centers as an opportunity to complete your observations.

Eclipse Studios

Thursday Planner

Objectives

- To produce a group of one to five objects
- To make a group equal in number to another group using one-to-one correspondence
- To count objects (or "steps" in a path) organized in a line up to 5
- To compare two groups to determine whether or not they have the same small number of objects
- To quickly recognize the number of objects in a small group when shown only briefly

Materials

- *Numeral Cards
- *counters
- *game board
- *game pieces
- *Number Cube

Math Throughout the Year

Review activity directions at the top of page 115, and complete each in class whenever appropriate.

Looking Ahead

For tomorrow, make copies of Family Letter Week 8 from the *Teacher's Resource Guide.*

*provided in Manipulative Kit

Thursday

1 Whole Group 5

Warm-Up: Listen and Copy

Clap one to five times, and tell children to clap the same number of times in the same way. Clap quickly, slowly, or with pauses to create patterns, such as clap, clap, pause, clap.

Where's My Number?

- Show a Numeral Card to children. Secretly put that many counters in one of your hands; put a different amount in your other hand.
- Hold out your closed hands, open them for two seconds, and then close them.
- Have children point to the hand with the number of counters that matches the Numeral Card.
- Repeat with new numerals.

2 Work Time 30

Small Group

Number Jump

- Show a number of fingers, and write that numeral for children to see. Tell children to jump safely that many times. Count the jumps in unison. Repeat with another appropriate number.
- As a variation, here is the subitizing version: hide your hands behind your back, tell children to jump only if you show three fingers, and show your fingers for only two seconds.

Monitoring Student Progress

If . . . children struggle during Number Jump,	**Then . . .** use smaller numbers, and say each number as you show your fingers.
If . . . children excel during Number Jump,	**Then . . .** use larger numbers, and show finger combinations using both hands.

Number Race

- Demonstrate the game again if needed before children begin. Remember the game has two players or two teams with two players, and is played with one game board, one Number Cube, and a game piece per player.
- Player One rolls the Number Cube and announces the number that was rolled. Player Two checks whether Player One said the correct number. Player One moves that many spaces.
- Player Two takes his or her turn, following the same steps. The game continues until all players reach the end. Review the directions with pairs or teams until they can proceed without guidance.
- To simplify the game, provide Number Cubes up to 3 only. For a challenge, provide a Number Cube with 3 to 8 (dots or numerals), or ask players to tell you how many it would take to land on a particular space or to finish the game.

Computer Center Building Blocks

Continue to provide each child with a chance to complete Numeral Train Game and Pizza Pizzazz 3.

Hands On Math Center

Based on what children continue to learn and benefit from most, choose from this week's Hands On Math Center activities: Compare Game, Get Just Enough, Find the Number, Places Scenes, and Pizza Game 1. Consult the Weekly Planner for corresponding materials and, if needed, previous days for activity directions.

3 Reflect

Ask children:

■ **What do you count when you play Number Race?**

Children might say: You count how many you rolled; or you count how many jumps to take, spaces to move, or the like.

4 Assess

During Small Group activities, use the Small Group Record Sheet from *Assessment* to observe and record children's progress.

Friday

Friday Planner

Objectives

- To produce a group of one to five objects
- To make a group equal in number to another group using one-to-one correspondence
- To count objects (or "steps" in a path) organized in a line up to 5
- To compare two groups to determine whether or not they have the same small number of objects
- To quickly recognize the number of objects in a small group when shown only briefly

Materials

- counting book
- coffee can
- marbles

Math Throughout the Year

Review activity directions at the top of page 115, and complete each in class whenever appropriate.

Looking Ahead

For next week, familiarize yourself with Memory Geometry 1: Exact Matches and Number Snapshots 2 from the ***Building Blocks*** software, and assemble the Shape Flip Book if you have not already.

1 Whole Group 15

Warm-Up: Number Jump

- Show a number of fingers, and write that numeral for children to see.
- Tell children to jump safely that many times, and count the jumps in unison.
- Repeat the activity with another appropriate numeral.

Listen and Count

- Read a book, such as *Blueberries for Sal* by Robert McCloskey, in which something specific is being counted. Remind children you are going to drop items into a can for them to count. Ask children to listen quietly as you slowly drop marbles (or counters) into a clean, empty coffee can.
- When you finish, have children hold up their fingers to show how many marbles they think are in the can. After you have observed their responses, ask children to say the number.
- Spill the items out of the can, and count them as a whole group to check.

Monitoring Student Progress

If . . . children struggle during Listen and Count,	**Then . . .** drop the items slower, encouraging children to count slowly as they hear an item make a single, distinct sound.
If . . . children excel during Listen and Count,	**Then . . .** increase the number of items you drop.

Eclipse Studios

Work Time

20

Computer Center

Building Blocks

Continue to provide each child with a chance to complete Numeral Train Game and Pizza Pizzazz 3.

Hands On Math Center

Based on what children continue to learn and benefit from most, choose from this week's Hands On Math Center activities: Compare Game, Get Just Enough, Find the Number, Places Scenes, and Pizza Game 1. Consult the Weekly Planner for corresponding materials and, if needed, previous days for activity directions.

Reflect

5

Briefly discuss with children what you have done in class this week, such as matching numerals to corresponding amounts, and quickly review numerals 1 through 6. Ask:

■ **What do numerals mean?**

Children might say: They are numbers; they are like letters, but they tell how many; or I use them to count.

Assess

Use the Weekly Record Sheet from ***Assessment*** to record children's progress. Use their time at the centers as an opportunity to complete your observations.

Home Connection

From the ***Teacher's Resource Guide,*** distribute to children copies of Family Letter Week 8 to share with their family. Each letter has an area for children to show families an example of what they have been doing in class.

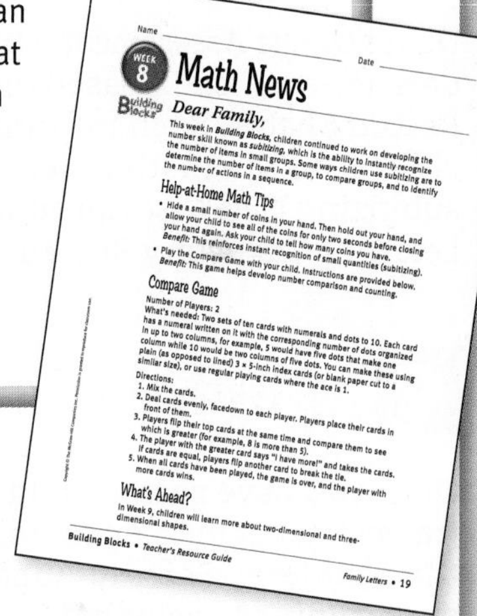

Name ______ Date ______

WEEK 8

Math News

Building Blocks

Dear Family,

This week in ***Building Blocks,*** children continued to work on developing the number skill known as ***subitizing,*** which is the ability to instantly recognize the number of items in small groups. Some ways children use subitizing are to determine the number of items in a group, to compare groups, and to identify the number of actions in a sequence.

Help-at-Home Math Tips

- Hide a small number of coins in your hand. Then hold out your hand, and allow your child to see all of the coins for only two seconds before closing your hand again. Ask your child to tell how many coins you have. *Benefit:* This reinforces instant recognition of small quantities (subitizing).
- Play the Compare Game with your child. Instructions are provided below. *Benefit:* This game helps develop number comparison and counting.

Compare Game

Number of Players: 2

What's needed: Two sets of ten cards with numerals and dots to 10. Each card has a numeral written on it with the corresponding number of dots organized in up to two columns, for example, 5 would have five dots that make one column while 10 would be two columns of five dots. You can make these using plain (as opposed to lined) 3 × 5-inch index cards (or blank paper cut to a similar size), or use regular playing cards where the ace is 1.

Directions:

1. Mix the cards.
2. Deal cards evenly, facedown to each player. Players place their cards in front of them.
3. Players flip their top cards at the same time and compare them to see which is greater (for example, 8 is more than 5).
4. The player with the greater card says "I have more!" and takes the cards. If cards are equal, players flip another card to break the tie.
5. When all cards have been played, the game is over, and the player with more cards wins.

What's Ahead?

In Week 9, children will learn more about two-dimensional and three-dimensional shapes.

Building Blocks • *Teacher's Resource Guide*

Family Letters • 19

Assess and Differentiate

A Gather Evidence

Review children's progress in mathematics by looking at the Weekly Record Sheets (Monday, Wednesday, Friday) and the Small Group Record Sheets (Tuesday, Thursday) from this past week.

B Summarize Findings

Using ***Online Assessment***, summarize and analyze assessment data for each child based on your weekly observations and Record Sheets. Such information helps determine where each child is on the math trajectory for counting, comparing, and subitizing. See ***Assessment*** for the print companion to each Learning Trajectory Record.

C Differentiate Instruction

Once you have seen a child exhibit specific levels of the trajectory, begin to encourage and work with that child toward the next level. Refer to Appendix A for individualized instruction opportunities, including Special Education concerns.

Object Counting

If . . . the child can count five items and tell how many with the last number counted,	**Then . . .** *Counter (Small Numbers)* Accurately counts objects in a line to 5, and answers "how many?" with the last number counted. When objects are visible, especially with small numbers, begins to understand cardinality.
If . . . the child can produce up to five items,	**Then . . .** *Producer (Small Numbers)* Counts out objects to 5.
If . . . the child can count groups up to 10 in a line,	**Then . . .** *Counter (10)* Counts structured arrangements of objects to 10; may be able to write or draw to represent 10.

Comparing and Ordering

If . . . the child can compare only similarly-sized items in small groups,	**Then . . .** *Counting Comparer (Same Size)* Accurate comparison via counting but only when objects are similar in size and in groups up to 5.
If . . . the child can compare various small groups, sometimes labeling one as having more or less,	**Then . . .** *Counting Comparer (5)* Compares with counting even when larger group's objects are smaller; figures out how many more or less.

Subitizing

If . . . the child can instantly identify groups up to 5,	**Then . . .** *Perceptual Subitizer to 5* Instantly recognizes collections of up to five when shown briefly and verbally names the number of items.
If . . . the child can label some arrangements up to 10,	**Then . . .** *Conceptual Subitizer to 5+* Verbally labels all arrangements to about five when shown briefly using groups: "I saw 2 and 2 so I said 4" or "I made 2 groups of 3 plus 1 in my mind so that is 7."

Overview

Big Ideas

- naming, describing, and matching shapes
- counting
- comparing numbers
- reading numerals

Teaching for Understanding

The unifying concept is that analyzing, comparing, and classifying shapes help create new knowledge of shapes and their relationships. Children can learn to match, name, and describe shapes, which builds mathematical and vocabulary skills. Children need repeated, consistent experience with number ideas so, even in a week that emphasizes geometry, number activities occur.

Shape Matching

Children match and name shapes in several settings, including a flip book, computer and print memory games, and box faces matched to shape outlines. These activities build a foundation for mathematical concepts, such as congruence (same size and same shape) and arrays (rows and columns). Memory skills are also developed.

Describing Shapes

Children determine whether or not geometric figures are members of a category. For example, with a rectangle, asking children why it is or is not the specified shape. This is a challenging task for young children but with patience, encouragement, and modeling, they can do it. As they learn and become more engaged with mathematical thinking, children gain confidence, understanding, and language proficiency.

Meaningful Connections

Shape-matching activities help children see the connection between two- and three-dimensional shapes. Such activities also make connections to other areas—again, congruence, arrays, and memory. In addition, counting the sides of shapes helps children connect number ideas to skills. They practice verbal and object counting, comparing numbers, naming numerals, and instantly recognizing small groups (subitizing).

What's Ahead?

In the weeks to come, children will learn more about naming the shapes they are matching. They will further develop their ability to figure out what shape someone is thinking about given clues or descriptions of the shape. Children will also extend their knowledge of number, such as learning how to represent amounts in various ways and seeing patterns in counting.

WEEK 9

Overview

English Learner

For children learning English, reinforce numerals with concrete representations of their values. Refer to page 85 of the *Teacher's Resource Guide*.

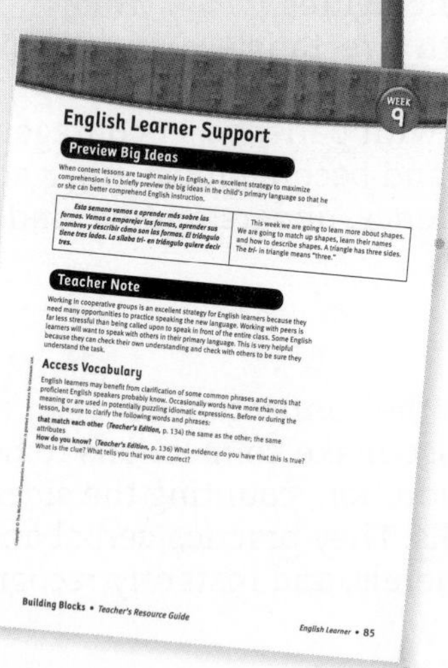
English Learner Support

Preview Big Ideas

Teacher Note

Access Vocabulary

Building Blocks • Teacher's Resource Guide

English Learner • 85

Technology Project

Children may get additional shape practice using Mystery Pictures Free Explore. Encourage children to drag as many shapes as they would like to make their pictures. Children may work in pairs, asking their partner to complete their mystery picture. As you visit children, ask them to describe what they are making, which shapes they are using, and why.

How Children Learn about Shapes and Number

Knowing where children are on the learning trajectory for geometry, counting, and number and what the next steps are helps facilitate their development. Children who easily surpass these trajectory levels might be challenged by larger groups and harder shapes and encouraged to assist other children.

Shapes: Matching

What to Look For What types of shapes can the child match?

Shape Matcher, Identical Matches familiar shapes (circle, square, typical triangle) of same size and orientation.

Shape Matcher, Sizes Matches familiar shapes of different sizes.

Shape Matcher, Orientations Matches familiar shapes of different orientations.

Shapes: Naming

What to Look For What types of shapes can the child recognize and name?

Shape Recognizer, All Rectangles Recognizes more rectangle sizes, shapes, and orientations.

Shape Recognizer, More Shapes Recognizes most familiar shapes and typical examples of other shapes, such as hexagon, rhombus (diamond), and trapezoid.

Shape Identifier Names most common shapes, including rhombuses, without making mistakes, such as calling ovals circles; implicitly recognizes right angles, thus can distinguish between rectangles and parallelograms without right angles.

Object Counting

What to Look For What size groups can the child label and/or produce?

Counter (Small Numbers) Accurately counts objects in a line to 5, and answers "how many?" with the last number counted. When objects are visible, especially with small numbers, begins to understand cardinality.

Producer (Small Numbers) Counts out objects to 5.

Counter (10) Counts structured arrangements of objects to 10; may be able to write or draw to represent 10.

Recognition of Number and Subitizing

What to Look For Does the child instantly recognize groups of at least 5?

Perceptual Subitizer to 5 Instantly recognizes collections of up to 5 when shown briefly and verbally names the number of items.

Conceptual Subitizer to 5+ Verbally labels all arrangements to about 5 when shown briefly using groups: "I saw 2 and 2 so I said 4," or "I made 2 groups of 3 plus 1 in my mind so that is 7."

Math Throughout the Year

Math Throughout the Year activities are recommended to build on the mathematical skills highlighted in each week. Here are suggested activities for **Week 9.**

Counting Jar

A counting jar holds a specified number of items for children to count. Use the same jar all year, changing its small amount of items weekly. Have children spill the items to count them.

Name Faces of Blocks

During circle or free play time, children name the faces (sides) of different building blocks. Tell children which classroom items are the same shape.

Numerals Every Day

Numerals are all around us. Help children notice and read numerals on common items throughout the day, such as clocks, food containers, street signs, room numbers, newspapers, and so on.

Shape Walk

Go for a walk outside of the classroom to search for one specific shape.

Center Preview

Computer Center Building Blocks

Get your classroom Computer Center ready for Memory Geometry 1: Exact Matches and Number Snapshots 2 from the *Building Blocks* software.

After you introduce Memory Geometry 1 and Number Snapshots 2, each child should complete the activities individually as you (or an assistant) monitor and guide him or her periodically. Ideally, each child will have at least ten minutes of computer time at least twice during the week. Use children's center time as an opportunity for assessment.

Hands On Math Center

This week's Hands On Math Center activities are Shape Flip Book, Is It or Not?, Match Shape Sets, Rectangles and Boxes, and Compare Game. Supply the center with these materials: Shape Flip Book, Shape Sets, various-sized rectangular boxes (some with square bottoms), large paper, and two or more Counting Card sets.

Literature Connections

These books help develop shape recognition.

- *Brown Rabbit's Shape Book* by Alan Baker
- *Bear in a Square* by Stella Blackstone
- *Up Goes the Skyscraper!* by Gail Gibbons
- *Into the Sky* by Ryan Ann Hunter
- *Manhattan Skyscrapers* by Eric Peter Nash and Norman McGrath

(t)Matt Meadows (b)Eclipse Studios

WEEK 9 Weekly Planner

Learning Trajectories

Week 9 Objectives

- To name and describe familiar two-dimensional shapes
- To distinguish between visually-similar non-examples of familiar two-dimensional shapes
- To match congruent shapes by memory
- To compare small numbers of objects after shown only briefly
- To produce small numbers of actions

	Developmental Path	Instructional Activities
Shapes: Matching	*Shape Matcher, Identical*	Shape Flip Book Memory Geometry 1
	Shape Matcher, Sizes	Shape Flip Book
	Shape Matcher, Orientations	Shape Flip Book Memory Geometry
	Shape Matcher, More Shapes	
Shapes: Naming	*Shape Recognizer, Circles, Squares, and Triangles+*	
	Shape Recognizer, All Rectangles	Rectangles and Boxes
	Shape Recognizer, More Shapes	"Three Straight Sides" Memory Geometry
	Shape Identifier	Is It or Not?
	Parts of Shapes Identifier	
Object Counting	*Corresponder*	
	Counter (Small Numbers)	Listen and Count
	Producer (Small Numbers)	Number Jump
	Counter (10)	Number Jump (Numerals)
	Counter and Producer (10+)	
Subitizing	*Perceptual Subitizer to 4*	
	Perceptual Subitizer to 5	Listen and Copy
	Conceptual Subitizer to 5+	Number Snapshots 2
	Conceptual Subitizer to 10	

Use this chart to plan for your specific class schedule. If you have your prekindergarteners for only three days, complete Monday, Tuesday, and Thursday of the week.

Work Time

Whole Group	Small Group	Computer	Hands On	Program Resources
Listen and Copy **Shape Flip Book**		**Memory Geometry 1**	**Shape Flip Book** **Is It or Not?** *Materials*: *Shape Sets **Match Shape Sets** *Materials*: *Shape Sets	***Teacher's Resource Guide*** **Shape Flip Book** Building Blocks ***Assessment*** **Weekly Record Sheet**
"Three Straight Sides" **Listen and Count** *Materials*: counting book coffee can marbles **Is It or Not?** *Materials*: *Shape Sets musical triangle	**Is It or Not?** *Materials*: *Shape Sets **Memory Geometry**	**Number Snapshots 2**	**Shape Flip Book** **Is It or Not?** **Match Shape Sets**	***Teacher's Resource Guide*** **Memory Geometry Cards Set A** Building Blocks ***Assessment*** **Small Group Record Sheet**
Number Jump **Rectangles and Boxes** *Materials*: rectangular boxes (some with square bottoms) large paper		• **Memory Geometry 1** • **Number Snapshots 2**	**Rectangles and Boxes** *Materials*: rectangular boxes (some with square bottoms) large paper **Compare Game** *Materials*: *Counting Cards	Building Blocks ***Assessment*** **Weekly Record Sheet**
Count and Move in Patterns **Number Jump (Numerals)** *Materials*: *Numeral Cards **Is It or Not?** *Materials*: *Shape Sets	**Is It or Not?** *Materials*: *Shape Sets **Memory Geometry**	• **Memory Geometry 1** • **Number Snapshots 2**	**Rectangles and Boxes** **Compare Game** **Shape Flip Book** **Is It or Not?** **Match Shape Sets**	***Teacher's Resource Guide*** **Memory Geometry Cards Set A** Building Blocks ***Assessment*** **Small Group Record Sheet**
"Three Straight Sides" **Number Jump (Numerals)** *Materials*: *Numeral Cards **Rectangles and Boxes** *Materials*: rectangular boxes large paper		• **Memory Geometry 1** • **Number Snapshots 2**	**Rectangles and Boxes** **Compare Game** **Shape Flip Book** **Is It or Not?** **Match Shape Sets**	Building Blocks ***Teacher's Resource Guide*** **Family Letter Week 9** ***Assessment*** **Weekly Record Sheet**

*provided in Manipulative Kit

Monday

Monday Planner

Objectives

- To produce small numbers of actions
- To distinguish between visually-similar non-examples of familiar two-dimensional shapes
- To name and describe familiar two-dimensional shapes
- To match congruent shapes by memory

Materials

*Shape Sets

Math Throughout the Year

Review activity directions at the top of page 131, and complete each in class whenever appropriate.

Looking Ahead

For today, if you have not already, prepare the Shape Flip Book from the ***Teacher's Resource Guide.*** For tomorrow, use the counting book and the clean, empty coffee can from Week 8's Listen and Count; provide a musical triangle for Is It or Not?; and make copies of Memory Geometry Cards Set A, which will also be used Thursday, from the ***Teacher's Resource Guide.***

*provided in Manipulative Kit

1 Whole Group

10

Warm-Up: Listen and Copy

Clap up to five times, and tell children to clap the same number of times in the same way. For example, clap quickly, slowly, or with pauses to create patterns, such as clap, clap, pause, clap. Repeat with different small numbers. This activity may also be used for transitions.

Shape Flip Book

- Display the Shape Flip Book. The three panels can be flipped separately: left panels show a shape, middle panels show an image in which the shape is clearly outlined, and right panels show the shape in another image. Model for children how to match the panels, emphasizing familiar shapes you have introduced, such as circle, square, triangle, and rectangle.
- Have all children name each shape with you. Ask whether they have anything at home with those shapes.
- Tell children they can "read" the book by matching panels and naming shapes.

2 Work Time

25

Computer Center

Building Blocks

Introduce Memory Geometry 1: Exact Matches from the **Building Blocks** software. This activity is based on the traditional concentration game; children will match geometric shapes, clicking *Yes* or *No* to confirm whether each pair is a match. Each child should have an opportunity to complete Memory Geometry 1 this week.

Monitoring Student Progress

If . . . a child seems bored or uninterested during his or her time using the computer,	**Then . . .** make sure he or she is being challenged. Increase motivation by having the child tutor another child in the current computer activity or introducing an activity's next level.

Hands On Math Center

After the Shape Flip Book, allow children to participate in the other listed activities as time permits.

Shape Flip Book

Children “read” and explore the Shape Flip Book by matching its panels and naming shapes.

Is It or Not?

Using a dry erase board or chalkboard, have children make and discuss triangles. If they decide a shape is not a triangle because one of its sides is not straight, it is not closed, or it has too many or too few sides, children can easily fix the shape to make it a triangle.

Match Shape Sets

From two complete Shape Sets, have children find matches that are the same size and the same shape.

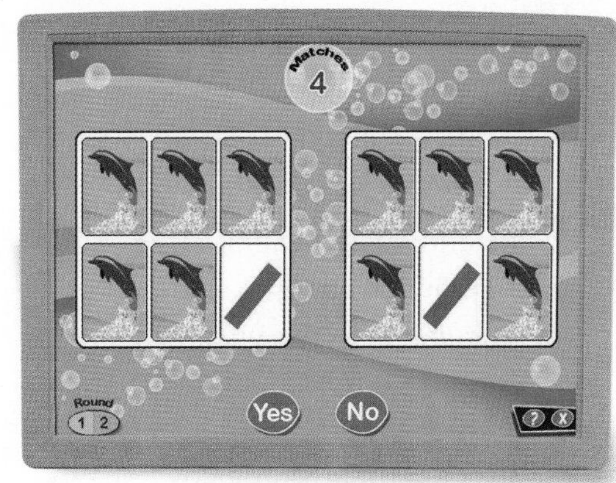

Memory Geometry 1

RESEARCH IN ACTION

The computer can teach and help children follow rules, such as flipping cards back over and waiting for their turn.

Reflect

Ask children:

- **How can you find shapes that match each other?**

Children might say: I found a rectangle in the Shape Flip Book and turned the pages until I found one just like it, or I placed Shape Sets on top of each other.

Assess

Use the Weekly Record Sheet from ***Assessment*** to record children’s progress. Use their time at the centers as an opportunity to complete your observations.

Tuesday Planner

Objectives

- To compare small numbers of objects after shown only briefly
- To name and describe familiar two-dimensional shapes
- To distinguish between visually-similar non-examples of familiar two-dimensional shapes
- To match congruent shapes by memory

Materials

- musical triangle
- counting book
- coffee can
- marbles
- *Shape Sets

Vocabulary

Triangles have three straight sides that connect at three corners. The *tri* in *triangle* means "three."

Math Throughout the Year

Review activity directions at the top of page 131, and complete each in class whenever appropriate.

Looking Ahead

For tomorrow, gather a variety of rectangular boxes, some with square bottoms, and have available large paper on which to trace the boxes.

*provided in Manipulative Kit

Tuesday

1 Whole Group

20

Warm-Up: "Three Straight Sides"

Teach the class to sing the lyrics below to the tune of "Three Blind Mice." You may wish to hit each side of a musical triangle while singing: "three (hit one side) straight (hit 2nd side) sides (hit 3rd side)." If so, explain that the instrument is not a true triangle if indeed it is not closed and has curved corners.

Three straight sides, three straight sides,
See how they meet, see how they meet,
They follow the path that a triangle makes.
Three straight sides and there's no mistakes.
Three sides and three corners, that's all it takes.
Three straight sides.

Listen and Count

- Read the same counting book from last week in which something specific is being counted. Remind children you are going to drop items into a can for them to count. Ask children to listen quietly as you slowly drop marbles into a clean, empty coffee can.
- When you finish, have children hold up their fingers to show how many marbles they think are in the can. After you have observed their responses, ask children to say the number.
- Spill the items out of the can, and count them as a whole group to check.
- If children are ready, try the adding version by pausing between drops. Start with up to four drops, and then add one or two more. Children show with their fingers how many items you dropped. Spill the items, and count them together.

Is It or Not?

- Showing a musical triangle, explain that it is not a true triangle because it is not closed and it has curved corners. Review that a triangle has to have three straight sides, which can be different lengths that are connected.
- Tell children you are going to try to fool them. Name and show a Shape Set triangle, and ask: Is this a triangle? Why or why not? Draw "foolers," shapes that look like but are not triangles, and draw unusually-shaped triangles. Ask which are true triangles. Summarize by reviewing triangle attributes (or sing "Three Straight Sides" again).
- Repeat with other, more difficult shapes as children are able.

Work Time

15

Small Group

Is It or Not?

Follow the Whole Group directions with triangles and rectangles.

Memory Geometry

Place two sets of Memory Geometry Cards Set A on a desktop with a strip of tape down the center, one set facedown in an array on each side of the tape. Players take turns exposing a card from each. Matching cards are kept; those that do not are replaced facedown. Players should discuss shapes.

Monitoring Student Progress

If . . .	Then . . .
If . . . children struggle during Memory Geometry,	**Then . . .** place one array faceup, or limit the cards used.
If . . . children excel during Memory Geometry,	**Then . . .** use Memory Geometry Cards Set B (rotated shapes) or D (smaller shapes) in one of the arrays.

Computer Center

Introduce Number Snapshots 2, in which children match a number to a different representation of that number. Each child should complete the activity.

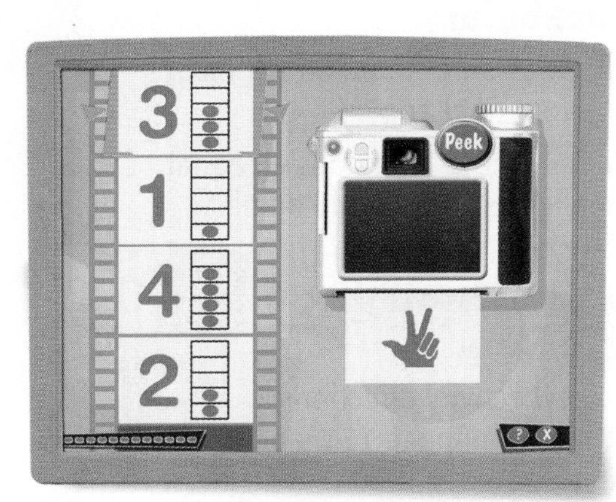

Number Snapshots 2

Hands On Math Center

Allow children to continue yesterday's Hands On Math Center activities.

Reflect

5

Ask children:

- **How do you know a shape is a triangle?**

Children might say: It has three straight sides with no holes, or the like.

Assess

During Small Group activities, use the Small Group Record Sheet from *Assessment* to observe and record children's progress.

Wednesday Planner

Objectives

- To name and describe familiar two-dimensional shapes
- To distinguish between visually-similar non-examples of familiar two-dimensional shapes
- To match congruent shapes by memory
- To compare small numbers of objects after shown only briefly
- To produce small numbers of actions

Materials

- rectangular boxes, (some with square bottoms)
- large paper
- *Counting Cards

Vocabulary

Rectangles have four straight sides and four right angles.

Squares have four sides all equal in length and four right angles.

Math Throughout the Year

Review activity directions at the top of page 131, and complete each in class whenever appropriate.

*provided in Manipulative Kit

Wednesday

1 Whole Group 15

Warm-Up: Number Jump

- Show a number of fingers, and write that numeral for children to see. Tell children to jump safely that many times, and count the jumps in unison.
- As a variation, hide your hands behind your back, and tell children to jump only if you show three fingers. Show your fingers for only two seconds.
- Repeat, changing to other appropriate numbers.

Monitoring Student Progress

If . . . children struggle during Number Jump,	**Then . . .** use smaller numerals, and say each numeral as you show your fingers.
If . . . children excel during Number Jump,	**Then . . .** use larger numerals, and show finger combinations using both hands.

Rectangles and Boxes

- Draw a large rectangle for the entire class to see, and trace it, counting each side as you go. Challenge children to draw a rectangle in the air as you count, reminding them that each side should be straight.
- Show a variety of boxes to children, such as toothpaste, pasta, and cereal boxes, and discuss their shape. Eventually focus on the faces of the boxes, which should mostly be rectangles. Talk about the sides and right angles.
- On large paper, place two boxes horizontally and trace their faces. Have children match the boxes to the traced rectangles. Trace more boxes and repeat.
- Help children consider other box face shapes, such as triangles (candy and food storage), octagons (hat and gift boxes), and circles/cylinders (toy and oats containers).

2 Work Time 20

Computer Center Building Blocks

Continue to provide each child with a chance to complete Memory Geometry 1 and Number Snapshots 2.

 Hands On Math Center

Based on what children learn and benefit from most, allow them to continue this week's previous Hands On Math Center activities, and add the following recurring activities.

Rectangles and Boxes

Have children help each other trace several boxes that have been placed horizontally on large sheets of paper. Children then match the boxes to the traced rectangles.

Compare Game

- For each pair of children playing the game, two or more sets of Counting Cards are needed. Children mix the cards and then deal them evenly facedown.
- Players simultaneously flip their top cards and compare them to find out which is greater. The player with the greater amount says "I have more!" and takes the opponent's card. If card amounts are equal, however, players each flip another card to determine a result.
- The game is over when all cards have been played, and the "winner" is the player with more cards. See Friday's Monitoring Student Progress for scaffolding.

RESEARCH IN ACTION

Boxes are solid (three-dimensional) shapes called *rectangular prisms*. Their faces are rectangles. As a square is a special type of rectangle with all four sides the same length, a cube is a special type of rectangular prism in which all faces are the same square.

3 Reflect

5

Ask children:

■ **Can you think of a box that has a face that is not a rectangle?**

Children might say: A candy box, an oatmeal box, a specially-shaped pencil case, and the like.

4 Assess

Use the Weekly Record Sheet from ***Assessment*** to record children's progress. Use their time at the centers as an opportunity to complete your observations.

Thursday Planner

Objectives

- To name and describe familiar two-dimensional shapes
- To distinguish between visually-similar non-examples of familiar two-dimensional shapes
- To match congruent shapes by memory
- To compare small numbers of objects after shown only briefly
- To produce small numbers of actions

Materials

- *Numeral Cards
- *Shape Sets

Math Throughout the Year

Review activity directions at the top of page 131, and complete each in class whenever appropriate.

Looking Ahead

For tomorrow, make copies of Family Letter Week 9 from the ***Teacher's Resource Guide.***

*provided in Manipulative Kit

Whole Group

15

Warm-Up: Count and Move in Patterns

Spend only a few minutes counting with children in patterns of 4 up to 16, or another appropriate even number. For more fun, get a drum or use the corners of a wooden block to tap along with the counting, tapping harder for emphasis at each fourth beat.

Number Jump (Numerals)

Show a Numeral Card 1 to 5. Have all children first say that numeral, and then safely jump that many times, counting the jumps in unison. Repeat with a different numeral. Change jumping to another safe movement (twirling, clapping, or hopping) your class would enjoy.

Is It or Not?

- Draw a rectangle, reviewing that rectangles have four closed, straight sides with four right angles. Explain also that two opposite sides of a rectangle must be the same length.
- Tell children you are going to try to fool them. Name and show a Shape Set rectangle, and ask: Is this a rectangle? Why or why not? Draw "foolers," shapes that look like but are not rectangles, and draw unusually-shaped rectangles. Ask which are rectangles.
- Summarize by reviewing rectangle attributes, and repeat with harder shapes as children are able.

Work Time

20

Small Group

Is It or Not?

Follow the Whole Group directions, making sure children discuss with you whether or not the shapes you show are true rectangles and why. Repeat with triangles (showing examples, drawing foolers, and so on), reviewing their attributes: three straight, closed sides.

Eclipse Studios

Monitoring Student Progress

If . . . children need a challenge during Is It or Not?	**Then . . .** require children to describe exactly why the shape is or is not a rectangle or triangle, or show a shape that is very close to the target shape but slightly incorrect, such as including a curved line.

Memory Geometry

- Place two sets of Memory Geometry Cards Set A on top of a flannel board or a desktop with a strip of tape down the center, one set facedown in an array on each side of the tape.
- Players take turns exposing one card from each array. Cards that do not match are replaced facedown; cards that match are kept by that player. Players should name and describe the shapes together.

Continue to provide each child with a chance to complete Memory Geometry 1 and Number Snapshots 2.

Based on what children continue to learn and benefit from most, choose from this week's Hands On Math Center activities: Rectangles and Boxes, Compare Game, Shape Flip Book, Is It or Not?, and Match Shape Sets. Consult the Weekly Planner for corresponding materials and, if needed, previous days for activity directions.

3 Reflect

5

Ask children:

■ **How is a rectangle different from a triangle?**

Children might say: Rectangles have four sides and right angles. Triangles have three sides.

During Small Group activities, use the Small Group Record Sheet from *Assessment* to observe and record children's progress.

RESEARCH IN ACTION

Activities like Is It or Not? develop sorting and classifying abilities. Sorting by mathematical properties is perhaps the most important type of classifying activity. Children learn to sort by mathematical attributes in the same way they sort by other attributes, such as color, thus deepening their understanding of mathematical attributes. Children also learn that classifications of such attributes are an important aspect of mathematics.

Friday Planner

Objectives

- To name and describe familiar two-dimensional shapes
- To distinguish between visually-similar non-examples of familiar two-dimensional shapes
- To match congruent shapes by memory
- To compare small numbers of objects after shown only briefly
- To produce small numbers of actions

Materials

- *Numeral Cards
- rectangular boxes (some with square bottoms)
- large paper

Math Throughout the Year

Review activity directions at the top of page 131, and complete each in class whenever appropriate.

Looking Ahead

For next week, familiarize yourself with Mystery Pictures 2: Name Shapes and Memory Geometry 2: Turned Shapes from the ***Building Blocks*** software.

*provided in Manipulative Kit

Friday

1 Whole Group

Warm-Up: "Three Straight Sides"

Sing the lyrics below to the tune of "Three Blind Mice."

Three straight sides, three straight sides,
See how they meet, see how they meet,
They follow the path that a triangle makes.
Three straight sides and there's no mistakes.
Three sides and three corners, that's all it takes.
Three straight sides.

Number Jump (Numerals)

Show a Numeral Card 1 to 5. Have all children first say that numeral, and then safely jump that many times, counting the jumps in unison. Repeat with a different numeral. Change jumping to another safe movement (twirling, clapping, or hopping) your class would enjoy.

Rectangles and Boxes

- Draw a large rectangle for the entire class to see, and trace it, counting each side as you go. Challenge children to draw a rectangle in the air as you count, reminding them that each side should be straight.
- Show children a variety of boxes, and discuss their shape. Eventually focus on the bases of the boxes; some should be squares, which are also rectangles. Discuss this, as well as sides and right angles.
- On large paper, place two boxes vertically and trace their bases (or box ends). Again, address those that are squares (four sides of equal length), emphasizing that they are rectangles as well.
- Set up several papers for children to trace boxes using the same color crayon or marker. If children need help tracing, lightly outline the boxes in pencil first, and then have them find a match to trace using crayon or marker. Otherwise, have children match the boxes to the traced rectangles and/or squares.

2 Work Time

Computer Center Building Blocks

Continue to provide each child with a chance to complete Memory Geometry 1 and Number Snapshots 2.

 Hands On Math Center

Based on what children continue to learn and benefit from most, choose from this week's Hands On Math Center activities: Rectangles and Boxes, Compare Game, Shape Flip Book, Is It or Not?, and Match Shape Sets. Consult the Weekly Planner for corresponding materials and, if needed, previous days for activity directions. Scaffolding for Compare Game follows.

Monitoring Student Progress

If . . . children struggle during Compare Game,	**Then . . .** use fewer cards.
If . . . children excel during Compare Game,	**Then . . .** use more cards, or have players count dot pairs instead of cards.

 Reflect

5 

Briefly discuss with children what you have done in class this week, such as comparing shapes in categories such as rectangle and triangle. Show children a rectangular "fooler," and ask:

- **Is this a rectangle?**

Children might say: No, it has an open space; its sides are not straight; or it does not have all right angles.

 Assess

Use the Weekly Record Sheet from ***Assessment*** to record children's progress. Use their time at the centers as an opportunity to complete your observations.

Home Connection

From the ***Teacher's Resource Guide,*** distribute to children copies of Family Letter Week 9, which includes the Memory Geometry: 2D Shapes game, to share with their family. Each letter has an area for children to show families an example of what they have been doing in class.

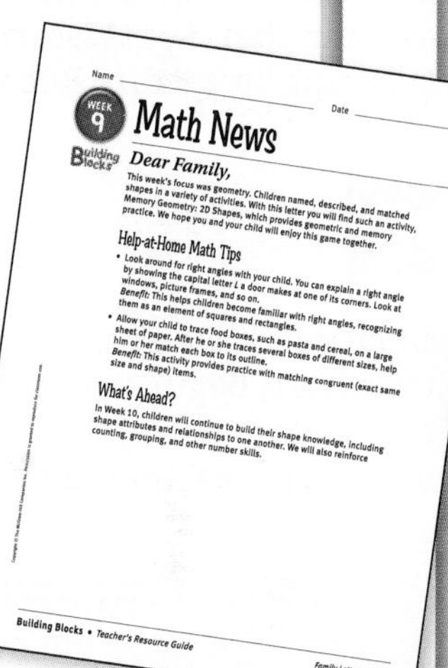

Name ______ Date ______

WEEK 9 **Math News**

Building Blocks

Dear Family,

This week's focus was geometry. Children named, described, and matched shapes in a variety of activities. With this letter you will find such an activity, Memory Geometry: 2D Shapes, which provides geometric and memory practice. We hope you and your child will enjoy this game together.

Help-at-Home Math Tips

- Look around for right angles with your child. You can explain a right angle by showing the capital letter *L* a door makes at one of its corners. Look at windows, picture frames, and so on.
  Benefit: This helps children become familiar with right angles, recognizing them as an element of squares and rectangles.
- Allow your child to trace food boxes, such as pasta and cereal, on a large sheet of paper. After he or she traces several boxes of different sizes, help him or her match each box to its outline.
  Benefit: This activity provides practice with matching congruent (exact same size and shape) items.

What's Ahead?

In Week 10, children will continue to build their shape knowledge, including shape attributes and relationships to one another. We will also reinforce counting, grouping, and other number skills.

Building Blocks • *Teacher's Resource Guide*

Family Letters • 21

WEEK 9 Wrap-Up

Assess and Differentiate

A Gather Evidence

Review children's progress in mathematics by looking at the Weekly Record Sheets (Monday, Wednesday, Friday) and the Small Group Record Sheets (Tuesday, Thursday) from this past week.

B Summarize Findings

Using ***Online Assessment,*** summarize and analyze assessment data for each student based on your weekly observations and Record Sheets. Such information helps determine where each child is on the math trajectory for geometry, counting, and number. See ***Assessment*** for the print companion to each Learning Trajectory Record.

C Differentiate Instruction

Once you have seen a child exhibit specific levels of the trajectory, begin to encourage and work with that child toward the next level. Refer to Appendix A for individualized instruction opportunities, including Special Education concerns.

Shapes: Matching

If . . .	Then . . .
If . . . the child can match shapes of same size and same orientation,	**Then . . .** *Shape Matcher, Identical* Matches familiar shapes (circle, square, typical triangle) of same size and orientation.
If . . . the child can match different-sized shapes,	**Then . . .** *Shape Matcher, Sizes* Matches familiar shapes of different sizes.
If . . . the child can match shapes of varied orientations,	**Then . . .** *Shape Matcher, Orientations* Matches familiar shapes of different orientations.

Shapes: Naming

If . . .	Then . . .
If . . . the child can recognize a large variety of rectangles,	**Then . . .** *Shape Recognizer, All Rectangles* Recognizes more rectangle sizes, shapes, and orientations.
If . . . the child can name most familiar shapes and some typical examples of less familiar shapes,	**Then . . .** *Shape Recognizer, More Shapes* Recognizes most familiar shapes and typical examples of other shapes, such as hexagon, rhombus (diamond), and trapezoid.
If . . . the child can consistently name common shapes, as well as recognize right angles,	**Then . . .** *Shape Identifier* Names most common shapes, including rhombuses, without making mistakes, such as calling ovals circles; implicitly recognizes right angles, thus can distinguish between rectangles and parallelograms without right angles.

Object Counting

If . . .	Then . . .
If . . . the child can count five items and tell how many with the last number counted,	**Then . . .** *Counter (Small Numbers)* Accurately counts objects in a line to 5, and answers "how many?" with the last number counted. When objects are visible, especially with small numbers, begins to understand cardinality.
If . . . the child can produce up to five items,	**Then . . .** *Producer (Small Numbers)* Counts out objects to 5.
If . . . the child can count groups up to 10 in a line,	**Then . . .** *Counter (10)* Counts structured arrangements of objects to 10; may be able to write or draw to represent 10.

Recognition of Number and Subitizing

If . . .	Then . . .
If . . . the child can instantly identify groups up to 5,	**Then . . .** *Perceptual Subitizer to 5* Instantly recognizes collections of up to 5 when shown briefly and verbally names the number of items.
If . . . the child can label some arrangements up to 10,	**Then . . .** *Conceptual Subitizer to 5+* Verbally labels all arrangements to about five when shown briefly using groups: "I saw 2 and 2 so I said 4," or "I made 2 groups of 3 plus 1 in my mind so that is 7."

WEEK 10

Overview

Big Ideas

- recognizing, naming, and sorting shapes
- putting together shapes
- counting
- comparing small numbers

Teaching for Understanding

Analyzing, comparing, and classifying shapes continue to help build children's knowledge of shapes and their relationships. Children find, name, and describe shapes in a variety of activities, which may lead to shape comparisons, as well as starting to see what happens when shapes are combined.

Shape Activities

After they have matched shapes, children learn to name and describe them. Guess My Rule is a game children play in order to figure out how someone else is sorting shapes. Hearing how others describe shapes and seeing similarities among shapes are beneficial to each child's own perception of shapes and their attributes.

Engaging in Dialogue

Children are asked: How did you figure out my rule? A question like this helps children learn geometric concepts and mathematical processes such as *inducing*, which is determining a general rule based on example. The process is not easy so gradually increase the difficulty of examples and questions, presenting simpler examples first and helping children respond. For example, if children say, "These are all circles," you may add, "Correct. These are all circles, and these are the shapes that are not circles." Follow up and ask, "How do you know?" Children might repeat their initial response so discuss the other shapes: "They are round like circles, or some are round a bit (pointing to an oval) but not even all around."

Meaningful Connections

When children participate in games like Guess My Rule, they are learning about attributes and sorting in addition to geometric shapes. Most of this week's activities also develop reasoning, problem solving, making connections, and communicating verbally. Children continue to practice counting and number comparing skills developed in previous weeks.

What's Ahead?

In the weeks to come, children's number knowledge grows as they discover how to represent numbers in various ways, such as with symbols or numerals, and how to see patterns while counting numbers. A simple yet powerful pattern is plus one. For example, when another item is added to a collection of five, the new amount of the collection is the next number word—six.

How Children Learn about Shapes

Knowing where children are on the learning trajectory for geometry and counting and what the next steps are helps facilitate their development. Children who easily surpass these trajectory levels might be challenged by harder shapes and larger groups and encouraged to assist other children.

Shapes: Matching

What to Look For What types of shapes can the child match?

Shape Matcher, Identical Matches familiar shapes (circle, square, typical triangle) of same size and orientation.

Shape Matcher, Sizes Matches familiar shapes of different sizes.

Shape Matcher, Orientations Matches familiar shapes of different orientations.

Shapes: Naming

What to Look For What types of shapes can the child recognize and name?

Shape Recognizer, All Rectangles Recognizes more rectangle sizes, shapes, and orientations.

Side Recognizer Identifies sides as distinct geometric objects.

Shape Recognizer, More Shapes Recognizes most familiar shapes and typical examples of other shapes, such as hexagon, rhombus (diamond), and trapezoid.

Shape Identifier Names most common shapes, including rhombuses, without making mistakes, such as calling ovals circles; implicitly recognizes right angles, thus can distinguish between rectangles and parallelograms without right angles.

Parts of Shapes Identifier Identifies shapes in terms of their components (no matter how skinny a triangle is, knows it is a triangle because it has three straight sides).

Shapes: Composing

What to Look For How does the child assemble simple outline puzzles?

Piece Assembler Makes pictures in which shapes touch and each represents a unique role, such as one shape for each body part, and fills simple outline puzzles using trial and error.

Picture Maker Puts several shapes together to make one part of a picture, such as two shapes for one arm. Uses trial and error, does not anticipate the creation of a new geometric shape, chooses shapes using general shape or side length, and fills simple outline puzzles that suggest placement of each shape.

Object Counting

What to Look For What size groups can the child label?

Counter (Small Numbers) Accurately counts objects in a line to 5, and answers "how many?" with the last number counted. When objects are visible, especially with small numbers, begins to understand cardinality.

English Learner

For children learning English, connect the shapes to the shapes of familiar objects. Review the following: *guess* and *piles*.

Refer to page 86 of the ***Teacher's Resource Guide.***

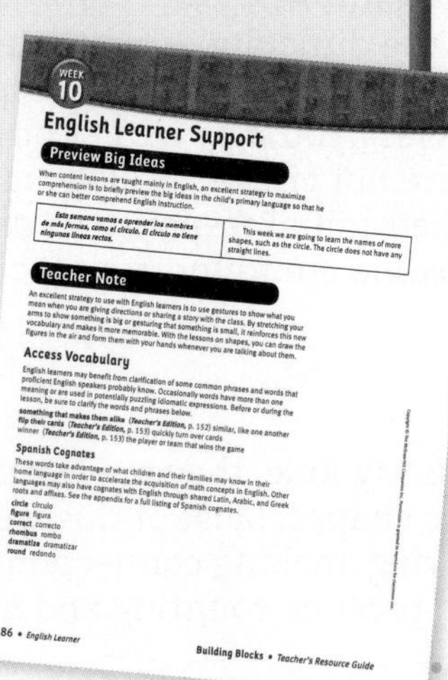

WEEK 10

English Learner Support

Preview Big Ideas

When content lessons are taught mainly in English, an excellent strategy to maximize comprehension is to briefly preview the big ideas in the child's primary language so that he or she can better comprehend English instruction.

Esta semana vamos a aprender los nombres de más formas, como el círculo. El círculo no tiene ningunas líneas rectas.	This week we are going to learn the names of more shapes, such as the circle. The circle does not have any straight lines.

Teacher Note

An excellent strategy to use with English learners is to use gestures to show what you mean when you are giving directions or sharing a story with the class. By stretching your arms to show something is big or gesturing that something is small, it reinforces this new vocabulary and makes it more memorable. With the lessons on shapes, you can draw the figures in the air and form them with your hands whenever you are talking about them.

Access Vocabulary

English learners may benefit from clarification of some common phrases and words that proficient English speakers probably know. Occasionally words have more than one meaning or are used in potentially puzzling idiomatic expressions. Before or during the lesson, be sure to clarify the words and phrases below.

something that makes them alike (*Teacher's Edition*, p. 152) similar, like one another
flip their cards (*Teacher's Edition*, p. 153) quickly turn over cards
winner (*Teacher's Edition*, p. 153) the player or team that wins the game

Spanish Cognates

These words take advantage of what children and their families may know in their home language in order to accelerate the acquisition of math concepts in English. Other languages may also have cognates with English through shared Latin, Arabic, and Greek roots and affixes. See the appendix for a full listing of Spanish cognates.

circle círculo
figure figura
correct correcto
rhombus rombo
dramatize dramatizar
round redondo

86 • *English Learner* Building Blocks • *Teacher's Resource Guide*

Technology Project

Children may get additional shape practice using Mystery Pictures Free Explore by making their own pictures. This enhances their shape knowledge and creativity. Technology Projects are great opportunities to increase positive social interactions around the computer. Have children invite their classmates to play too!

Math Throughout the Year

Math Throughout the Year activities are recommended to build on the mathematical skills highlighted in each week. Here are suggested activities for **Week 10.**

Counting Jar

A counting jar holds a specified number of items for children to count. Use the same jar all year, changing its small amount of items weekly. Have children spill the items to count them.

Numerals Every Day

Numerals are all around us. Help children notice and read numerals on common items throughout the day, such as clocks, food containers, street signs, room numbers, newspapers, and so on.

Shape Walk

Go for a walk outside of the classroom to search for one specific shape.

Center Preview

Computer Center Building Blocks

Get your classroom Computer Center ready for Mystery Pictures 2: Name Shapes and Memory Geometry 2: Turned Shapes from the ***Building Blocks*** software.

After you introduce Mystery Pictures 2 and Memory Geometry 2, each child should complete the activities individually as you (or an assistant) monitor and guide him or her periodically. Ideally, each child will have at least ten minutes of computer time at least twice during the week. Use children's center time as an opportunity for assessment.

Hands On Math Center

This week's Hands On Math Center activities are Shape Pictures, Memory Geometry, Rectangles and Boxes, Shape Flip Book, and Compare Game. Supply the center with these materials: Shape Sets, Pattern Blocks, Memory Geometry Cards Set A, various-sized rectangular boxes (some with square bottoms), large paper, Shape Flip Book, and two or more Numeral Card sets.

Literature Connections

These books help develop shape recognition.

- *Brown Rabbit's Shape Book* by Alan Baker
- *Bear in a Square* by Stella Blackstone
- *What Is Square?* by Rebecca Kai Dotlich
- *Village of Round and Square Houses* by Ann Grifalconi
- *There's a Square: A Book about Shapes* by Mary Serfozo

Weekly Planner

Learning Trajectories

Week 10 Objectives

- To name and describe familiar two-dimensional shapes
- To distinguish between visually-similar non-examples of familiar two-dimensional shapes
- To match congruent shapes by memory
- To compare small numbers of objects after shown only briefly
- To produce small numbers of actions

	Developmental Path	Instructional Activities
Shapes: Matching	***Shape Matcher, Identical***	Shape Flip Book
	Shape Matcher, Sizes	Shape Flip Book
	Shape Matcher, Orientations	Shape Flip Book Memory Geometry (print) Memory Geometry 2 (computer)
	Shape Matcher, More Shapes	
Shapes: Naming	*Shape Recognizer, Circles, Squares, and Triangles+*	
	Shape Recognizer, All Rectangles	I Spy Rectangles and Boxes Mystery Pictures 2
	Side Recognizer	I Spy Shape Step Guess My Rule
	Shape Recognizer, More Shapes	I Spy Shape Step Guess My Rule Memory Geometry Mystery Pictures 2
	Shape Identifier	I Spy
	Parts of Shapes Identifier	Guess My Rule I Spy
	Shape Class Identifier	
Shapes: Composing	*Pre-Composer*	
	Piece Assembler	Shape Pictures
	Picture Maker	Shape Pictures
	Shape Composer	
Counting	*Corresponder*	
	Counter (Small Numbers)	"Five Little Fingers"
	Producer (Small Numbers)	

Use this chart to plan for your specific class schedule. If you have your prekindergarteners for only three days, complete Monday, Tuesday, and Thursday of the week.

Work Time

Whole Group	Small Group	Computer	Hands On	Program Resources
I Spy *Materials:* *Shape Sets **Shape Step** *Materials:* large shapes to step on		**Mystery Pictures 2**	**Shape Pictures** *Materials:* *Shape Sets *Pattern Blocks **Memory Geometry** **Rectangles and Boxes** *Materials:* rectangular boxes large paper	Building Blocks ***Teacher's Resource Guide*** Memory Geometry Cards Sets A, B, and D ***Assessment*** Weekly Record Sheet
"Five Little Fingers" **Guess My Rule** *Materials:* *Shape Sets	**Guess My Rule** *Materials:* *Shape Sets **Shape Step** *Materials:* large shapes to step on	**Memory Geometry 2**	**Shape Flip Book** **Compare Game** *Materials:* *Counting Cards **Shape Pictures** **Memory Geometry** **Rectangles and Boxes**	Building Blocks ***Assessment*** Small Group Record Sheet
Building Shapes **I Spy** *Materials:* *Shape Sets **Shape Step** *Materials:* large shapes to step on		• **Mystery Pictures 2** • **Memory Geometry 2**	**Shape Pictures** **Memory Geometry** **Rectangles and Boxes** **Shape Flip Book** **Compare Game**	***Big Book Building Shapes*** Building Blocks ***Assessment*** Weekly Record Sheet
"Five Little Fingers" **Guess My Rule** *Materials:* *Shape Sets	**Guess My Rule** *Materials:* *Shape Sets **Shape Step** *Materials:* large shapes to step on	• **Mystery Pictures 2** • **Memory Geometry 2**	**Shape Pictures** **Memory Geometry** **Rectangles and Boxes** **Shape Flip Book** **Compare Game**	Building Blocks ***Assessment*** Small Group Record Sheet
I Spy *Materials:* *Shape Sets **Shape Step** *Materials:* large shapes to step on		• **Mystery Pictures 2** • **Memory Geometry 2**	**Shape Pictures** **Memory Geometry** **Rectangles and Boxes** **Shape Flip Book** **Compare Game**	Building Blocks ***Teacher's Resource Guide*** Family Letter Week 10 ***Assessment*** Weekly Record Sheet

*provided in Manipulative Kit

Monday Planner

Objectives

- To name and describe familiar two-dimensional shapes
- To distinguish between visually-similar non-examples of familiar two-dimensional shapes
- To match congruent shapes by memory
- To compare small numbers of objects after shown only briefly

Materials

- *Shape Sets
- large shapes to step on (See directions.)
- *Pattern Blocks
- rectangular boxes, some with square bottoms
- large paper

Math Throughout the Year

Review activity directions at the top of page 147, and complete each in class whenever appropriate.

Looking Ahead

For today, recurring from last week, you will need Memory Geometry Cards Set A (from the ***Teacher's Resource Guide***), rectangular boxes (some with square bottoms), and large paper. For tomorrow, also from last week, you will need the Shape Flip Book.

*provided in Manipulative Kit

Monday

1 Whole Group

15

Warm-Up: I Spy

- Beforehand, place various Shape Set shapes throughout the classroom in plain view.
- Name the shape of something in the room. You may wish to start with something easily recognizable, such as "three sides." Have children guess the item or shape you are thinking about. If able, have the child who guessed correctly think of the next item or shape for you and the class to guess.
- As a variation, try the properties version: describe a shape's attributes and see whether children can guess which item or shape you mean. This can also be done with Shape Sets, actual objects in the room, and/or other shape manipulatives.

Shape Step

- If you do not already have large shapes on which children can step, do one of the following: make shapes on the floor with masking or colored tape; make chalk shapes outdoors; tape laminated paper shapes to the floor; or use a copier to enlarge Shape Sets from the ***Teacher's Resource Guide***.
- Show several triangles. Mix up all those triangles with plenty of other shapes on the floor.
- Tell children to step on triangles only. Have a group of five children step on the triangles. Ask the rest of the class to watch carefully to make sure the group steps on all triangles.
- Whenever possible, ask children to explain why the shape they stepped on was the correct shape ("How do you know that was a triangle?").
- Repeat the activity until all groups have stepped on shapes. Children enjoy playing Shape Step repeatedly so dramatize it in different ways, such as pretending that shapes are rocks in a pond and children will get their feet wet if they step on anything that is not a "rock" (the chosen shape).
- Repeat with another shape, such as rectangles.

Monitoring Student Progress

If . . .	Then . . .
If . . . children need more help during Shape Step,	**Then . . .** have them feel and count a chosen shape's sides, draw it in the air, and review its attributes (for example, a triangle has three straight sides). If children miss a shape they should have stepped on, you step on it and ask whether or not it is correct.
If . . . children need a challenge during Shape Step,	**Then . . .** incorporate less familiar shapes.

Work Time

20

Computer Center

Introduce Mystery Pictures 2: Name Shapes from the ***Building Blocks*** software. In this activity, children identify shapes to construct each mystery picture, yet they will still enjoy guessing what each picture will be. Each child should have an opportunity to complete Mystery Pictures 2 this week.

Mystery Pictures 2

Hands On Math Center

Before children go to this center, formally yet briefly introduce Pattern Blocks, especially naming any new or difficult shapes like rhombuses. After Shape Pictures, allow children to participate in the other listed activities as time permits. These activities, two of which continue from last week, occur all week.

Shape Pictures

Place Shape Sets and Pattern Blocks on different tables, and have children play with them, creating pictures, designs, and patterns.

Memory Geometry

Place two sets of Memory Geometry Cards Set A facedown, each in an array. Players take turns exposing one card from each array. Cards that do not match are replaced facedown; cards that match are kept by that player. Players should name and describe the shapes together. See Monitoring Student Progress on page 137 for ways to support children who struggle or excel during Memory Geometry.

Rectangles and Boxes

Have children help each other trace several boxes that have been placed horizontally on large sheets of paper. Children then match the boxes to the traced rectangles.

RESEARCH IN ACTION

All squares are rhombuses—and not just when they are tilted—because a square has four sides of equal length. Squares are also rectangles due to their right angles. However, not all rhombuses are squares because a rhombus does not always have right angles.

Reflect

5

Show children two different familiar shapes, and ask:

- **How are these shapes different?**

Children might say: One is bigger, one has four sides, it is a triangle, or the like.

Assess

Use the Weekly Record Sheet from ***Assessment*** to record children's progress. Use their time at the centers as an opportunity to complete your observations.

Tuesday Planner

Objectives

- To name and describe familiar two-dimensional shapes
- To distinguish between visually-similar non-examples of familiar two-dimensional shapes
- To match congruent shapes by memory
- To compare small numbers of objects after shown only briefly
- To produce small numbers of actions

Materials

- *Shape Sets
- large shapes to step on
- *Counting Cards

Math Throughout the Year

Review activity directions at the top of page 147, and complete each in class whenever appropriate.

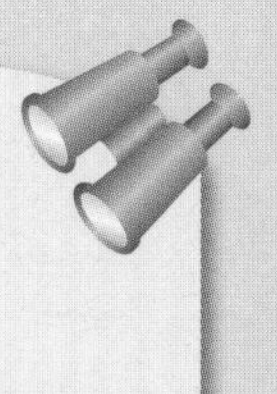

Looking Ahead

If needed, preview ***Big Book Building Shapes*** for tomorrow.

*provided in Manipulative Kit

Tuesday

1 Whole Group 10

Warm-Up: "Five Little Fingers"

Help children suit their actions to the finger play's words.

> One little finger standing on its own, (*Show index finger.*)
> Two little fingers, now they're not alone. (*Add middle finger.*)
> Three little fingers, happy as can be. (*Add ring finger.*)
> Four little fingers go walking down the street. (*Add pinkie; wiggle all four fingers.*)
> Five little fingers, this one is a thumb. (*Show all fingers, emphasizing the thumb.*)
> Wave good-bye 'cause now we're done. (*Wave hand.*)

Guess My Rule

- Tell children to watch carefully as you sort Shape Set shapes into piles based on something that makes them alike. Ask children to silently guess your sorting rule, such as circles versus squares or four-sided shapes versus other shapes.
- Sort shapes one at a time, continuing until there are at least two shapes in each pile.
- Signal "shhh," and pick up a new shape. With a look of confusion, gesture to children to encourage all of them to point quietly to which pile the shape belongs. Place the shape in its pile.
- After all shapes are sorted, ask children what they think the sorting rule is. Repeat with other shapes and new rules. A complete sorting rule suggestion list appears on page A11 of this guide.

2 Work Time 25

Small Group

Guess My Rule

Help children follow the Whole Group directions. Once all shapes are sorted, ask children: What is my rule? How do you know? Say: Show me how the shapes fit the rule.

Eclipse Studios

Monitoring Student Progress

If . . . children struggle with Guess My Rule,	**Then . . .** use simpler rules, or state rules as choices, having children say *yes* or *no*.
If . . . children excel at Guess My Rule,	**Then . . .** use harder rules, such as circles versus ovals.

RESEARCH IN ACTION

Alternating ways of making Shape Step shapes (tape, chalk, cut, or enlarged) on different days helps children generalize their shape knowledge by seeing different representations of the same geometric ideas.

Shape Step

Using a large-shape method from Monday's Whole Group, show the shapes on the floor to children. Tell them on which shape to step, and observe and discuss whether they step on only the target shape.

Computer Center Building Blocks

Introduce Memory Geometry 2: Turned Shapes from the ***Building Blocks*** software, in which children match turned shapes and click yes or no to ensure comprehension. Each child should complete the activity this week.

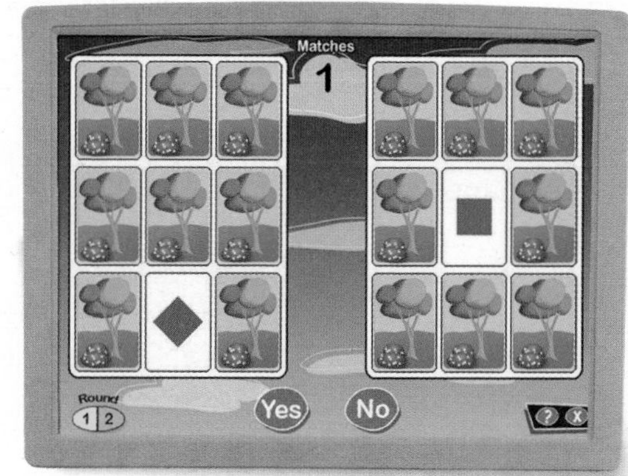

Memory Geometry 2

Hands On Math Center

After Shape Flip Book and Compare Game, both of which recur from Week 9, allow children to continue yesterday's activities as time permits.

Shape Flip Book

Children "read" and explore by matching panels and naming shapes.

Compare Game

- For each pair of children playing, two or more sets of Counting Cards are needed. Children mix the cards and then deal them evenly facedown.
- Players simultaneously flip their top cards and compare them to find out which is greater. The player with the greater amount says "I have more!" and takes the opponent's card. If equal, each player each flips a card to break the tie. Once all cards are played, the "winner" has more.

3 Reflect

Ask children:

■ **How did you figure out my sorting rule?**

Children might say: By looking at the piles, by comparing, or the like.

4 Assess

During Small Group activities, use the Small Group Record Sheet from *Assessment* to observe and record children's progress.

Wednesday Planner

Objectives
- To name and describe familiar two-dimensional shapes
- To distinguish between visually-similar non-examples of familiar two-dimensional shapes
- To match congruent shapes by memory
- To compare small numbers of objects after shown only briefly
- To produce small numbers of actions

Materials
- *Shape Sets
- large shapes to step on

Math Throughout the Year

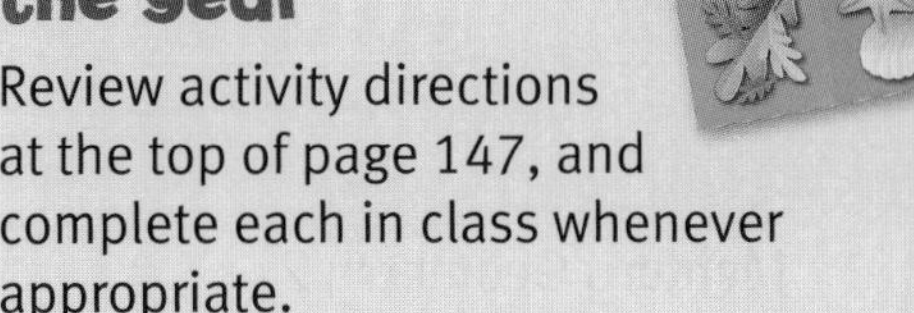

Review activity directions at the top of page 147, and complete each in class whenever appropriate.

*provided in Manipulative Kit

Wednesday

1 Whole Group 15

Warm-Up: *Building Shapes*

Read ***Big Book Building Shapes*** to the class, and briefly review any less familiar shapes. Together identify similar shapes in the classroom.

I Spy

- Beforehand, place various Shape Set shapes throughout the classroom in plain view.
- Name the shape of something in the room ("I spy something that is a big rectangle."). Have all children guess the item or shape you are thinking about. If able, have the child who guessed correctly think of the next item or shape for you and the class to guess.

Monitoring Student Progress

If . . . children struggle during I Spy,	**Then . . .** place fewer Shape Set shapes and refer to only those.
If . . . children excel during I Spy,	**Then . . .** consider harder items or shapes for them to guess.

Shape Step

- If you do not already have large shapes on which children can step, do one of the following: make shapes on the floor with masking or colored tape; make chalk shapes outdoors; tape laminated paper shapes to the floor; or use a copy machine to enlarge Shape Set shapes from the ***Teacher's Resource Guide*** for laminating.
- Show several rectangles. Mix up all those rectangles with plenty of other shapes on the floor.
- Tell children to step on rectangles only. Have a group of five children step on the rectangles. Ask the rest of the class to watch carefully to make sure the group steps on all rectangles. Whenever possible, ask children to explain why the shape they stepped on was the correct shape.
- Repeat until all groups have stepped on shapes. Remember to pretend during the activity.

Eclipse Studios

2 Work Time

20

Computer Center

Building Blocks

Continue to provide each child with a chance to complete Mystery Pictures 2 and Memory Geometry 2.

Hands On Math Center

Based on what children learn and benefit from most, allow them to continue this week's previous Hands On Math Center activities.

3 Reflect

5

Ask children:

- **How do you know a shape is a rectangle when I ask you to step on a rectangle?**

Children might say: It has four sides and right corners; two sides are long and two are short; or it is not a square because the sides are not the same.

4 Assess

Use the Weekly Record Sheet from *Assessment* to record children's progress. Use their time at the centers as an opportunity to complete your observations.

RESEARCH IN ACTION

Rectangles and rectangular prisms (box-like shapes) are found nearly everywhere—windows, doors, photo frames, boards, light fixtures, floors, walls, and ceilings. Show and discuss examples of these with children. Remember that all squares are rectangles but not all rectangles are squares—only those with all four sides equal in length.

Thursday Planner

Objectives

- To name and describe familiar two-dimensional shapes
- To distinguish between visually-similar non-examples of familiar two-dimensional shapes
- To match congruent shapes by memory
- To compare small numbers of objects after shown only briefly
- To produce small numbers of actions

Materials

- *Shape Sets
- large shapes to step on

Math Throughout the Year

Review activity directions at the top of page 147, and complete each in class whenever appropriate.

Looking Ahead

For tomorrow, make copies of Family Letter Week 10 from the ***Teacher's Resource Guide.***

*provided in Manipulative Kit

Thursday

1 Whole Group

Warm-Up: "Five Little Fingers"

Help children suit their actions to the finger play's words.

> One little finger standing on its own, (*Show index finger.*)
> Two little fingers, now they're not alone. (*Add middle finger.*)
> Three little fingers, happy as can be. (*Add ring finger.*)
> Four little fingers go walking down the street. (*Add pinkie; wiggle all four fingers.*)
> Five little fingers, this one is a thumb. (*Show all fingers, emphasizing the thumb.*)
> Wave good-bye 'cause now we're done. (*Wave hand.*)

Guess My Rule

- Tell children to watch carefully as you sort Shape Set shapes into piles based on something that makes them alike. Ask children to silently guess your sorting rule, such as four-sided shapes versus other shapes.
- Sort shapes one at a time, continuing until there are at least two shapes in each pile.
- Signal "shhh," and pick up a new shape. With a look of confusion, gesture to children to encourage all of them to point quietly to which pile the shape belongs. Place the shape in its pile.
- After all shapes are sorted, ask children what they think the sorting rule is. Repeat with other shapes and new rules, such as squares versus non-squares. A complete sorting rule suggestion list appears on page A11 of this guide.
- If able, allow children to "play teacher" by choosing the sorting rule.

2 Work Time

Small Group

Guess My Rule

- Help children follow the Whole Group directions for this activity. After shapes are sorted, ask children: Why did you guess that rule? How do you know? Say: Show me how the shapes fit your rule.
- To simplify the activity, use simpler rules, such as triangles versus non-triangles, or state rules as choices (shapes with three straight sides versus all other shapes), and have children say *yes* to agree or *no* to disagree. For a challenge, use harder rules, such as squares versus rectangles.

Shape Step

Using large shapes to step on (tape, chalk, cut, or enlarged), specify which shape children should step on, and observe whether they step on only the target shape. Encourage children to discuss why the shapes they stepped on are the correct shape.

Monitoring Student Progress

If . . . children struggle during Shape Step,	**Then . . .** have them feel and count a chosen shape's sides, draw it in the air, and review its attributes. If children miss a shape they should have stepped on, you step on it and ask whether or not it is correct.
If . . . children excel during Shape Step,	**Then . . .** incorporate less familiar shapes.

Computer Center Building Blocks

Continue to provide each child with a chance to complete Mystery Pictures 2 and Memory Geometry 2.

Hands On Math Center

Based on what children continue to learn and benefit from most, choose from this week's Hands On Math Center activities: Shape Pictures, Memory Geometry, Rectangles and Boxes, Shape Flip Book, and Compare Game. Consult the Weekly Planner for corresponding materials and, if needed, Monday and Tuesday for activity directions.

3 Reflect

5

Ask children:

- **What picture did you make on the computer?**
- **What shapes were in your picture?**

Based on what they made, children might say: I made a building of squares and a triangle on top for a roof, I made a face with circle eyes and rectangles in the mouth for teeth, or the like.

4 Assess

During Small Group activities, use the Small Group Record Sheet from *Assessment* to observe and record children's progress.

Eclipse Studios

Friday Planner

Objectives
- To name and describe familiar two-dimensional shapes
- To distinguish between visually-similar non-examples of familiar two-dimensional shapes
- To match congruent shapes by memory
- To compare small numbers of objects after shown only briefly
- To produce small numbers of actions

Materials
- *Shape Sets
- large shapes to step on

Math Throughout the Year

Review activity directions at the top of page 147, and complete each in class whenever appropriate.

Looking Ahead

For next week, familiarize yourself with Party Time 2: Count Placemats and Memory Number 1: Counting Cards from the ***Building Blocks*** software.

*provided in Manipulative Kit

Friday

1 Whole Group 15

Warm-Up: I Spy

- Beforehand, place various Shape Set shapes throughout the classroom in plain view.
- Name the shape of something in the room. Have children guess the item or shape you are thinking about. If able, have the child who guessed correctly think of the next item or shape for you and the class to guess.
- As a variation, try the properties version: describe a shape's attributes and see whether children can guess which item or shape you mean.

Shape Step

- If you do not already have large shapes on which children can step, do one of the following: make shapes on the floor with masking or colored tape; make chalk shapes outdoors; tape laminated paper shapes to the floor; or use a copy machine to enlarge Shape Set shapes from the ***Teacher's Resource Guide*** for laminating.
- Show several of a shape of your choice. Mix them up with plenty of other shapes on the floor. In groups of 5, have children step on that shape only.
- Ask the rest of the class to watch carefully to make sure each group steps on the target shape. Whenever possible, ask children to explain why the shape they stepped on was the correct shape.
- Repeat until all groups have stepped on target shapes. Remember to pretend during the activity.

2 Work Time 20

Computer Center

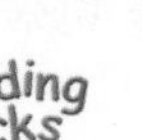

Continue to provide each child with a chance to complete Mystery Pictures 2 and Memory Geometry 2.

Eclipse Studios

 Hands On Math Center

Based on what children continue to learn and benefit from most, choose from this week's Hands On Math Center activities: Shape Pictures, Memory Geometry, Rectangles and Boxes, Shape Flip Book, and Compare Game. Consult the Weekly Planner for corresponding materials and, if needed, Monday and Tuesday for activity directions. Scaffolding for Compare Game follows.

Monitoring Student Progress

If . . . children struggle during Compare Game,	**Then . . .** use fewer cards.
If . . . children excel during Compare Game,	**Then . . .** use more cards, or have players count dot pairs instead of cards.

 Reflect 

5

Briefly discuss with children what you have done in class this week, such as combining shapes to make pictures, and ask:

■ **What did you learn about shapes this week?**

Children might say: How they are the same, how they are different, how a square's sides are the same, how rectangles are not squares, what color they are in the Pattern Blocks, or the like.

 Assess

Use the Weekly Record Sheet from ***Assessment*** to record children's progress. Use their time at the centers as an opportunity to complete your observations.

Home Connection

From the ***Teacher's Resource Guide,*** distribute to children copies of Family Letter Week 10 to share with their family. Each letter has an area for children to show families an example of what they have been doing in class.

Name

WEEK 10 Math News

Building Blocks

Dear Family,

This week in ***Building Blocks,*** we focused on shapes and spatial relationships. Children named, described, and sorted shapes in activities that emphasized the attributes of familiar shapes. As they continued to build their knowledge of circles, squares, rectangles, and triangles, children also experimented with combining these shapes to make pictures.

With this letter you will find a reference sheet of shapes the children will learn about throughout the year. Please keep the sheet in a convenient place, such as on the door of your refrigerator.

Help-at-Home Math Tips

- Play I Spy with shapes. For example, say, "I spy something that is a rectangle." Have your child guess the item, and then take turns being the guesser and the spy. As an alternative to naming the shape, give clues instead. For example, for a square item, say, "I spy something that has four sides the same length." *Benefit:* This activity provides practice with identifying shapes and their attributes.
- Ask your child which shape he or she would like a sandwich, pancake, or the like cut into. Take time to look for shapes in all sorts of food. *Benefit:* This helps children name and describe shapes.

What's Ahead?

In Week 11, the main focus returns to building children's number knowledge. Computer games and other classroom activities help reinforce skills that have already been taught, as well as introduce simple addition.

Don't forget . . .

Review your child's work to discover what he or she has been doing in class!

24 • *Family Letters*

Building Blocks • *Teacher's Resource Guide*

Assess and Differentiate

A Gather Evidence

Review children's progress in mathematics by looking at the Weekly Record Sheets (Monday, Wednesday, Friday) and the Small Group Record Sheets (Tuesday, Thursday) from this past week.

B Summarize Findings

Using ***Online Assessment***, summarize and analyze assessment data for each child based on your weekly observations and Record Sheets. Such information helps determine where each child is on the math trajectory for geometry and counting. See ***Assessment*** for the print companion to each Learning Trajectory Record.

C Differentiate Instruction

Once you have seen a child exhibit specific levels of the trajectory, begin to encourage and work with that child toward the next level. Refer to Appendix A for individualized instruction opportunities, including Special Education concerns.

Shapes: Matching

If . . .	Then . . .
If . . . the child can match shapes of same size and same orientation,	**Then . . .** *Shape Matcher, Identical* Matches familiar shapes (circle, square, typical triangle) of same size and orientation.
If . . . the child can match different-sized shapes,	**Then . . .** *Shape Matcher, Sizes* Matches familiar shapes of different sizes.
If . . . the child can match shapes of varied orientations,	**Then . . .** *Shape Matcher, Orientations* Matches familiar shapes of different orientations.

Shapes: Naming

If . . .	Then . . .
If . . . the child can recognize a large variety of rectangles,	**Then . . .** *Shape Recognizer, All Rectangles* Recognizes more rectangle sizes, shapes, and orientations.
If . . . the child can recognize sides of shapes as parts of the whole,	**Then . . .** *Side Recognizer* Identifies sides as distinct geometric objects.
If . . . the child can name most familiar shapes and some typical examples of less familiar shapes,	**Then . . .** *Shape Recognizer, More Shapes* Recognizes most familiar shapes and typical examples of other shapes, such as hexagon, rhombus (diamond), and trapezoid.
If . . . the child can consistently name common shapes, as well as recognize right angles,	**Then . . .** *Shape Identifier* Names most common shapes, including rhombuses, without making mistakes, such as calling ovals circles; implicitly recognizes right angles, thus can distinguish between rectangles and parallelograms without right angles.
If . . . the child can recognize some shapes by their attributes when atypical in appearance,	**Then . . .** *Parts of Shapes Identifier* Identifies shapes in terms of their components (no matter how skinny a triangle is, knows it is a triangle because it has three straight sides).

Shapes: Composing

If . . .	Then . . .
If . . . the child can complete simple outline puzzles,	**Then . . .** *Piece Assembler* Makes pictures in which shapes touch and each represents a unique role, such as one shape for each body part, and fills simple outline puzzles using trial and error.
If . . . the child can complete simple outline puzzles, putting multiple shapes together to make one,	**Then . . .** *Picture Maker* Puts several shapes together to make one part of a picture, such as two shapes for one arm. Fills simple outline puzzles that suggest placement of each shape.

Object Counting

If . . .	Then . . .
If . . . the child can accurately count to 10 and knows what "how many?" means,	**Then . . .** *Counter (Small Numbers)* Accurately counts objects in a line to 5, and answers "how many?" with the last number counted.

Overview

Big Ideas

- counting
- reading numerals
- connecting numerals to quantities
- comparing amounts and numbers

Teaching for Understanding

We want children to understand that numbers tell how many. The more ways children can express numbers, the stronger their number knowledge is. This week children play games with numbers, figure out simple number problems, and represent numbers with body motions, manipulatives, and symbols. Experiencing a variety of number activities should help children realize the connection between numbers and everyday situations (the role that math takes on in other areas and contexts), as well as keep their number interest stimulated.

Number Activities

Number activities deepen children's knowledge of the meaning of counting. They learn when another object is added to a collection that the number of that collection increases by one. Children also learn to read more written numerals and match numerals to numbers of objects. They continue to practice these abilities in meaningful and motivational settings.

Computer Activities

Children continue to move forward mathematically through this program's computer activities and games, hopefully gaining confidence on the computer itself while motivating a genuine curiosity. Much of the software has a print or hands-on version that correlates to it, which builds a multidimensional connection for children. Children should begin to realize that math serves a purpose and inundates many areas of their everyday lives.

Meaningful Connections

As children are formally introduced to more numerals and learn to read them, literacy is being developed, thus linking it to mathematics. With that link made, a child is better equipped to excel in both areas. Using a numeral to represent a group is a key step toward abstract mathematical thinking. Games enhance the link between language and number through comparison. For example, using words to describe one item's position to another.

What's Ahead?

In the weeks to come, children will learn to count to larger numbers, as well as learn more about how counting and adding are related. Next week specifically, children will engage in dramatic play activities to further their number and counting knowledge. They will also be introduced to the numeral 10.

Eclipse Studios

Overview

English Learner

Articulate the counting numbers carefully and frequently for English learners. Review the following: counting jar, whole group, and faceup. Refer to page 87 of the ***Teacher's Resource Guide.***

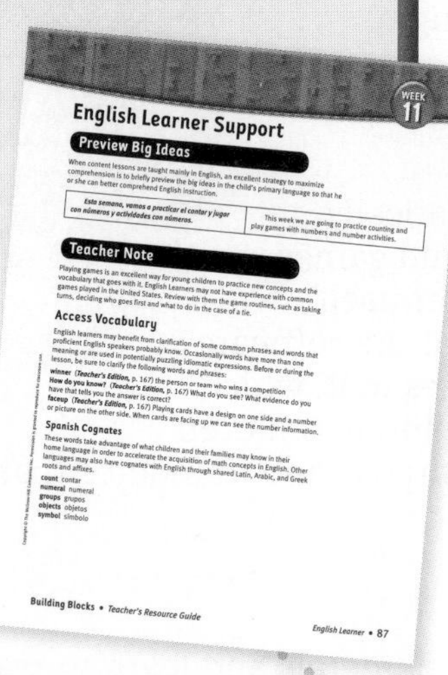

WEEK 11

English Learner Support

Preview Big Ideas

When content lessons are taught mainly in English, an excellent strategy to maximize comprehension is to briefly preview the big ideas in the child's primary language so that he or she can better comprehend English instruction.

Esta semana, vamos a practicar el contar y jugar con números y actividades con números.	This week we are going to practice counting and play games with numbers and number activities.

Teacher Note

Playing games is an excellent way for young children to practice new concepts and the vocabulary that goes with it. English Learners may not have experience with common games played in the United States. Review with them the game routines, such as taking turns, deciding who goes first and what to do in the case of a tie.

Access Vocabulary

English learners may benefit from clarification of some common phrases and words that proficient English speakers probably know. Occasionally words have more than one meaning or are used in potentially puzzling idiomatic expressions. Before or during the lesson, be sure to clarify the following words and phrases:

winner (***Teacher's Edition,*** p. 167) the person or team who wins a competition
How do you know? (***Teacher's Edition,*** p. 167) What do you see? What evidence do you have that tells you the answer is correct?
faceup (***Teacher's Edition,*** p. 167) Playing cards have a design on one side and a number or picture on the other side. When cards are facing up we can see the number information.

Spanish Cognates

These words take advantage of what children and their families may know in their home language in order to accelerate the acquisition of math concepts in English. Other languages may also have cognates with English through shared Latin, Arabic, and Greek roots and affixes.

count contar
numeral numeral
groups grupos
objects objetos
symbol símbolo

Building Blocks • *Teacher's Resource Guide*

English Learner • 87

Technology Project

Children may get additional problem-posing and problem-solving experience using Party Time Free Explore. Encourage children to make up their own problems together and then solve them.

A child might set out a certain number of placemats and ask his or her partner to place one or two items on each placemat; they could explore how many items they could get on each placemat or the entire table!

How Children Count, Add, and Order

Knowing where children are on the learning trajectory for counting, number, and comparing and what the next steps are helps facilitate their development. Children who easily surpass these trajectory levels might be challenged by larger amounts and encouraged to assist other children.

Object Counting

What to Look For What types of groups can the child count and/or produce?

Counter (Small Numbers) Accurately counts objects in a line to 5, and answers "how many?" with the last number counted. When objects are visible, especially with small numbers, begins to understand cardinality.

Producer (Small Numbers) Counts out objects to 5.

Counter (10) Counts structured arrangements of objects to 10, may be able to write numerals or draw collections to represent 1–10.

Adding and Subtracting

What to Look For Can the child add and/or subtract small numbers or groups?

Small Number +/– Finds sums for joining problems up to 3+2 by counting objects. For example: There are two crayons. One more crayon is added. Ask the child how many crayons there are in all. The child counts two, counts one more, and then counts "1, 2, 3... 3!"

Comparing and Ordering

What to Look For What types of groups can the child accurately compare?

Counting Comparer (5) Compares with counting even when larger group's objects are smaller; figures out how many more or less.

Counting Comparer (10) Compares with counting up to 10 even when larger group's objects are smaller.

Numerals

What to Look For Which numerals can the child read?

Numeral Recognizer Reads single-digit numerals, recognizing them as symbolizing number words.

Math Throughout the Year

Math Throughout the Year activities are recommended to build on the mathematical skills highlighted in each week. Here are suggested activities for **Week 11**.

Cleanup (Pick a Number)

Say a number during cleanup, and each child picks up that many items.

Count Motions (Transitioning)

While waiting during transitions, have children count how many times you put up your hand, turn around, or perform some other motion. Then have them do the motion a specified number of times, such as "Jump five times" or "Clap six times." Initially count the actions with children. Later, do the motions but model and explain how to count silently. Children who understand how many motions will stop, but others will continue doing the motion.

Counting Jar

A counting jar holds a specified number of items for children to count. Use the same jar all year, changing its small amount of items weekly. Have children spill the items to count them.

Numerals Every Day

Numerals are all around us. Help children notice and read numerals on common items.

Center Preview

Computer Center Building Blocks

Get your classroom Computer Center ready for Party Time 2: Count Placemats and Memory Number 1: Counting Cards from the *Building Blocks* software.

After you introduce Party Time 2 and Memory Number 1, each child should complete the activities individually as you (or an assistant) monitor and guide him or her periodically. Ideally, each child will have at least ten minutes of computer time at least twice during the week. Use children's center time as an opportunity for assessment.

Hands On Math Center

This week's Hands On Math Center activities are Memory Number, Compare Game, Places Scenes, and Pizza Game 1. Supply the center with these materials: Dot Cards, Numeral Cards, Places Scenes, various counters (including round), copies of Pizza Game 1 activity sheet, and a Number Cube labeled 3–8.

Literature Connections

These books help develop counting.

- *Ten, Nine, Eight* by Molly Bang
- *Two Ways to Count to Ten* by Ruby Dee
- *10 Bears in My Bed: A Goodnight Countdown* by Stanley Mack
- *One Duck Stuck* by Phyllis Root
- *My Love for You* by Susan L. Roth

(t)Matt Meadows (b)Eclipse Studios

Learning Trajectories

Week 11 Objectives

- To count objects up to 5 or 10
- To recognize numerals and the quantities they represent
- To compare small amounts
- To connect counting to simple addition

	Developmental Path	Instructional Activities
Object Counting	*Corresponder*	
	Counter (Small Numbers)	Pizza Game 1 "Little Bird"
	Producer (Small Numbers)	Number Jump (Numerals) Pizza Game 1 Places Scenes
	Counter (10)	Number Jump (Numerals) Party Time 2 Pizza Game 1
	Counter and Producer (10+)	
Adding and Subtracting	*Nonverbal +/–*	
	Small Number +/–	How Many Now? Number Choice How Many Now? (Hidden version)
	Find Result +/–	
Comparing and Ordering	*Counting Comparer*	
	Counting Comparer (5)	Memory Number (print) Compare Game
	Counting Comparer (10)	Memory Number 1
	Mental Number Line to 10	

Use this chart to plan for your specific class schedule. If you have your prekindergarteners for only three days, complete Monday, Tuesday, and Thursday of the week.

Work Time

Whole Group	Small Group	Computer	Hands On	Program Resources
Number Jump (Numerals) How Many Now? *Materials:* *counters *Numeral Cards		Party Time 2	Memory Number *Materials*: *Dot Cards *Numeral Cards Places Scenes *Materials*: *counters *Numeral Cards Compare Game *Materials*: *Dot Cards	Building Blocks ***Teacher's Resource Guide*** Places Scenes ***Assessment*** Weekly Record Sheet
"Little Bird" **Numeral 7**	**How Many Now?** *Materials*: *counters **Number Choice** *Materials*: *game board *game pieces *Number Cubes	**Memory Number 1**	**Memory Number** **Places Scenes** **Compare Game**	***Big Book Where's One?*** Building Blocks ***Teacher's Resource Guide*** • Counting Cards • Places Scenes ***Assessment*** Small Group Record Sheet
"Little Bird" **Number Jump (Numerals)** *Materials*: *Numeral Cards		• **Party Time 2** • **Memory Number 1**	Pizza Game 1 *Materials*: *round counters *Number Cube 3–8 paper plates **Memory Number** **Places Scenes** **Compare Game**	Building Blocks ***Teacher's Resource Guide*** • Pizza Game 1 • Places Scenes ***Assessment*** Weekly Record Sheet
"Little Bird" Numeral 8	**How Many Now?** *Materials*: *counters **Number Choice** *Materials*: *game board *game pieces *Number Cubes	• **Party Time 2** • **Memory Number 1**	**Pizza Game 1** **Memory Number** **Places Scenes** **Compare Game**	***Big Book Where's One?*** Building Blocks ***Teacher's Resource Guide*** • Counting Cards • Pizza Game 1 • Places Scenes ***Assessment*** Small Group Record Sheet
Number Jump (Numerals) *Materials*: *Numeral Cards **How Many Now? (Hidden version)** *Materials*: *counters *Numeral Cards dark cloth		• **Party Time 2** • **Memory Number 1**	**Pizza Game 1** **Memory Number** **Places Scenes** **Compare Game**	Building Blocks ***Teacher's Resource Guide*** • Family Letter Week 11 • Pizza Game 1 • Places Scenes ***Assessment*** Weekly Record Sheet

*provided in Manipulative Kit

Monday Planner

Objectives

- To count objects up to 5 or 10
- To recognize numerals and the quantities they represent
- To compare small amounts

Materials

- *Numeral Cards
- *counters
- *Dot Cards

Math Throughout the Year

Review activity directions at the top of page 163, and complete each in class whenever appropriate.

Looking Ahead

For today, make sure Places Scenes are at the Hands On Math Center and, for tomorrow's Small Group, make one copy of the Counting Cards for each child, both from the *Teacher's Resource Guide.*

*provided in Manipulative Kit

Monday

Whole Group

10

Warm-Up: Number Jump (Numerals)

- Hold up a Numeral Card 1–8, and have all children first say that numeral, safely jump that number of times, and then, together, count the jumps.
- Repeat with a different numeral. Change jumping to another movement the children would enjoy, such as twirling, clapping, or squatting.

How Many Now?

- Show the children three counters, and count them together.
- Ask children to tell how many counters there are, and then show the matching Numeral Card.
- Add one counter, and ask children to tell how many there are now. To check, count the counters together. Show the matching Numeral Card.
- Repeat the process; however, alternate adding and removing a counter, eventually doing the same with two counters.

Work Time

25

Computer Center Building Blocks

Introduce Party Time 2: Count Placemats from the ***Building Blocks*** software. In this activity, children count the settings at a table. Have children help you solve a couple problems, showing them how you click on the numeral to tell how many items are needed. Ask how children could figure out which numeral to click on if they are unsure, such as by counting from 1. Each child should have an opportunity to complete Party Time 2 this week.

Hands On Math Center

These activities, two of which recur, may be set up at the center all week.

Memory Number

- For each pair of children, one set of Dot Cards and one set of Numeral Cards are needed. If more are needed, make copies from the *Teacher's Resource Guide*.
- Place card sets facedown in two separate arrays. At different times, each player chooses, flips, and shows a card from each array.

- If the cards do not match, they are returned facedown to their arrays. If the cards match, that player keeps them.

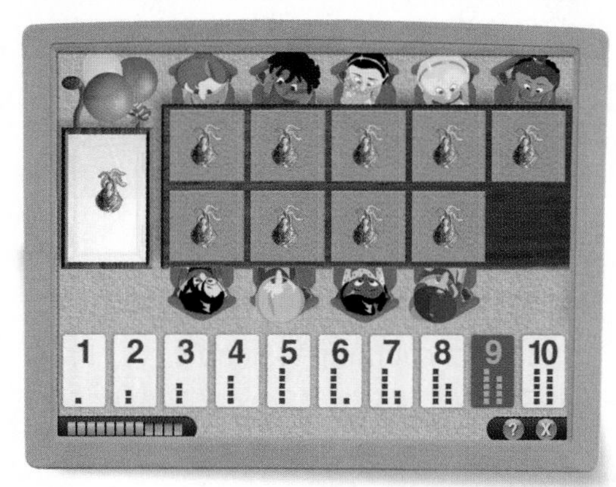

Party Time 2

Monitoring Student Progress

If . . . children struggle during Memory Number,	**Then . . .** place one array face up; use simpler matches (copies from the same set); or limit the number of cards.
If . . . children excel during Memory Number,	**Then . . .** use more cards in each set, or make and use cards for numerals beyond 10.

Places Scenes

Children choose several Places Scenes backgrounds, and put counters on each scene to match the number of the Numeral Card, which has been placed on the table. Children then tell stories about their scene. See Wednesday's Monitoring Student Progress for scaffolding.

Compare Game

- For each pair of children playing the game, two sets of Dot Cards are needed.
- Mix the cards, and deal them evenly to each player. Players place cards facedown in front of them.
- Players simultaneously flip their top cards to compare which has more dots. The player with more says "I have more!" and takes the opponent's card. If dot amounts are equal, players each flip another card to break the tie.
- The game ends when all cards have been played, and the "winner" is the player with more cards. This game can be played without a winner by not allowing players to collect opponent's cards.
- See Thursday's Monitoring Student Progress for scaffolding.

Reflect

5

Show children a numeral, and ask:

■ **What is this? How do you know what it means?**

Children might say: The numeral's name itself; "it is this" (showing that many fingers); or they might just count to that numeral.

Assess

Use the Weekly Record Sheet from ***Assessment*** to record children's progress. Use their time at the centers as an opportunity to complete your observations.

Tuesday Planner

Objectives

- To count objects up to 5 or 10
- To recognize numerals and the quantities they represent
- To compare small amounts
- To connect counting to simple addition

Materials

- *counters
- *game board
- *game pieces
- *Number Cubes

Math Throughout the Year

Review activity directions at the top of page 163, and complete each in class whenever appropriate.

*provided in Manipulative Kit

Tuesday

1 Whole Group

10 

Warm-Up: "Little Bird"

Hold up a finger for each new bird. Here are the finger play's words:

> One little bird with lovely feathers blue (*Show first finger.*)
> Sat beside another one, then there were two. (*Show second finger.*)
> Two little birds singing in the tree, another came to join them
> Then there were three. (*Show third finger.*)
> Three little birds wishing there were more, along came another bird
> Then there were four. (*Show fourth finger.*)
> Four little birds glad to be alive, found a lonely friend
> Then there were five. (*Show fifth finger.*)
> Five little birds picking up sticks, along came a helper
> Then there were six. (*Show one hand and a finger on the other.*)
> Six little birds looking up to heaven, another bird joined them
> Then there were seven. (*Show one hand and two fingers on the other.*)
> Seven little birds just as happy as can be,
> Seven little birds singing songs for you and me.

Numeral 7

- Read *Big Book Where's One?*. Return to the page with the numeral 7; show and point to the 7, and have children say 7.
- Count together how many things there are only seven of on the page. Explain that the number of things matches the numeral.
- Model and explain how the numeral 7 is formed. It has two straight parts; at the top from the left is a straight line across and from the end of that line down is a slanted line to the bottom. Practice with children forming the numeral 7 in the air with fingers.

2 Work Time

25 

Small Group

How Many Now?

- Give each child a copy of a Counting Card set from the ***Teacher's Resource Guide***. Set out three counters, and count them with children. Tell children, "Look!" and add another counter.
- Ask each child to show the card that tells how many counters there are. Repeat the process, adding or subtracting a counter or two.

Number Choice

- Teach the game by playing it with children. One game board, two Number Cubes, and a game piece per player are needed.
- Player One rolls both cubes, decides which to choose, and announces his or her choice. Help children understand which number is the better choice based on whether it lands them on a positive action space (green circle) or avoids a negative action space (red circle). After landing on a positive space, players roll again to move forward that many; on a negative space, they roll again to move backward that many. A positive action space moves a player forward; a negative one moves him or her backward. Player Two checks whether Player One has said a correct number. If so, Player One moves that many spaces. Players switch roles, and the game continues until both players reach the end.

Monitoring Student Progress

If . . . children struggle during Number Choice,	**Then . . .** provide Number Cubes labeled 1–3 only.
If . . . children excel during Number Choice,	**Then . . .** use one Number Cube labeled 5–10 and/or ask players to say beforehand what number they would need to roll to land on a particular space, finish the game, or the like.

Computer Center Building Blocks

Introduce Memory Number 1: Counting Cards from the *Building Blocks* software, in which children play the traditional concentration game by matching cards with both numerals and dots. Each child should have an opportunity to complete Memory Number 1 this week.

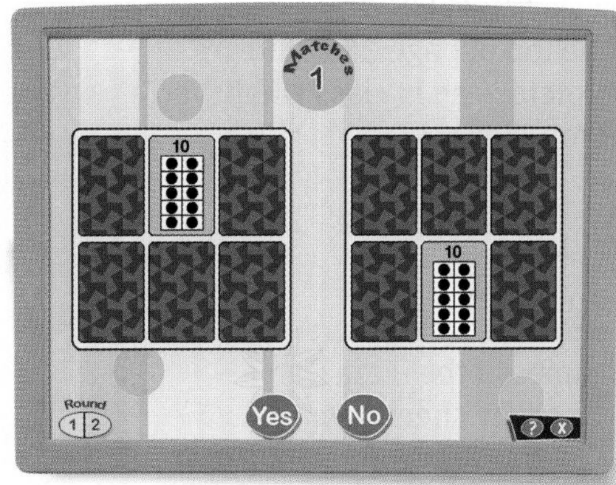

Memory Number 1

Hands On Math Center

Allow children to continue yesterday's Hands On Math Center activities.

3 Reflect

Ask children:

- **When there were three counters and one was added, how did you know the new number?**

Children might say: I counted, I know that four comes after three, or the like.

4 Assess

During Small Group activities, use the Small Group Record Sheet from *Assessment* to observe and record children's progress.

Wednesday Planner

Objectives

- To count objects up to 5 or 10
- To recognize numerals and the quantities they represent
- To compare small amounts
- To connect counting to simple addition

Materials

- *Numeral Cards
- *round counters
- *Number Cube 3–8
- paper plates

Math Throughout the Year

Review activity directions at the top of page 163, and complete each in class whenever appropriate.

Looking Ahead

For today, make sure copies of the Pizza Game 1 activity sheet from the ***Teacher's Resource Guide*** are at the Hands On Math Center.

*provided in Manipulative Kit

Wednesday

1 Whole Group

10

Warm-Up: "Little Bird"

Hold up a finger for each new bird. Here are the finger play's words:

One little bird with lovely feathers blue (*Show first finger.*)
Sat beside another one, then there were two. (*Show second finger.*)
Two little birds singing in the tree, another came to join them
Then there were three. (*Show third finger.*)
Three little birds wishing there were more, along came another bird
Then there were four. (*Show fourth finger.*)
Four little birds glad to be alive, found a lonely friend
Then there were five. (*Show fifth finger.*)
Five little birds picking up sticks, along came a helper
Then there were six. (*Show one hand and a finger on the other.*)
Six little birds looking up to heaven, another bird joined them
Then there were seven. (*Show one hand and two fingers on the other.*)
Seven little birds just as happy as can be,
Seven little birds singing songs for you and me.

Number Jump (Numerals)

- Hold up a Numeral Card 1–8, and have all children first say that numeral, safely jump that number of times, and then, together, count the jumps.
- Repeat with a different numeral. Change jumping to another movement the children would enjoy, such as twirling, clapping, or squatting.

2 Work Time

25

Computer Center Building Blocks

Continue to provide each child with a chance to complete Party Time 2 and Memory Number 1.

Hands On Math Center

After reviewing Pizza Game 1, children may play it and/or continue this week's previous Hands On Math Center activities based on what they learn and benefit from most. Scaffolding for Places Scenes follows.

Monitoring Student Progress

If . . . children need help with the Places Scenes activity,	**Then . . .** use Numeral Cards of lesser amounts; have children use Counting Cards and place counters on the card's dots before placing them on their scenes; or have children use the beach or space scene, which guides them to put one to five counters in specified locations (placement spots).
If . . . children need a challenge during the Places Scenes activity,	**Then . . .** use or make Numeral Cards of greater amounts.

Pizza Game 1

- Introduce the Number Cube with three to eight dots to children. Each player should have a copy of the Pizza Game 1 activity sheet.
- Player One rolls the cube, and puts that many round counters on his or her paper plate. Player One asks Player Two, "Is this correct?" Player Two agrees or disagrees with Player One (but Player One has to be correct before proceeding). Once correct, Player One moves the counters to the pizza on his or her activity sheet. Players take turns until all pizzas have "toppings."

3 Reflect

5

Ask children:

■ **On the computer, how did you know how many toppings were hidden?**

Children might say: I counted, I saw it in my head, I knew 1 plus another 1 is 2, or the like.

It is expected that some children will still offer answers like "I'm smart," "I just know," or "My (mom/dad/other adult) told me." In such cases, help them think mathematically about the question(s) and apply the skills they are currently learning.

4 Assess

Use the Weekly Record Sheet from *Assessment* to record children's progress. Use their time at the centers as an opportunity to complete your observations.

Eclipse Studios

Thursday Planner

Objectives

- To count objects up to 5 or 10
- To recognize numerals and the quantities they represent
- To compare small amounts
- To connect counting to simple addition

Materials

- *counters
- *game board
- *game pieces
- *Number Cubes

Math Throughout the Year

Review activity directions at the top of page 163, and complete each in class whenever appropriate.

Looking Ahead

For tomorrow, make copies of Family Letter Week 11 from the *Teacher's Resource Guide.*

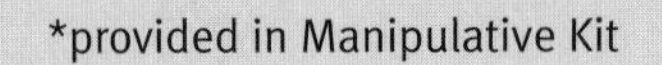

*provided in Manipulative Kit

Thursday

1 Whole Group

10

Warm-Up: "Little Bird"

- See yesterday's Whole Group for the finger play's beginning lines.
- Because you will be counting to eight, replace the last line with the following:

  Seven little birds think they are just great, one more came
  And now there are eight. *(Show one hand and three fingers on the other.)*
  Eight little birds just as happy as can be,
  Eight little birds singing songs for you and me.

Numeral 8

- Review *Big Book Where's One?*. Return to the page with the numeral 8; show and point to the 8, and have children say 8.
- Count together how many things there are only eight of on the page. Explain that the number of things matches the numeral.
- Model and explain how the numeral 8 is formed. It looks like two circles, one on top of the other, by writing the capital letter S, and then curving back to the start. Practice with children forming the numeral 8 in the air with fingers.

2 Work Time

25

Small Group

How Many Now?

- Give each child a copy of a Counting Card set from the ***Teacher's Resource Guide***. Set out three counters, and count them with children. Tell children, "Look!" and add another counter.
- Ask each child to show the Counting Card that tells how many counters there are.
- Repeat the process several times, adding a counter each time (eventually adding or subtracting two).

Number Choice

- Remind children how to play the game. One game board, two Number Cubes, and a game piece per player are needed. Player One rolls both cubes, decides which to choose, and announces his or her choice.

Review how it might be better to choose the smaller number if it lands a player on a positive (move forward) instead of a negative (move backward) action space.
- Player Two checks whether Player One has said a correct number. If so, Player One moves that many spaces. Players switch roles, and the game continues until both players reach the end.

 Computer Center Building Blocks

Continue to provide each child with a chance to complete Party Time 2 and Memory Number 1.

 Hands On Math Center

Based on what children continue to learn and benefit from most, choose from this week's Hands On Math Center activities: Pizza Game 1, Memory Number, Places Scenes, and Compare Game. Consult the Weekly Planner for corresponding materials and, if needed, previous days for activity directions.

Monitoring Student Progress

If . . . children struggle during Compare Game,	**Then . . .** use cards with fewer dots.
If . . . children excel during Compare Game,	**Then . . .** have them count dot pairs instead of just dots; use cards with more dots; or use Numeral Cards.

 Reflect  5

Ask children:
- **When there were three counters and one was added, how did you know the new number?**

Children might say: I counted, I know that four comes after three, or the like.

 Assess

During Small Group activities, use the Small Group Record Sheet from *Assessment* to observe and record children's progress.

Eclipse Studios

Friday Planner

Objectives

- To count objects up to 5 or 10
- To recognize numerals and the quantities they represent
- To compare small amounts
- To connect counting to simple addition

Materials

- *Numeral Cards
- *counters
- dark cloth

Math Throughout the Year

Review activity directions at the top of page 163, and complete each in class whenever appropriate.

Looking Ahead

For next week, familiarize yourself with Dinosaur Shop 1: Label Boxes and Space Race: Number Choice from the ***Building Blocks*** software.

*provided in Manipulative Kit

Friday

1 Whole Group

15 

Warm-Up: Number Jump (Numerals)

- Hold up a Numeral Card 1–8, and have all children first say that numeral, safely jump that number of times, and then, together, count the jumps.
- Repeat with a different numeral. Change jumping to another movement the children would enjoy, such as twirling, clapping, or squatting.

Monitoring Student Progress

If . . . children need help during Number Jump,	**Then . . .** use smaller numbers, and say the numeral when you show your fingers.
If . . . children need a challenge during Number Jump,	**Then . . .** use larger numbers, or show finger combinations, such as two on the left and three on the right.

How Many Now? (Hidden version)

- Show the children three counters, and count them together. Then cover the counters with a dark cloth.
- Ask children to tell how many counters there are. Uncover the counters to count to check, and then show the matching Numeral Card.
- Cover the counters again, and add or subtract another counter. Ask children to tell how many counters there are now. Uncover them to count together to check, and then show the matching Numeral Card.
- Repeat the process, eventually adding and subtracting two counters.

Work Time

Computer Center 

Continue to provide each child with a chance to complete Party Time 2 and Memory Number 1.

Hands On Math Center

Based on what children continue to learn and benefit from most, choose from this week's Hands On Math Center activities: Pizza Game 1, Memory Number, Places Scenes, and Compare Game. Consult the Weekly Planner for corresponding materials and, if needed, Monday and Tuesday for activity directions.

Reflect

Briefly discuss with children what you have done in class this week, such as adding small numbers, and ask:

- **When you have two blocks and add one more block, how many do you have?**

Children might say: Three, I count to three, I know three is more than two, or the like.

Assess

Use the Weekly Record Sheet from ***Assessment*** to record children's progress. Use their time at the centers as an opportunity to complete your observations.

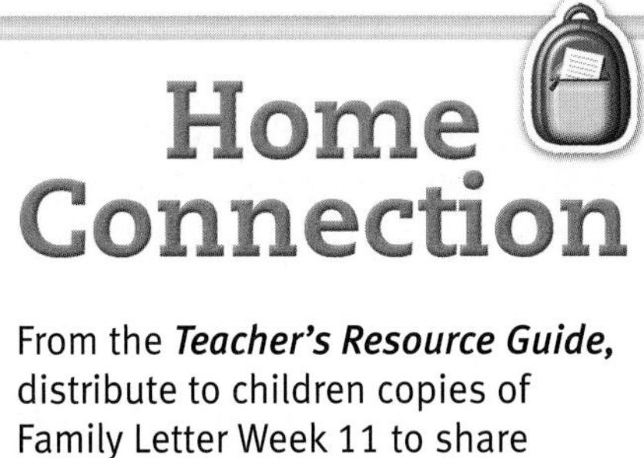

Home Connection

From the ***Teacher's Resource Guide,*** distribute to children copies of Family Letter Week 11 to share with their family. Each letter has an area for children to show families an example of what they have been doing in class.

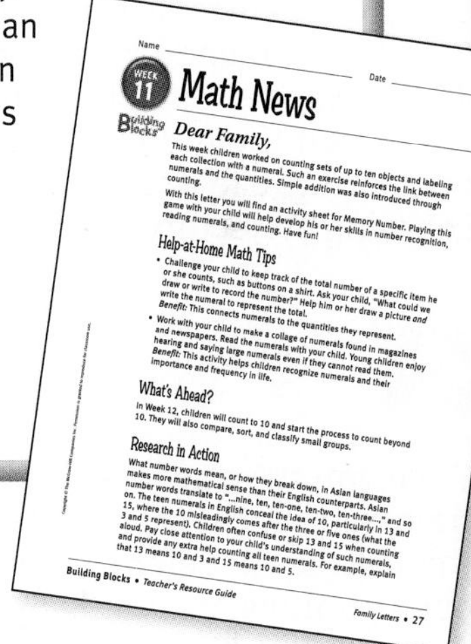

Name ______ Date ______

WEEK 11 Building Blocks

Math News

Dear Family,

This week children worked on counting sets of up to ten objects and labeling each collection with a numeral. Such an exercise reinforces the link between numerals and the quantities. Simple addition was also introduced through counting.

With this letter you will find an activity sheet for Memory Number. Playing this game with your child will help develop his or her skills in number recognition, reading numerals, and counting. Have fun!

Help-at-Home Math Tips

- Challenge your child to keep track of the total number of a specific item he or she counts, such as buttons on a shirt. Ask your child, "What could we draw or write to record the number?" Help him or her draw a picture *and* write the numeral to represent the total. *Benefit:* This connects numerals to the quantities they represent.
- Work with your child to make a collage of numerals found in magazines and newspapers. Read the numerals with your child. Young children enjoy hearing and saying large numerals even if they cannot read them. *Benefit:* This activity helps children recognize numerals and their importance and frequency in life.

What's Ahead?

In Week 12, children will count to 10 and start the process to count beyond 10. They will also compare, sort, and classify small groups.

Research in Action

What number words mean, or how they break down, in Asian languages makes more mathematical sense than their English counterparts. Asian number words translate to "...nine, ten, ten-one, ten-two, ten-three...," and so on. The teen numerals in English conceal the idea of 10, particularly in 13 and 15, where the 10 misleadingly comes after the three or five ones (what the 3 and 5 represent). Children often confuse or skip 13 and 15 when counting aloud. Pay close attention to your child's understanding of such numerals, and provide any extra help counting all teen numerals. For example, explain that 13 means 10 and 3 and 15 means 10 and 5.

Building Blocks • *Teacher's Resource Guide* — *Family Letters* • 27

Assess and Differentiate

A Gather Evidence

Review children's progress in mathematics by looking at the Weekly Record Sheets (Monday, Wednesday, Friday) and the Small Group Record Sheets (Tuesday, Thursday) from this past week.

B Summarize Findings

Using *Online Assessment,* summarize and analyze assessment data for each child based on your weekly observations and Record Sheets. Such information helps determine where each child is on the math trajectory for counting, number, and comparing. See ***Assessment*** for the print companion to each Learning Trajectory Record.

C Differentiate Instruction

Once you have seen a child exhibit specific levels of the trajectory, begin to encourage and work with that child toward the next level. Refer to Appendix A for individualized instruction opportunities, including Special Education concerns.

Object Counting

If . . .	Then . . .
If . . . the child can accurately count to 10 and knows what "how many?" means,	**Then . . .** *Counter (Small Numbers)* Accurately counts objects in a line to 5, and answers "how many?" with the last number counted. When objects are visible, especially with small numbers, begins to understand cardinality.
If . . . the child can produce at least five items,	**Then . . .** *Producer (Small Numbers)* Counts out objects to 5.
If . . . the child can count set arrangements of 10 and attempts to represent numbers in writing,	**Then . . .** *Counter (10)* Counts structured arrangements of objects to 10, may be able to write numerals or draw collections to represent 1–10.

Adding and Subtracting

If . . .	Then . . .
If . . . the child can add and/or subtract small numbers,	**Then . . .** *Small Number +/–* Finds sums for joining problems up to 3+2 by counting objects. For example: There are two crayons. One more crayon is added. Ask the child how many crayons there are in all. The child counts two, counts one more, and then counts "1, 2, 3... 3!".

Comparing and Ordering

If . . .	Then . . .
If . . . the child can compare small groups by counting,	**Then . . .** *Counting Comparer (5)* Compares with counting even when larger group's objects are smaller; figures out how many more or less.
If . . . the child can compare groups of at least 10 by counting,	**Then . . .** *Counting Comparer (10)* Compares with counting up to 10 even when larger group's objects are smaller.

WEEK 12

Overview

Big Ideas

- counting objects to 10
- numeral recognition
- sorting and classifying

Teaching for Understanding

The idea we are trying to convey to children is that counting can tell us how many are in a collection. Children build their number knowledge by connecting quantities to words and symbols like 9 and 10 and by using numbers in new ways, such as counting only part of a large group. These skills are reinforced as children truly consider their actions—actually discussing *how* to count correctly.

Counting Activities

This week a puppet is used to help children identify, discuss, and correct counting errors. Although some people often wonder and worry whether talking about mistakes will actually teach those mistakes to children, studies show that the opposite is true; when children help identify and correct errors, they develop more reliable skills themselves. Children can often identify errors others make before they can recognize theirs, and doing so helps children prevent their own errors.

Numerals

Children continue to study early numerals, completing through 10 this week. They will have opportunities to simply recognize and sort numerals, as well as count through literature and dramatic play. Many children enjoy reading large numerals that they might not even understand; if you see such curiosity or enthusiasm, encourage it by counting beyond 10 with the entire class.

Meaningful Connections

Many activities develop several important mathematical skills, such as counting, sorting, and classifying. As children sort dinosaurs and then tell how many of each category, for example, type of dinosaur or color, they are learning to classify, count, and count *only* a subset of objects, such as just the triceratops. This is a challenging and essential counting skill which forges a connection between quantifying and classifying, providing the beginnings of part-whole understanding.

What's Ahead?

In the weeks to come, children will learn to attach the meanings of different stair lengths or heights to counting numbers. The "stairs" are actually made of cubes that link together. Children begin to order numbers, for example, placing numerals in proper sequence, and gain deeper knowledge of shapes.

Eclipse Studios

Overview

How Children Learn to Count and Compare

Knowing where children are on the learning trajectory for counting and comparing and what the next steps are helps facilitate their development. Children who easily surpass these trajectory levels might be challenged by larger numbers and encouraged to assist other children.

Object Counting

What to Look For What size groups can the child count accurately?

Counter (10) Counts structured arrangements of objects to 10, may be able to write numerals or draw collections to represent 1–10.

Counter and Producer (10+) Accurately counts and produces objects to 10, then beyond to 30; has explicit understanding of cardinality (that numbers tell "how many?") and keeps track of objects that have and have not been counted even in different arrangements. Writes or draws to represent 1–10 (then 20 and 30), and gives next number to 20s or 30s; recognizes errors in others' counting and can eliminate most errors in own counting if asked to try hard.

Comparing and Ordering

What to Look For What type of number relationships does the child recognize through comparing?

Counting Comparer (10) Compares with counting up to 10 even when larger group's objects are smaller.

Mental Number Line to 10 Uses internal images and knowledge of number relationships to determine relative size and position. For example, in determining which number, 4 or 6, is closer to 9.

English Learner

For children learning English, write number words with the actual numerals to reinforce their connection. Refer to page 88 of the ***Teacher's Resource Guide.***

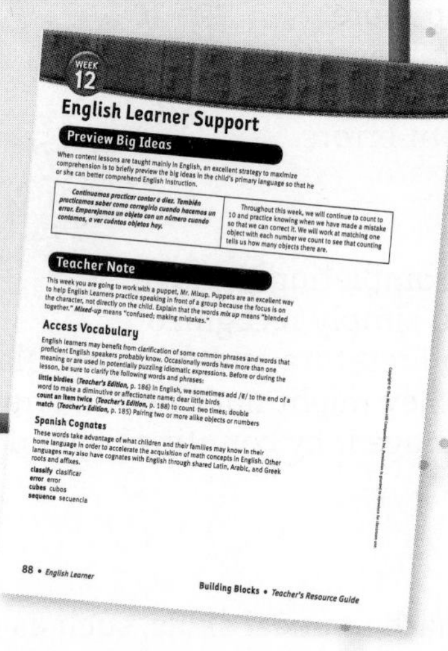

WEEK 12

English Learner Support

Preview Big Ideas

When content lessons are taught mainly in English, an excellent strategy to maximize comprehension is to briefly preview the big ideas in the child's primary language so that he or she can better comprehend English instruction.

Continuamos practicar contar a diez. También practicamos saber como corregirlo cuando hacemos un error. Emparejamos un objeto con un número cuando contamos, a ver cuántos objetos hay.	Throughout this week, we will continue to count to 10 and practice knowing when we have made a mistake so that we can correct it. We will work at matching one object with each number we count to see that counting tells us how many objects there are.

Teacher Note

This week you are going to work with a puppet, Mr. Mixup. Puppets are an excellent way to help English Learners practice speaking in front of a group because the focus is on the character, not directly on the child. Explain that the words *mix up* means "blended together." *Mixed-up* means "confused; making mistakes."

Access Vocabulary

English learners may benefit from clarification of some common phrases and words that proficient English speakers probably know. Occasionally words have more than one meaning or are used in potentially puzzling idiomatic expressions. Before or during the lesson, be sure to clarify the following words and phrases:

little birdies (*Teacher's Edition*, p. 186) In English, we sometimes add /ē/ to the end of a word to make a diminutive or affectionate name; dear little birds
count an item twice (*Teacher's Edition*, p. 188) to count two times; double
match (*Teacher's Edition*, p. 185) Pairing two or more alike objects or numbers

Spanish Cognates

These words take advantage of what children and their families may know in their home language in order to accelerate the acquisition of math concepts in English. Other languages may also have cognates with English through shared Latin, Arabic, and Greek roots and affixes.

classify clasificar
error error
cubes cubos
sequence secuencia

88 • *English Learner* Building Blocks • *Teacher's Resource Guide*

Technology Project

Using Dinosaur Shop Free Explore, children can count along with the program as they place dinosaurs. Or, working cooperatively, one child names a number, and the other puts that many dinosaurs in a box. Challenge them to think of other ways to explore the program. Ask children to think about the link between a numeral and the number of dinosaurs in a box. For more involved mathematical play, put Counting Cards and a toy telephone next to the computer so children can truly mimic filling orders.

Math Throughout the Year

Math Throughout the Year activities are recommended to build on the mathematical skills highlighted in each week. Here are suggested activities for **Week 12.**

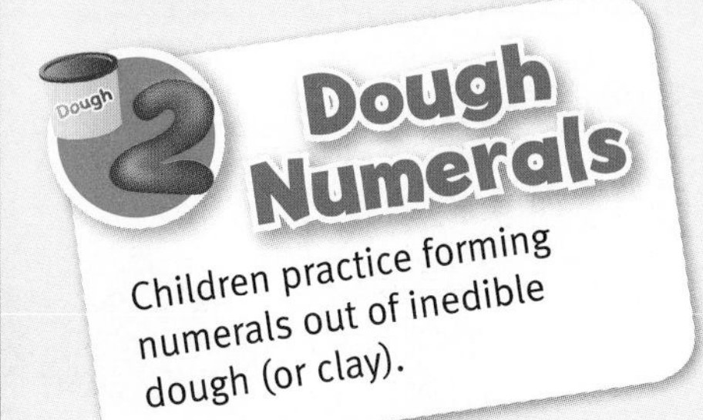

Dough Numerals

Children practice forming numerals out of inedible dough (or clay).

Dinosaur Shop (Dramatic Play)

Children set up and run a toy dinosaur shop, applying the mathematical concepts of sorting, counting, and adding. First, children sort and classify toy dinosaurs (or dinosaur counters) by color, size, type, or other criterion. Some children are customers who place orders; others take turns being salespeople. The "orders" are put into boxes, which are then labeled with the corresponding numeral (using a self-sticking note or other label). Play money may be used to add realism and challenge to the activity. Thematic images—such as dinosaur, fossil, and appropriate habitat images—may be used to decorate the "shop."

Numerals Every Day

Numerals are all around us. Help children notice and read numerals on common items throughout the day around school.

Center Preview

Computer Center Building Blocks

Get your classroom Computer Center ready for Dinosaur Shop 1: Label Boxes and Space Race: Number Choice from the *Building Blocks* software.

After you introduce Dinosaur Shop 1 and Space Race, each child should complete the activities individually as you (or an assistant) monitor and guide him or her periodically. Ideally, each child will have at least ten minutes of computer time at least twice during the week. Use children's center time as an opportunity for assessment.

Hands On Math Center

This week's Hands On Math Center activities are Dinosaur Shop, Places Scenes, Memory Number, and Numeral Sorting. Supply the center with these materials: toy dinosaurs (and/or dinosaur counters), boxes, labels (or self-sticking notes), Places Scenes, counters, Numeral Cards, Dot Cards, tactile numerals (magnetic, plastic, foam, or the like), and muffin pans.

Literature Connections

These books help develop counting.

- *My Visit to the Dinosaurs* by Aliki
- *Count-a-saurus* by Nancy Blumenthal
- *Too Many Eggs: A Counting Book* by M. Christina Butler
- *Two Ways to Count to Ten* by Ruby Dee
- *March of the Dinosaurs: A Prehistoric Counting Book* by Jakki Wood

(t)Matt Meadows (b)Eclipse Studios

WEEK 12

Weekly Planner

Use this chart to plan for your specific class schedule. If you have your prekindergarteners for only three days, complete Monday, Tuesday, and Thursday of the week.

Learning Trajectories

Week 12 Objectives

- To recognize numerals and the quantities they represent
- To compare small amounts
- To sort and classify small groups
- To count objects to 10 and beyond

	Developmental Path	Instructional Activities
Object Counting	*Producer (Small Numbers)*	
	Counter (10)	Places Scenes Memory Number
	Counter and Producer (10+)	Number Jump (Numerals) Dinosaur Shop 1 (computer) *Makayla's Magnificent Machine* Dinosaur Shop (print) Mr. Mixup Places Scenes
	Counter Backward from 10	
Comparing and Ordering	*Counting Comparer (5)*	
	Counting Comparer (10)	Dinosaur Shop 1 (computer)
	Mental Number Line to 10	Space Race: Number Choice
	Serial Orderer to 6+	

Work Time

Whole Group	Small Group	Computer	Hands On	Program Resources
Numeral 9 **Number Jump (Numerals)** *Materials:* *Numeral Cards		**Dinosaur Shop 1**	**Dinosaur Shop** *Materials:* dinosaurs boxes labels **Places Scenes** *Materials:* *counters *Numeral Cards	Building Blocks ***Teacher's Resource Guide*** Places Scenes ***Assessment*** Weekly Record Sheet
Makayla's Magnificent Machine ***Mr. Mixup**	***Mr. Mixup** **Dinosaur Shop (Sort and Label)** *Materials:* dinosaurs boxes labels	**Space Race: Number Choice**	**Memory Number** *Materials:* *Dot Cards *Numeral Cards **Numeral Sorting** *Materials:* numerals muffin pans	***Big Book Makayla's Magnificent Machine*** Building Blocks ***Teacher's Resource Guide*** Places Scenes ***Assessment*** Small Group Record Sheet
"Ten Little Birdies" **Numeral 10** ***Mr. Mixup**		• **Dinosaur Shop 1** • **Space Race: Number Choice**	**Memory Number** **Numeral Sorting** **Dinosaur Shop** **Places Scenes**	Building Blocks ***Teacher's Resource Guide*** Places Scenes ***Assessment*** Weekly Record Sheet
Number Jump (Numerals) *Materials:* *Numeral Cards ***Mr. Mixup**	***Mr. Mixup** **Dinosaur Shop (Sort and Label)** *Materials:* dinosaurs boxes labels	• **Dinosaur Shop 1** • **Space Race: Number Choice**	Memory Number Numeral Sorting Dinosaur Shop Places Scenes	Building Blocks ***Teacher's Resource Guide*** Places Scenes ***Assessment*** Small Group Record Sheet
"Ten Little Birdies" ***Mr. Mixup**		• **Dinosaur Shop 1** • **Space Race: Number Choice**	Memory Number Numeral Sorting Dinosaur Shop Places Scenes	Building Blocks ***Teacher's Resource Guide*** • Family Letter Week 12 • Places Scenes ***Assessment*** Weekly Record Sheet

*provided in Manipulative Kit

Monday Planner

Objectives
- To recognize numerals and the quantities they represent
- To compare small amounts
- To sort and classify small groups

Materials
- *Numeral Cards
- dinosaurs
- boxes
- labels
- *counters

Math Throughout the Year

Review activity directions at the top of page 179, and complete each in class whenever appropriate.

Looking Ahead

For today's Hands On Math Center, provide boxes, labels (or self-sticking notes) with numerals 1–10 written on them (one numeral per label), and Places Scenes, which were used last week from the ***Teacher's Resource Guide.*** For tomorrow, preview ***Big Book Makayla's Magnificent Machine*** and provide muffin pans.

*provided in Manipulative Kit

Monday

1 Whole Group 20

Warm-Up: Numeral 9
- Write the numeral 9 where everyone can see it.
- Teach how the numeral 9 is formed. Explain that its parts are a small circle on top and an up-and-down (vertical) line. You may need to clarify right and left with any strategy you find useful, such as "Start on the side where my desk is." Use this rhyme if you would like: "A circle on top and a straight down line. 9 is fine!"
- Children should practice forming the numeral 9 in the air with their fingers. If you have a tactile numeral 9 for children to feel, have them do so.

Number Jump (Numerals)
- Hold up a Numeral Card 1–9, and have all children first say that numeral, safely jump that number of times, and then, together, count the jumps.
- Repeat with a different numeral. Change jumping to another movement the children would enjoy, such as twirling, clapping, or squatting.

2 Work Time 15

Computer Center Building Blocks

Introduce Dinosaur Shop 1: Label Boxes from the ***Building Blocks*** software. In this activity, children work in a store that sells toy dinosaurs, and they help label boxes with numerals to tell how many dinosaurs are in each box. Each child should have an opportunity to complete Dinosaur Shop 1 this week.

Hands On Math Center

These activities may be set up at the center all week. (Places Scenes recurs from last week).

Dinosaur Shop

Using toy dinosaurs (or dinosaur counters), show children how to sort dinosaurs into boxes and how to label each box with the numeral that tells how many dinosaurs there are. Allow children to then count, sort, and label on their own.

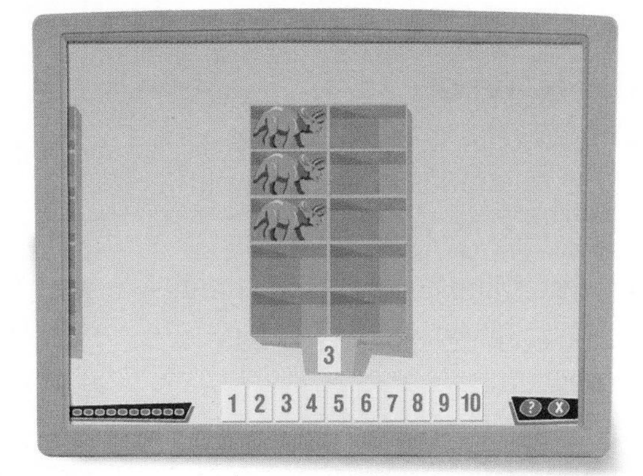

Dinosaur Shop 1

Monitoring Student Progress

If . . . children struggle during Dinosaur Shop,	**Then . . .** use smaller numerals (fewer counters).
If . . . children excel during Dinosaur Shop,	**Then . . .** use larger numerals (more counters), or encourage children to sort by attributes (color, type, and so on).

Places Scenes

Children choose several Places Scenes, and put counters on each scene to match the number of the Numeral Card, which has been placed on the table. To correlate to this week's computer activity, encourage children to use dinosaur-themed scenes and items. Children then tell stories about their scene. See Wednesday's Monitoring Student Progress for scaffolding.

3 Reflect

5

Show children a numeral, and ask:

- **What is this? How do you know what it means?**

Children might say: The numeral's name itself; they might just count to that numeral; "It is this" (showing that many fingers); or "It is like a letter but it helps us count."

4 Assess

Use the Weekly Record Sheet from ***Assessment*** to record children's progress. Use their time at the centers as an opportunity to complete your observations.

Tuesday Planner

Objectives

- To recognize numerals and the quantities they represent
- To compare small amounts
- To sort and classify small groups
- To count objects to 10 and beyond

Materials

- *Mr. Mixup
- dinosaurs
- boxes
- labels
- *Dot Cards
- *Numeral Cards
- numerals
- muffin pans

Math Throughout the Year

Review activity directions at the top of page 179, and complete each in class whenever appropriate.

*provided in Manipulative Kit

Tuesday

Whole Group

15 

Warm-Up: *Makayla's Magnificent Machine*

Read the story to children, and help children consider how they would use each number of objects. When possible, act out their solutions as a group.

Mr. Mixup

- Introduce Mr. Mixup, a puppet that often makes mistakes, to children. Use a silly voice, and have fun! Tell children they are going to help Mr. Mixup count. If he makes a mistake, they need to stop him right then and correct him.
- Make mistakes in verbal counting: wrong order (1, 2, 3, 4, 6, 5), skipped number (11, 13, 14, 16, 17), repeated number (4, 5, 6, 7, 7, 8), and substituted number (12, 19, 14, 19, 16, 17, 18, 19).

Work Time

20 

Small Group

Mr. Mixup

Follow the Whole Group directions, making additional mistakes and observing which children can identify and correct those mistakes. Then have children work in pairs, taking turns being Mr. Mixup, making counting mistakes for their partner to correct.

Monitoring Student Progress

If . . . children struggle correcting Mr. Mixup,	**Then . . .** clearly exaggerate his mistakes using numbers less than 5.
If . . . children excel correcting Mr. Mixup,	**Then . . .** make his mistakes more subtle using greater numbers.

Dinosaur Shop (Sort and Label)

Have children work in pairs to sort and count dinosaurs. First, they sort dinosaurs into boxes *consistently* (using the same criterion as they sort) and *fully* (sorting each dinosaur by that criterion) into boxes. Children then count the dinosaurs in the boxes, labeling each with the correct numeral to tell how many there are.

 Computer Center Building Blocks

Introduce Space Race: Number Choice from the ***Building Blocks*** software, where children click on the numeral that is the better choice determined by which space they will land on (some spaces move players ahead, others send them back). Discuss with children directions and strategies as appropriate. Each child should complete the activity this week.

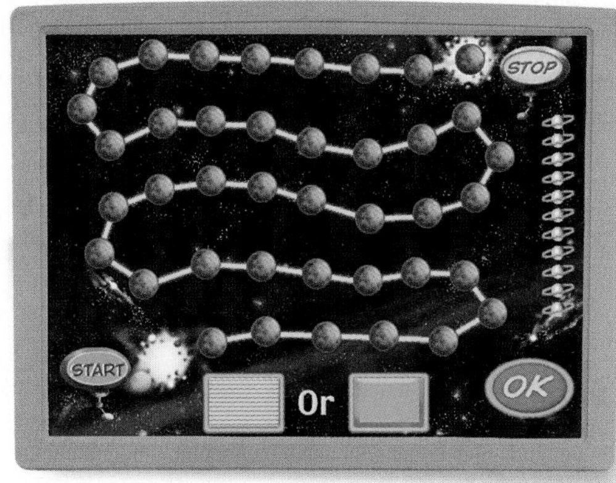

Space Race: Number Choice

 Hands On Math Center

After reviewing these activities, children may play them and/or continue yesterday's.

Memory Number

- For each pair of children, one set of Dot Cards and one set of Numeral Cards are needed. If more are needed, make copies from the ***Teacher's Resource Guide.***
- Place card sets facedown in two separate arrays. At different times, each player chooses, flips, and shows a card from each array. If cards do not match, they are replaced facedown in their arrays. If cards match, that player keeps them.
- See Thursday's Monitoring Student Progress for scaffolding.

Numeral Sorting

- Provide several sets of tactile numerals 0–10 (foam, magnetic, other), a list of numerals 0–10, and muffin pans for sorting. Children sort the numerals into the pans in numerical order.
- To simplify, use paper cups, each labeled with a numeral 0–10. For a challenge, have children sort from highest to lowest instead.

RESEARCH IN ACTION

Use any opportunity, especially Mr. Mixup activities, to discuss the emotional aspect of making mistakes. Ask children, for example, "How do you think Mr. Mixup feels when he makes a mistake? How could we help him? We could tell him, 'It is okay, everybody makes mistakes, and you can learn from them.'"

 Reflect 5 

Ask children:

■ **How did you know Mr. Mixup made a mistake?**

Children might say: He counted wrong, he forgot a number, or the like.

 Assess

During Small Group activities, use the Small Group Record Sheet from *Assessment* to observe and record children's progress.

Wednesday Planner

Objectives

- To recognize numerals and the quantities they represent
- To compare small amounts
- To sort and classify small groups
- To count objects to 10 and beyond

Materials

*Mr. Mixup

Math Throughout the Year

Review activity directions at the top of page 179, and complete each in class whenever appropriate.

*provided in Manipulative Kit

Wednesday

1 Whole Group 25

Warm-Up: "Ten Little Birdies"

Hold up both hands, showing all ten fingers. As each "bird" leaves, "fly" both hands away and bring them back showing one less finger.

Ten little birdies chirping just fine,
One flew away, then there were nine.
Nine little birdies wait, wait, wait,
One flew away, then there were eight.
Eight little birdies, not quite eleven,
One flew away, then there were seven.
Seven little birdies in a nest they fix,
One flew away, then there were six.
Six little birdies see a busy hive,
One flew away, then there were five.
Five little birdies watching others soar,
One flew away, then there were four.
Four little birdies sitting in a tree,
One flew away, then there were three.
Three little birdies love a sky of blue,
One flew away, then there were two.
Two little birdies warm in the sun,
One flew away, then there was one.
One little birdie having little fun,
The birdie flew away, then there was none.

Numeral 10

Write the numeral 10 where everyone can see it. Emphasize that it is a 1 next to a 0. Review that 1 is an up-and-down (vertical) line, and 0 is an oval. Children should practice forming the numeral 10 in the air with their fingers.

Mr. Mixup

- Tell children they are going to help Mr. Mixup count. If he makes a mistake, they need to stop him right then and correct him. Give Mr. Mixup a silly voice, and have fun!
- Continue making mistakes in verbal counting, working with greater numbers as children are able: wrong order (1, 2, 3, 4, 6, 5), skipped number (11, 13, 14, 16, 17), repeated number (4, 5, 6, 7, 7, 8), and substituted number (12, 19, 14, 19, 16, 17, 18, 19).

Work Time

10

RESEARCH IN ACTION

Activities like Numeral Sorting and Dinosaur Shop develop the important mathematical skills of sorting and classifying.

Computer Center Building Blocks

Continue to provide each child with a chance to complete Dinosaur Shop 1 and Space Race: Number Choice.

Hands On Math Center

Based on what children learn and benefit from most, choose from this week's Hands On Math Center activities: Memory Number, Numeral Sorting, Dinosaur Shop, and Places Scenes. Consult the Weekly Planner for corresponding materials and, if needed, previous days for activity directions. Scaffolding for Places Scenes follows.

Monitoring Student Progress

If . . .	Then . . .
If . . . children struggle during Places Scenes,	**Then . . .** use Counting Cards, or have children use the beach or space scene, which guides them to put one to five counters in specified locations, and then help them label each with its numeral.
If . . . children excel during Places Scenes,	**Then . . .** make and use Numeral Cards of greater amounts.

Reflect

Ask children:

- **Where have you seen the numerals 9 and 10?**

Children might say: I watch channel 10; on my clock; on calendars; at the store; and the like.

Assess

Use the Weekly Record Sheet from *Assessment* to record children's progress. Use their time at the centers as an opportunity to complete your observations.

Thursday Planner

Objectives

- To recognize numerals and the quantities they represent
- To compare small amounts
- To sort and classify small groups
- To count objects to 10 and beyond

Materials

- *Numeral Cards
- *Mr. Mixup
- dinosaurs
- boxes
- labels

Math Throughout the Year

Review activity directions at the top of page 179, and complete each in class whenever appropriate.

Looking Ahead

For tomorrow, make copies of Family Letter Week 12 from the ***Teacher's Resource Guide.***

*provided in Manipulative Kit

Thursday

1 Whole Group

10

Warm-Up: Number Jump (Numerals)

- Hold up a Numeral Card 1–9, and have all children first say that numeral, safely jump that number of times, and then, together, count the jumps.
- Repeat with a different numeral. Change jumping to another movement the children would enjoy, such as twirling, clapping, or squatting.

Mr. Mixup

- Tell children they are going to help Mr. Mixup count. Remind them if he makes a mistake, they need to stop him right then and correct him. Give Mr. Mixup a silly voice, and have fun!
- If children have moved beyond verbal counting errors, make object counting mistakes, such as skipping an item or counting an item twice.

2 Work Time

25

Small Group

Mr. Mixup

Follow the Whole Group directions, making additional mistakes and observing which children can identify and correct those mistakes. Then have children work in pairs, taking turns being Mr. Mixup, making counting mistakes for their partner to correct.

Dinosaur Shop (Sort and Label)

Have children work in pairs to sort and count dinosaurs. First, they sort dinosaurs into boxes *consistently* (using the same criterion as they sort) and *fully* (sorting each dinosaur by that criterion). Children then count the dinosaurs in the boxes, labeling each with the correct numeral to tell how many there are.

Eclipse Studios

Computer Center Building Blocks

Continue to provide each child with a chance to complete Dinosaur Shop 1 and Space Race: Number Choice.

Hands On Math Center

Based on what children learn and benefit from most, choose from this week's Hands On Math Center activities: Memory Number, Numeral Sorting, Dinosaur Shop, and Places Scenes. Consult the Weekly Planner for corresponding materials and, if needed, Monday and Tuesday for activity directions. Scaffolding for Memory Number follows.

Monitoring Student Progress

If . . . children struggle during Memory Number,	**Then . . .** place one array face up; use simpler matches (copies from the same set); or limit the number of cards.
If . . . children excel during Memory Number,	**Then . . .** use more cards in each set, or make and use cards for numerals beyond 10.

3 Reflect

5 

Show children this type of 4 and this type of **4**, and ask:

- **How do you know to sort them together?**

Children might say: They look different but both are 4, or the like.

4 Assess

During Small Group activities, use the Small Group Record Sheet from *Assessment* to observe and record children's progress.

Friday Planner

Objectives

- To recognize numerals and the quantities they represent
- To compare small amounts
- To sort and classify small groups
- To count objects to 10 and beyond

Materials

*Mr. Mixup

Math Throughout the Year

Review activity directions at the top of page 179, and complete each in class whenever appropriate.

Looking Ahead

For next week, familiarize yourself with Build Stairs 1: Count Steps and Build Stairs 2: Order Steps from the ***Building Blocks*** software.

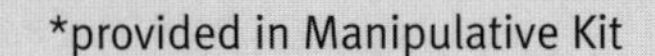

*provided in Manipulative Kit

Friday

1 Whole Group 15

Warm-Up: "Ten Little Birdies"

Hold up both hands, showing all ten fingers. As each "bird" leaves, "fly" both hands away and bring them back showing one less finger.

Ten little birdies chirping just fine,
One flew away, then there were nine.
Nine little birdies wait, wait, wait,
One flew away, then there were eight.
Eight little birdies, not quite eleven,
One flew away, then there were seven.
Seven little birdies in a nest they fix,
One flew away, then there were six.
Six little birdies see a busy hive,
One flew away, then there were five.
Five little birdies watching others soar,
One flew away, then there were four.
Four little birdies sitting in a tree,
One flew away, then there were three.
Three little birdies love a sky of blue,
One flew away, then there were two.
Two little birdies warm in the sun,
One flew away, then there was one.
One little birdie having little fun,
The birdie flew away, then there was none.

Mr. Mixup

- Tell children they are going to help Mr. Mixup count. Remind them if he makes a mistake, they need to stop him right then and correct him.
- If children have moved beyond verbal counting errors, make object counting mistakes, such as skipping an item or counting an item twice.

Monitoring Student Progress

If . . .	Then . . .
If . . . children struggle correcting Mr. Mixup,	**Then . . .** clearly exaggerate his mistakes using numbers less than 5.
If . . . children excel correcting Mr. Mixup,	**Then . . .** make his mistakes more subtle using greater numbers.

Work Time

20

Computer Center Building Blocks

Continue to provide each child with a chance to complete Dinosaur Shop 1 and Space Race: Number Choice.

Hands On Math Center

Based on what children learn and benefit from most, choose from this week's Hands On Math Center activities: Memory Number, Numeral Sorting, Dinosaur Shop, and Places Scenes. Consult the Weekly Planner for corresponding materials and, if needed, Monday and Tuesday for activity directions.

Reflect

5

Briefly discuss with children what you have done in class this week, such as correcting counting errors, and ask:

■ **How would you teach Mr. Mixup to count better?**

Children might say: Keep numbers in order; do not forget any numbers; or the like.

It is expected that some children will still offer answers like "I'm smart," "I just know," or "My (mom/dad/other adult) told me." In such cases, help them think mathematically about the question(s) and apply the skills they are currently learning.

Assess

Use the Weekly Record Sheet from ***Assessment*** to record children's progress. Use their time at the centers as an opportunity to complete your observations.

Home Connection

From the ***Teacher's Resource Guide,*** distribute to children copies of Family Letter Week 12 to share with their family. Each letter has an area for children to show families an example of what they have been doing in class.

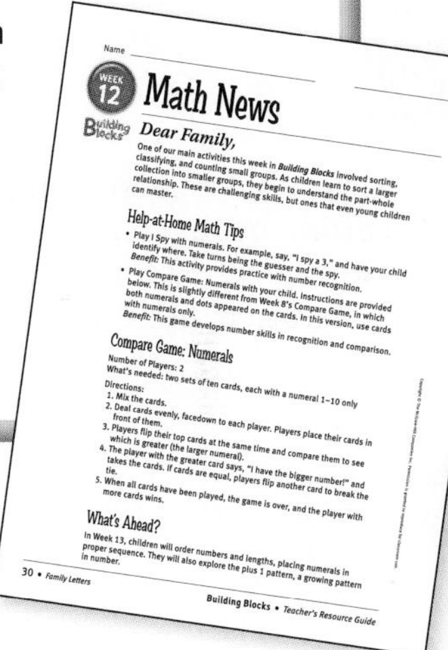

Name

WEEK 12

Math News

Building Blocks

Dear Family,

One of our main activities this week in ***Building Blocks*** involved sorting, classifying, and counting small groups. As children learn to sort a larger collection into smaller groups, they begin to understand the part-whole relationship. These are challenging skills, but ones that even young children can master.

Help-at-Home Math Tips

- Play I Spy with numerals. For example, say, "I spy a 3," and have your child identify where. Take turns being the guesser and the spy.
  Benefit: This activity provides practice with number recognition.
- Play Compare Game: Numerals with your child. Instructions are provided below. This is slightly different from Week 8's Compare Game, in which both numerals and dots appeared on the cards. In this version, use cards with numerals only.
  Benefit: This game develops number skills in recognition and comparison.

Compare Game: Numerals

Number of Players: 2

What's needed: two sets of ten cards, each with a numeral 1–10 only

Directions:

1. Mix the cards.
2. Deal cards evenly, facedown to each player. Players place their cards in front of them.
3. Players flip their top cards at the same time and compare them to see which is greater (the larger numeral).
4. The player with the greater card says, "I have the bigger number!" and takes the cards. If cards are equal, players flip another card to break the tie.
5. When all cards have been played, the game is over, and the player with more cards wins.

What's Ahead?

In Week 13, children will order numbers and lengths, placing numerals in proper sequence. They will also explore the plus 1 pattern, a growing pattern in number.

30 • *Family Letters*

Building Blocks • *Teacher's Resource Guide*

Assess and Differentiate

A Gather Evidence

Review children's progress in mathematics by looking at the Weekly Record Sheets (Monday, Wednesday, Friday) and the Small Group Record Sheets (Tuesday, Thursday) from this past week.

B Summarize Findings

Using ***Online Assessment,*** summarize and analyze assessment data for each child based on your weekly observations and Record Sheets. Such information helps determine where each child is on the math trajectory for counting and comparing. See ***Assessment*** for the print companion to each Learning Trajectory Record.

C Differentiate Instruction

Once you have seen a child exhibit specific levels of the trajectory, begin to encourage and work with that child toward the next level. Refer to Appendix A for individualized instruction opportunities, including Special Education concerns.

Object Counting

If . . .	Then . . .
If . . . the child can consistently count ten items in a set arrangement,	**Then . . .** *Counter (10)* Counts structured arrangements of objects to 10, may be able to write numerals or draw collections to represent 1–10.
If . . . the child can accurately count to 10 and attempts to count beyond 10 in varied arrangements,	**Then . . .** *Counter and Producer (10+)* Accurately counts and produces objects to 10, then beyond to 30; has explicit understanding of cardinality (that numbers tell "how many?") and keeps track of objects that have and have not been counted even in different arrangements. Writes or draws to represent 1–10 (then 20 and 30), and gives next number to 20s or 30s; recognizes errors in others' counting and can eliminate most errors in own counting if asked to try hard.

Comparing and Ordering

If . . .	Then . . .
If . . . the child can compare groups of at least 10 by counting,	**Then . . .** *Counting Comparer (10)* Compares with counting up to 10 even when larger group's objects are smaller.
If . . . the child can use prior knowledge to decide a number's size and position relative to another,	**Then . . .** *Mental Number Line to 10* Uses internal images and knowledge of number relationships to determine relative size and position. For example, in determining which number, 4 or 6, is closer to 9.

WEEK 13

Overview

Big Ideas

- counting
- ordering numbers and lengths
- patterning

Teaching for Understanding

Patterns can be used to recognize relationships, and patterning is at the heart of mathematics. Some define mathematics as the science of patterns in number and shape. Children have been working with number, including rhythmic and geometric (visual) patterns, since the first weeks. Here we emphasize the most important mathematical pattern of all: the growing pattern of plus 1. That is, when we count, we see that we are always adding one more.

Number: Ordering and Patterns

Children who can count still need to develop ordering skills regarding number. They need to be able to take collections of various amounts and put them in order from fewest to most items. Children also need to be able to order actual numerals and lengths. Research indicates that these important abilities are frequently *not* taught and thus not learned. Children without these skills are at risk for later academic failure.

Early Addition

Even very young children can figure out how many there will be if only one or two items are added to a small group of objects. Researchers believe children first do this by storing images of the objects in their minds. Our activities build on this initial ability, discussing numbers so children form a foundation for later arithmetic.

Meaningful Connections

Patterns are not just simple repeating patterns. Most mathematics relies on the plus 1 pattern, which is embedded in the counting sequence, and it is perhaps the simplest growing pattern. The plus 1 pattern connects repeating patterns to numerical patterns, which are the foundation of algebra. You are teaching children their first prealgebra lessons!

What's Ahead?

In the weeks to come, children will learn more about shapes and their attributes. They will name new shapes and describe all the shapes they know using mathematical vocabulary. We will eventually return to patterns with more typical repeating patterns.

WEEK 13 Overview

How Children Learn to Count and Pattern

Knowing where children are on the learning trajectory for counting, comparing, and patterning and what the next steps are helps facilitate their development. Children who easily surpass these trajectory levels might be challenged by larger amounts and encouraged to assist other children.

Counting

What to Look For Which numerals can the child count to and from accurately?

Counter (10) Counts structured arrangements of objects to 10, may be able to write numerals or draw collections to represent 1–10.

Counter and Producer (10+) Accurately counts and produces objects to 10 and then beyond to 30; has explicit understanding of cardinality (that numbers tell "how many?") and keeps track of objects that have and have not been counted even in different arrangements. Writes or draws to represent 1–10 (then 20 and 30), and gives next number to 20s or 30s; recognizes errors in others' counting and can eliminate most errors in own counting if asked to try hard.

Counter Backward from 10 Counts back verbally from 10 or when removing items from a group.

Comparing and Ordering

What to Look For How many units can the child order by length and/or number?

Serial Orderer to 6+ Orders lengths marked into units 1–6 and then beyond. For example, given a cube tower, puts cubes in order from 1 to 6; can later order collections, such as cards with one to six dots.

Adding and Subtracting

What to Look For What is the largest sum the child can calculate, and how does the child calculate?

Small Number (+/–) Finds sums for joining problems up to 3+2 by counting all objects. For example: There are two crayons. One more crayon is added. How many crayons are there in all? Child counts 2, counts one more, and then counts "1, 2, 3...3!"

Patterning

What to Look For What types of patterns does the child recognize?

Pattern Recognizer Recognizes a simple pattern ("This pattern is like steps going up higher each time.").

English Learner

For children learning English, show concrete representations of numerals, such as writing the numeral and showing multiple examples of that many items. Refer to page 89 of the ***Teacher's Resource Guide.***

WEEK 13

English Learner Support

Preview Big Ideas

When content lessons are taught mainly in English, an excellent strategy to maximize comprehension is to briefly preview the big ideas in the child's primary language so that he or she can better comprehend English instruction.

Esta semana, vamos a trabajar con un patrón importante, el más uno. Cuando recitamos los números contadores, es un ejemplo del patrón de más uno. En el patrón más uno, cada número agrega uno más para hallar el siguiente número en el patrón.	This week we are going to work with an important pattern called plus one. When we recite the counting numbers, this is an example of the plus one pattern. In the plus 1 pattern, each number adds one more to find the next number in the pattern.

Teacher Note

This week we learn the finger play "Five Little Monkeys" and use hands and fingers to reinforce counting and taking away or counting down. For English learners, review the name of each finger and the parts of the hand, such as the palm, back of the hand, knuckles, wrist, nails and each of the fingers (thumb, index finger, middle finger, ring finger and pinky).

Access Vocabulary

English learners may benefit from clarification of some common phrases and words that proficient English speakers probably know. Occasionally words have more than one meaning or are used in potentially puzzling idiomatic expressions. Before or during the lesson, be sure to clarify the following words and phrases:

most (*Teacher's Edition,* p. 193) the set with more than any other set has
sum (*Teacher's Edition,* p. 193) the total when adding numbers
stairs (*Teacher's Edition,* p. 198) the steps we climb to go up in a building

Spanish Cognates

These words take advantage of what children and their families may know in their home language in order to accelerate the acquisition of math concepts in English. Other languages may also have cognates with English through shared Latin, Arabic, and Greek word roots and affixes.

vertical vertical
horizontal horizontal
order orden
describe describir

Building Blocks • *Teacher's Resource Guide* — *English Learner* • 89

Technology Project

Children may get additional number-patterning practice using Build Stairs Free Explore by building traditional stairs and stairs "up and then down." Or, working cooperatively, one child places a step, and another places the next. They can also build stairs that go in "jumps" of two: 1, 3, 5, 7, 9, 7, 5, 3, 1. Challenge children to think of other ways to explore, and help them discuss what they do, as well as the relationship between numbers and step heights. Later, we will encourage children to practice addition by putting one step on top of another. When the sum is 10 or less, the steps combine to make a new step.

Math Throughout the Year

Math Throughout the Year activities are recommended to build on the mathematical skills highlighted in each week. Here are suggested activities for **Week 13.**

Dinosaur Shop (Dramatic Play)

Children set up and run a toy dinosaur shop, applying the mathematical concepts of sorting, counting, and adding. First, children sort and classify toy dinosaurs (or dinosaur counters) by color, size, type, or other criterion. Some children are customers who place orders; others take turns being salespeople. The "orders" are put into boxes, which are then labeled with the corresponding numeral (using a self-sticking note or other label). Play money may be used to add realism and challenge to the activity. Thematic images—such as dinosaur, fossil, and appropriate habitat images—may be used to decorate the "shop."

Cleanup

Say a number and, during the cleanup of an activity or for the day, each child picks up that many objects to put away. Repeat as necessary.

Numerals Every Day

Numerals are all around us. Help children notice and read numerals on common items throughout the day around school.

Center Preview

Computer Center Building Blocks

Get your classroom Computer Center ready for Build Stairs 1: Count Steps and Build Stairs 2: Order Steps from the ***Building Blocks*** software.

After you introduce both levels of Build Stairs, each child should complete the activities individually as you (or an assistant) monitor and guide him or her periodically. Ideally, each child will have at least ten minutes of computer time at least twice during the week. Use children's center time as an opportunity for assessment.

Hands On Math Center

This week's Hands On Math Center activities are Build Cube Stairs, Dinosaur Shop, Places Scenes, and Order Cards. Supply the center with these materials: Connecting Cubes, small toy figures, toy dinosaurs (and/or dinosaur counters), boxes, labels (or self-sticking notes), Places Scenes, counters, Numeral Cards, and Dot Cards.

Literature Connections

These books help develop counting.

- *One Was Johnny: A Counting Book* by Maurice Sendak
- *Rooster's Off to See the World* by Eric Carle
- *Animal Orchestra* by Scott Gustafson
- *Five Little Monkeys Jumping on the Bed* by Eileen Christelow
- *On the Stairs* by Julie Hofstrand Larios

(t)Matt Meadows (b)Eclipse Studios

WEEK 13

Weekly Planner

Learning Trajectories

Week 13 Objectives

- To order numbers and lengths
- To count objects to 10 and beyond
- To understand the plus 1 pattern in the counting sequence

	Developmental Path	Instructional Activities
Counting	*Producer (Small Numbers)*	
	Counter (10)	Build Cube Stairs Build Stairs 1 ***Victor Diego Seahawk's Big Red Wagon*** Order Cards
	Counter and Producer (10+)	Build Stairs 1 and 2 Dinosaur Shop Places Scenes
	Counter Backward from 10	Count and Move (Forward and Back) "Five Little Monkeys"
	Counter from N (N + 1, N – 1)	
Comparing and Ordering	*Mental Number Line to 10*	
	Serial Orderer to 6+	Build Cube Stairs Build Stairs 2 Order Cards
	Place Value Comparer	
Adding and Subtracting	*Nonverbal (+/–)*	
	Small Number (+/–)	How Many Now? (Hidden version)
	Find Result (+/–)	
Patterning	*Pre-Patterner*	
	Pattern Recognizer	Build Cube Stairs
	Pattern Fixer	

Use this chart to plan for your specific class schedule. If you have your prekindergarteners for only three days, complete Monday, Tuesday, and Thursday of the week.

Work Time

Whole Group	Small Group	Computer	Hands On	Program Resources
Count and Move **Build Cube Stairs** *Materials:* *Connecting Cubes		**Build Stairs 1**	**Build Cube Stairs** *Materials:* *Connecting Cubes small toy figures **Dinosaur Shop** *Materials:* dinosaurs boxes labels	Building Blocks ***Assessment*** Weekly Record Sheet
Count and Move (Forward and Back) **Build Cube Stairs** *Materials:* *Connecting Cubes	**Build Cube Stairs** *Materials:* *Connecting Cubes **How Many Now? (Hidden version)** *Materials:* *counters dark cloth	**Build Stairs 2**	**Places Scenes** *Materials:* *counters *Numeral Cards **Build Cube Stairs** **Dinosaur Shop**	***Teacher's Resource Guide*** • Numeral Cards • Places Scenes Building Blocks ***Assessment*** Small Group Record Sheet
Victor Diego Seahawk's Big Red Wagon **Order Cards** *Materials:* *Dot Cards		• **Build Stairs 1** • **Build Stairs 2**	**Order Cards** *Materials:* *Dot Cards **Places Scenes** **Build Cube Stairs** **Dinosaur Shop**	***Big Book Victor Diego Seahawk's Big Red Wagon*** Building Blocks ***Teacher's Resource Guide*** Places Scenes ***Assessment*** Weekly Record Sheet
Count and Move (Forward and Back) **Order Cards** *Materials:* *Dot Cards	**Build Cube Stairs** *Materials:* *Connecting Cubes **How Many Now? (Hidden version)** *Materials:* *counters dark cloth	• **Build Stairs 1** • **Build Stairs 2**	Order Cards Places Scenes Build Cube Stairs Dinosaur Shop	***Teacher's Resource Guide*** • Numeral Cards • Places Scenes Building Blocks ***Assessment*** Small Group Record Sheet
"Five Little Monkeys" ***Victor Diego Seahawk's Big Red Wagon***		• **Build Stairs 1** • **Build Stairs 2**	Order Cards Places Scenes Build Cube Stairs Dinosaur Shop	***Big Book Victor Diego Seahawk's Big Red Wagon*** Building Blocks ***Teacher's Resource Guide*** • Family Letter Week 13 • Places Scenes ***Assessment*** Weekly Record Sheet

*provided in Manipulative Kit

Monday Planner

Objectives

- To order numbers and lengths
- To count objects to 10 and beyond
- To understand the plus 1 pattern in the counting sequence

Materials

- *Connecting Cubes
- small toy figures
- dinosaurs
- boxes
- labels

Math Throughout the Year

Review activity directions at the top of page 195, and complete each in class whenever appropriate.

Looking Ahead

Continuing from last week, keep boxes and labels (or self-sticking notes) with numerals 1–10 at the Hands On Math Center. For tomorrow, make sure Places Scenes (also from last week) are at the Hands On Math Center.

*provided in Manipulative Kit

Monday

Whole Group

15 

Warm-Up: Count and Move

- Before doing this activity, have children walk up stairs with you and count as you do. Talk about the stairs and walking up them one step at a time.
- In a circle, outdoors, or wherever children are able to march in place, help children pretend to climb ten steps, counting aloud each step.

Build Cube Stairs

- Talk about the height of stairs and how we walk up or down them one step at a time.
- Show children how to link connecting cubes, and then show them two steps using the cubes (one cube for the first step, two cubes for the second).
- Ask children how many cubes it will take to make the next step (three). "Walk" your fingers up the steps, asking children to count "1, 2, 3" with you.
- Tell children they will make stairs of connecting cubes at the Hands On Math Center.

Work Time

20 

Computer Center Building Blocks

Introduce Build Stairs 1: Count Steps from the ***Building Blocks*** software. In Build Stairs activities, characters need help climbing steps to get to a higher level, thus children build or fix the stairs. In this first level, children click on numerals to build the stairs. Each child should have an opportunity to complete Build Stairs 1 this week.

Eclipse Studios

Hands On Math Center

These activities may be set up at the center all week (Dinosaur Shop recurs from last week).

Build Cube Stairs

Children build stairs with connecting cubes. For additional motivation, provide small toy figurines for children to use to climb their completed stairs.

Monitoring Student Progress

If . . . children struggle building stairs,	**Then . . .** have them build stairs up to 5 only, or you make stairs to 5 (optionally labeled with each step's corresponding numeral), and have children order them before building their own.
If . . . children excel building stairs,	**Then . . .** have them build stairs to 10, 12, or 20.

Dinosaur Shop

Using toy dinosaurs (or dinosaur counters), remind children how to sort dinosaurs into boxes and how to label each box with the numeral that tells how many dinosaurs there are. Allow children to then count, sort, and label on their own. See Friday's Monitoring Student Progress for scaffolding.

3 Reflect

5

Show children a numeral, and ask:

- **How did you figure out which step to start with and which came next?**

Children might say: I started with the smallest step, I started with one, or the like; and I counted to find the next step, I made a step that was one bigger, or the like.

4 Assess

Use the Weekly Record Sheet from *Assessment* to record children's progress. Use their time at the centers as an opportunity to complete your observations.

Build Stairs 1

RESEARCH IN ACTION

If children need help completing Build Stairs on the computer, keep connecting cube towers 1 to 10 at the Computer Center. You may even label the cubes with corresponding numerals. Encourage children to recreate the steps that are needed on the computer screen using the connecting cubes.

Tuesday Planner

Objectives

- To order numbers and lengths
- To count objects to 10 and beyond
- To understand the plus 1 pattern in the counting sequence

Materials

- *connecting cubes
- *counters
- dark cloth
- *Numeral Cards

Vocabulary

Horizontally is side to side.

Vertically is up and down.

Math Throughout the Year

Review activity directions at the top of page 195, and complete each in class whenever appropriate.

Looking Ahead

Preview ***Big Book Victor Diego Seahawk's Big Red Wagon*** for tomorrow's Warm-Up.

*provided in Manipulative Kit

Tuesday

1 Whole Group

10

Warm-Up: Count and Move (Forward and Back)

Everyone starts in a crouched position, and slowly rises to a standing position while counting aloud to 10. Then, while counting backward from 10, everyone slowly sinks back down. Repeat as time permits.

Build Cube Stairs

- Display steps one to three made from connecting cubes. Steps four and five should be made ahead of time, but initially hidden from children's view.
- Ask children to describe the pattern. They might say, "It gets bigger each time." Follow up by asking how many cubes are needed to make the next step. If needed, chant the counting pattern with children, pointing to each step as you say, for example, "1, 2, 3...4!"
- Demonstrate that, if you count up one step (count how many cubes make up a step), it is the same number as counting side to side (horizontally). For example, the fourth step is made of four cubes.

2 Work Time

25

Small Group

Build Cube Stairs

- Children, individually or with a partner, build stairs up to at least 10 with connecting cubes. As they work, discuss the plus 1 pattern; that is, when we count, the next number is one more than the one before. In stairs, each step has one more cube than the one before. Review how counting the number of steps (horizontally) and the number of cubes in a step (vertically) equal the same number.
- Like many of this week's activities, Build Cube Stairs integrates many skills. Children put numbers in order, count from one number to the next, and recognize the step-like visual pattern.

How Many Now? (Hidden version)

- Give each participating child a copy of the Numeral Cards from the ***Teacher's Resource Guide.*** Have children put their cards in order. (You can limit the range of the Numeral Cards the children have.) Then set out two counters, and ask children to show the card that tells how many counters there are. Cover the counters with a dark cloth, and then add another counter. Have children show the card that tells how many counters there are now.

- One child lifts the cloth and the group counts the counters to check their answers. The cloth is then replaced. Add or subtract a counter and, again, ask all children to show the card that tells how many counters there are. Lift the cloth to count to check. Repeat as children's interest and time allow.

Computer Center Building Blocks

Introduce Build Stairs 2: Order Steps from the ***Building Blocks*** software, where children are given a group of steps in random order to put in proper order to build stairs. Each child should complete the activity this week.

Build Stairs 2

Hands On Math Center

After reviewing this activity, children may play it and/or continue yesterday's.

Places Scenes

Children choose several Places Scenes, and put counters on each scene to match the number of the Numeral Card, which has been placed on the table. Children then tell stories about their scene.

Monitoring Student Progress

If . . .	Then . . .
If . . . children struggle during Places Scenes,	**Then . . .** use Counting Cards, or have children use the beach or space scene, which guides them to put 1–5 counters on specified locations, and then label each with its numeral on a self-sticking note.
If . . . children excel during Places Scenes,	**Then . . .** make and use Numeral Cards of greater amounts.

RESEARCH IN ACTION

Patterning is at the heart of mathematics. Some people define math as the science of patterns in number and shape. Children have been working with number (rhythmic) and geometric (visual) patterns since Week 1, but this week focuses on the growing pattern of plus 1.

3 Reflect

5

While counting, pretend with children they are walking up five steps, and ask:

- **What step comes next? How do you know?**

Children might say: I know 6 is next because it comes after 5, or I counted from 1 to get to 6.

4 Assess

During Small Group activities, use the Small Group Record Sheet from *Assessment* to observe and record children's progress.

Wednesday Planner

Objectives
- To order numbers and lengths
- To count objects to 10 and beyond
- To understand the plus 1 pattern in the counting sequence

Materials
*Dot Cards

Math Throughout the Year
Review activity directions at the top of page 195, and complete each in class whenever appropriate.

*provided in Manipulative Kit

Wednesday

1 Whole Group

Warm-Up: *Victor Diego Seahawk's Big Red Wagon*

- Read ***Big Book Victor Diego Seahawk's Big Red Wagon***, focusing on the plus 1 pattern. Point to the numeral 1, and ask children: What is the next numeral? Point to the numeral 2, and express delight that children knew the answer. Count the two bats on the page to check.
- Continue to have children predict the next numeral (the plus 1 pattern) to 10, and repeat the steps above. Conclude by asking children to describe the plus 1 pattern.

Order Cards

- Place Dot Cards 1–5 so they are left to right from the children's perspective. Ask children to describe the pattern.
- Tell children to count aloud, predicting the next number as you continue to place the next Dot Card in the pattern. Explain that they will put these cards in order on their own at the Hands On Math Center.

Monitoring Student Progress

If . . . children struggle during Order Cards,	**Then . . .** model counting from 1 to generate the next number, or use cards 1–5 only, increasing the number as children succeed.
If . . . children excel during Order Cards,	**Then . . .** make and use Numeral Cards to an appropriate number beyond 10 based on children's success.

Eclipse Studios

2 Work Time 15

Computer Center Building Blocks

Continue to provide each child with a chance to complete Build Stairs 1 and 2.

Hands On Math Center

After reviewing Order Cards below, allow children to complete it and/or this week's previous Hands On Math Center activities based on what they learn and benefit from most.

Order Cards

Each child shuffles a set of Dot Cards (1–10), and then puts them in order, applying the plus 1 pattern. When children work in pairs, each child takes a turn placing the next card. If more sets of Dot Cards are needed, make copies from the ***Teacher's Resource Guide.***

3 Reflect 5

Ask children:

- **When you put the cards in order, how do you know what comes next?** Children might say: I count them all; I start from 1 and keep counting; I add 1 to the number I see; or the like.

4 Assess

Use the Weekly Record Sheet from **Assessment** to record children's progress. Use their time at the centers as an opportunity to complete your observations.

RESEARCH IN ACTION

The fives and tens formats on the Dot Cards highlight the plus 1 growing pattern. Each card has one more square filled in. When using Dot Cards, emphasize the squares more than the numerals.

Thursday Planner

Objectives
- To order numbers and lengths
- To count objects to 10 and beyond
- To understand the plus 1 pattern in the counting sequence

Materials
- *Dot Cards
- *connecting cubes
- *counters
- dark cloth

Math Throughout the Year
Review activity directions at the top of page 195, and complete each in class whenever appropriate.

Looking Ahead
For tomorrow, make copies of Family Letter Week 13 from the *Teacher's Resource Guide.*

*provided in Manipulative Kit

Thursday

1 Whole Group

Warm-Up: Count and Move (Forward and Back)
Everyone starts in a crouched position, and slowly rises to a standing position while counting aloud to 10. Then, while counting backward from 10, everyone slowly sinks back down. Repeat as time permits.

Order Cards
- Place Dot Cards 1–5 so they are left to right from the children's perspective. Ask children to describe the pattern.
- Tell children to count aloud, predicting the next number as you continue to place the next Dot Card in the pattern.

2 Work Time

Small Group

Build Cube Stairs

Children build stairs up to at least 10 with connecting cubes. As they work, review the plus 1 pattern—when we count, the next number is one more than the one before, and review how counting the number of steps (horizontally) and the number of cubes in a step (vertically) are the same number.

Monitoring Student Progress

If . . . children struggle building stairs,	**Then . . .** have them build stairs up to 5 only, or you make stairs to 5 (optionally labeled with each step's corresponding numeral) and have children order them before building their own.
If . . . children excel building stairs,	**Then . . .** have them build stairs to 10, 12, or 20.

How Many Now? (Hidden version)

- Give each participating child a copy of the Numeral Cards from the ***Teacher's Resource Guide.*** Have children put their cards in order. (You can limit the range of the Numeral Cards the children have.) Then set out two counters, and ask children to show the card that tells how many counters there are. Cover the counters with a dark cloth, and then add another counter. Have children show the card that tells how many counters there are now.
- One child lifts the cloth and the group counts the counters to check their answers. The cloth is then replaced. Add or subtract a counter and, again, ask all children to show the card that tells how many counters there are. Lift the cloth to count to check. Repeat as children's interest and time allow.

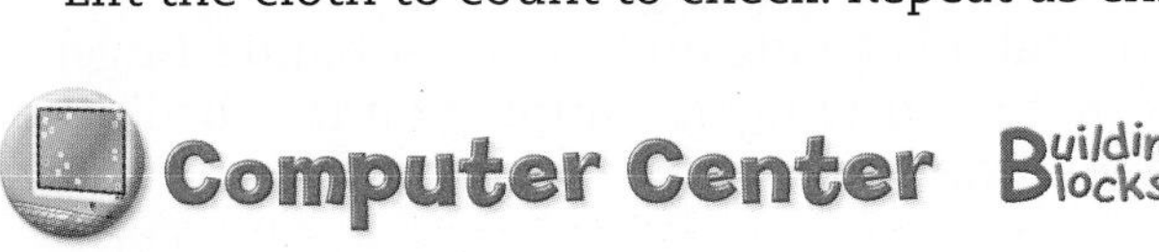

Continue to provide each child with a chance to complete Build Stairs 1 and 2.

Based on what children learn and benefit from most, choose from this week's Hands On Math Center activities: Order Cards, Places Scenes, Build Cube Stairs, and Dinosaur Shop. Consult the Weekly Planner for corresponding materials and, if needed, previous days for activity directions.

Ask children:

■ **When you put the cards in order, how do you know what comes next?**

Children might say: I count them all; I start from 1 and keep counting; I add 1 to the number I see; or the like.

During Small Group activities, use the Small Group Record Sheet from ***Assessment*** to observe and record children's progress.

Eclipse Studios

Friday Planner

Objectives
- To order numbers and lengths
- To count objects to 10 and beyond
- To understand the plus 1 pattern in the counting sequence

Materials
no new materials

Math Throughout the Year

Review activity directions at the top of page 195, and complete each in class whenever appropriate.

Looking Ahead

For next week, familiarize yourself with Mystery Pictures 3: Match New Shapes and Memory Geometry 3: Shapes-A-Round from the ***Building Blocks*** software.

Friday

1 Whole Group 20

Warm-Up: "Five Little Monkeys"
- Everyone opens one hand so the palm is facing up. The other hand's fingers are the "monkeys;" show all of that hand's fingers jumping on the "bed" (the other hand's palm). As each monkey falls off the bed, a finger folds (to show fewer fingers).
- Here are the words and additional actions:

  Five little monkeys jumping on the bed, one fell off and bumped his head. *(Lightly tap head.)*
  We called for the doctor, and the doctor said, *(Hold pretend phone to ear.)*
  No more monkeys jumping on the bed! *(Shake index finger in a "that's a no-no" way.)*
- Counting backward, repeat the above lines for numerals 4 to 2. For one monkey, say:

  One little monkey jumping on the bed, he fell off and bumped his head. *(Lightly tap head.)*
  We called for the doctor, and the doctor said, *(Hold pretend phone to ear.)*
  (Loudly) That's what you get for jumping on the bed! *(Shake index finger in a "that's a no-no" way.)*

Victor Diego Seahawk's Big Red Wagon
- Read ***Big Book Victor Diego Seahawk's Big Red Wagon***, focusing on the plus 1 pattern. Point to the numeral 1, and ask children: What is the next numeral? Point to the numeral 2, and express delight that children knew the answer. Count the two bats on the page to check.
- Continue to have children predict the next numeral (the plus 1 pattern) to 10, and repeat the steps above. Conclude by asking children to describe the plus 1 pattern.

2 Work Time

15

Computer Center

Continue to provide each child with a chance to complete Build Stairs 1 and 2.

Hands On Math Center

Based on what children learn and benefit from most, choose from this week's Hands On Math Center activities: Order Cards, Places Scenes, Build Cube Stairs, and Dinosaur Shop. Consult the Weekly Planner for corresponding materials and, if needed, previous days for activity directions. Scaffolding for Dinosaur Shop follows.

Monitoring Student Progress

If . . . children struggle during Dinosaur Shop,	**Then . . .** use smaller numerals.
If . . . children excel during Dinosaur Shop,	**Then . . .** use larger numerals, or encourage children to sort by attributes (color, type, and so on).

Home Connection

From the ***Teacher's Resource Guide,*** distribute to children copies of Family Letter Week 13 to share with their family. Each letter has an area for children to show families an example of what they have been doing in class.

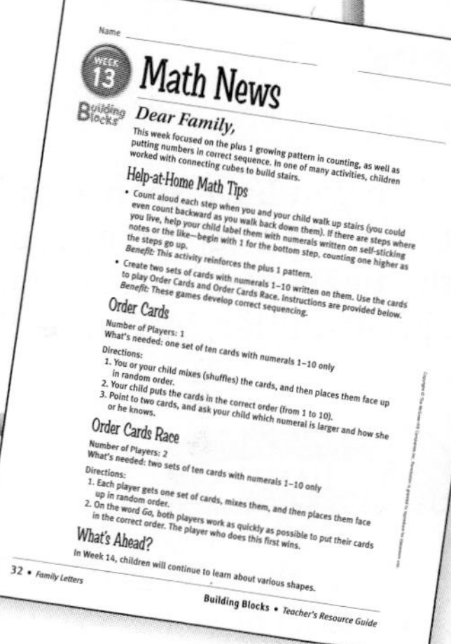

Name

Week 13

Math News

Dear Family,

This week focused on the plus 1 growing pattern in counting, as well as putting numbers in correct sequence. In one of many activities, children worked with connecting cubes to build stairs.

Help-at-Home Math Tips

- Count aloud each step when you and your child walk up stairs (you could even count backward as you walk back down them). If there are steps where you live, help your child label them with numerals written on self-sticking notes or the like—begin with 1 for the bottom step, counting one higher as the steps go up.
  Benefit: This activity reinforces the plus 1 pattern.
- Create two sets of cards with numerals 1–10 written on them. Use the cards to play Order Cards and Order Cards Race. Instructions are provided below.
  Benefit: These games develop correct sequencing.

Order Cards

Number of Players: 1
What's needed: one set of ten cards with numerals 1–10 only

Directions:
1. You or your child mixes (shuffles) the cards, and then places them face up in random order.
2. Your child puts the cards in the correct order (from 1 to 10).
3. Point to two cards, and ask your child which numeral is larger and how she or he knows.

Order Cards Race

Number of Players: 2
What's needed: two sets of ten cards with numerals 1–10 only

Directions:
1. Each player gets one set of cards, mixes them, and then places them face up in random order.
2. On the word *Go,* both players work as quickly as possible to put their cards in the correct order. The player who does this first wins.

What's Ahead?

In Week 14, children will continue to learn about various shapes.

32 • *Family Letters* Building Blocks • *Teacher's Resource Guide*

3 Reflect

5

Briefly discuss with children what you have done in class this week, such as patterning.

While counting, pretend with children they are walking up five steps, and ask:

- **What step comes next? How do you know?**

Children might say: I know 6 is next because it comes after 5; I counted from 1 to get to 6; I know the next number is one bigger; or the like.

4 Assess

Use the Weekly Record Sheet from ***Assessment*** to record children's progress. Use their time at the centers as an opportunity to complete your observations.

WEEK 13 Wrap-Up

Assess and Differentiate

A Gather Evidence

Review children's progress in mathematics by looking at the Weekly Record Sheets (Monday, Wednesday, Friday) and the Small Group Record Sheets (Tuesday, Thursday) from this past week.

B Summarize Findings

Using *Online Assessment,* summarize and analyze assessment data for each child based on your weekly observations and Record Sheets. Such information helps determine where each child is on the math trajectory for counting, comparing, and patterning. See *Assessment* for the print companion to each Learning Trajectory Record.

C Differentiate Instruction

Once you have seen a child exhibit specific levels of the trajectory, begin to encourage and work with that child toward the next level. Refer to Appendix A for individualized instruction opportunities, including Special Education concerns.

Counting

If . . .	Then . . .
If . . . the child can consistently count ten items in a set arrangement,	**Then . . .** *Counter (10)* Counts structured arrangements of objects to 10 and may be able to write numerals or draw collections to represent 1–10.
If . . . the child can accurately count to 10 and attempts to count beyond 10 in varied arrangements,	**Then . . .** *Counter and Producer (10+)* Accurately counts and produces objects to 10 and then beyond to 30; has explicit understanding of cardinality (that numbers tell "how many?") and keeps track of objects that have and have not been counted even in different arrangements. Writes or draws to represent 1–10 (then 20 and 30), and gives next number to 20s or 30s; recognizes errors in others' counting and can eliminate most errors in own counting if asked to try hard.
If . . . the child can accurately and consistently count backward from 10,	**Then . . .** *Counter Backward from 10* Counts back verbally from 10 or when removing items from a group.

Comparing and Ordering

If . . .	Then . . .
If . . . the child can order items from 1 to at least 6,	**Then . . .** *Serial Orderer to 6+* Orders lengths marked into units 1–6 and then beyond. For example, given a cube tower, puts cubes in order from 1 to 6; can later order collections, such as cards with one to six dots.

Adding and Subtracting

If . . .	Then . . .
If . . . the child can add or subtract small amounts by counting,	**Then . . .** *Small Number (+/–)* Finds sums for joining problems up to 3+2 by counting all objects. For example: There are two crayons. One more crayon is added. How many crayons are there in all? Child counts two, counts one more, and then counts "1, 2, 3...3!"

Patterning

If . . .	Then . . .
If . . . the child can identify when he or she sees a true pattern,	**Then . . .** *Pattern Recognizer* Recognizes a simple pattern ("This pattern is like steps going up higher each time.").

Overview

Big Ideas

- shape identification
- shape matching
- shapes in the environment

Teaching for Understanding

Geometric shapes can be used to represent and understand objects in the world around us. When children find and see shapes in the world, they learn more about how shapes make up the world. This week focuses on such skills by allowing children to truly feel the attributes of a shape and seeing how individual shapes combine to form an entirely new thing.

Shape Activities

Shape activities help children learn how to match and identify new categories of shapes and, more importantly, how to describe shapes in terms of their parts or attributes, not just by shape names. Children should start to consider automatically, for example, that your desk is a rectangle and not a square because the lengths of all its sides are not equal. Some will not be able to articulate this, but they will be thinking about what makes a rectangle different from a square.

Mathematical Discussion

"How do you know a shape is a rhombus?" People are often surprised to hear such a sophisticated question in an early childhood class. Preschoolers can discuss such ideas in their own words with impressive results. Some begin by saying, "I'm smart." However, given opportunity, repeated experience, and sound examples, children gradually develop the ability to respond mathematically: "They have four sides the same. They are diamonds. They are like squares squished." We use the term *rhombus*, but allow children to say *diamond*.

Meaningful Connections

We link geometry to number whenever possible. For example, children talk about the number of sides and angles (corners) in shapes, find shapes with six sides, and so on. The concept that number knowledge is used even when focusing on shapes, and shape knowledge is applied even when studying number should excite children about progressing in mathematics; at the very least, it enables them to see the usefulness of what they learn.

What's Ahead?

In the weeks to come, children will learn to sort shapes and identify the rule for sorting and will begin in-depth work on repeating patterns.

Eclipse Studios

Overview

How Children Identify Shapes

Knowing where children are on the learning trajectory for geometry and counting and what the next steps are helps facilitate their development. Children who easily surpass these trajectory levels might be challenged by less familiar shapes and encouraged to assist other children.

Shapes: Matching

What to Look For What types of shapes can the child match?

Shape Matcher, More Shapes Matches a wider variety of shapes of same size and orientation.

Shape Matcher, Sizes and Orientations Matches a wider variety of shapes of different sizes and orientations. For example, matches the same triangle even if the examples are turned differently.

Shapes: Naming

What to Look For What types of shapes and shape parts can the child identify?

Side Recognizer Parts: Identifies sides as distinct objects. For example, the child explains that a shape is a triangle because it has three sides and he or she runs his or her finger along the length of each side.

Shape Recognizer, More Shapes Recognizes most familiar shapes and typical examples of other shapes, such as hexagon, rhombus (diamond), and trapezoid.

Shape Identifier Names most common shapes, including rhombuses, without making mistakes, such as calling ovals circles; implicitly recognizes right angles, thus can distinguish between rectangles and parallelograms without right angles.

Counting

What to Look For Which numerals can the child count to and from accurately?

Counter and Producer (10+) Accurately counts and produces objects to 10, and then beyond to 30; has explicit understanding of cardinality (that numbers tell "how many?") and keeps track of objects that have and have not been counted even in different arrangements. Writes or draws to represent 1–10 (then 20 and 30), and gives next number to 20s or 30s; recognizes errors in others' counting and can eliminate most errors in own counting if asked to try hard.

Counter Backward from 10 Counts back verbally from 10 or when removing items from a group.

English Learner

Connect shape words to the shape words in the English learner's primary language. Review the following: *match shapes, guess, sort,* and *hide the blocks*. Refer to page 90 of the *Teacher's Resource Guide.*

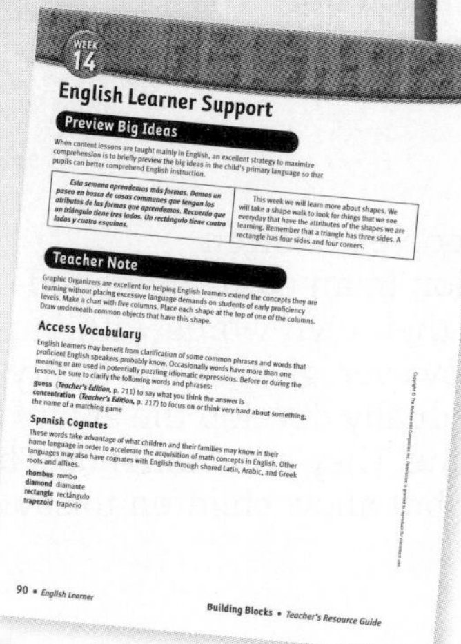
WEEK 14

English Learner Support

Preview Big Ideas

When content lessons are taught mainly in English, an excellent strategy to maximize comprehension is to briefly preview the big ideas in the child's primary language so that pupils can better comprehend English instruction.

Esta semana aprendemos más formas. Damos un paseo en busca de cosas communes que tengan los atributos de las formas que aprendemos. Recuerda que un triángulo tiene tres lados. Un rectángulo tiene cuatro lados y cuatro esquinas.	This week we will learn more about shapes. We will take a shape walk to look for things that we see everyday that have the attributes of the shapes we are learning. Remember that a triangle has three sides. A rectangle has four sides and four corners.

Teacher Note

Graphic Organizers are excellent for helping English learners extend the concepts they are learning without placing excessive language demands on students of early proficiency levels. Make a chart with five columns. Place each shape at the top of one of the columns. Draw underneath common objects that have this shape.

Access Vocabulary

English learners may benefit from clarification of some common phrases and words that proficient English speakers probably know. Occasionally words have more than one meaning or are used in potentially puzzling idiomatic expressions. Before or during the lesson, be sure to clarify the following words and phrases:

guess (*Teacher's Edition*, p. 211) to say what you think the answer is
concentration (*Teacher's Edition*, p. 217) to focus on or think very hard about something; the name of a matching game

Spanish Cognates

These words take advantage of what children and their families may know in their home language in order to accelerate the acquisition of math concepts in English. Other languages may also have cognates with English through shared Latin, Arabic, and Greek roots and affixes.

rhombus rombo
diamond diamante
rectangle rectángulo
trapezoid trapecio

90 • *English Learner*   **Building Blocks** • *Teacher's Resource Guide*

Technology Project

Children may get additional practice with shapes using Mystery Pictures Free Explore. Encourage children to drag as many shapes as they would like to make their own pictures on the computer. If they click the play button, children can invite a classmate to complete their mystery picture, which is challenging because the correct shape in the correct orientation must be selected. As you visit children at the computer, help them describe what they are making, what shapes they are using, and why.

Math Throughout the Year

Math Throughout the Year activities are recommended to build on the mathematical skills highlighted in each week. Here are suggested activities for **Week 14.**

Counting Jar

Using the same jar from previous weeks, place larger items in it this week so fewer will fill the jar than the last time. Children spill the jar's items to count them.

Guessing Bag

Engage children in this integrated science, math, and language activity before the week's Feely Box activities. With a mystery item already hidden in a paper bag, challenge children to feel it over the next few days to guess what it is by shape, weight, and so on. Children *secretly* tell classmates their guesses after everyone has felt the item. Start with easily identifiable items, such as a comb or block. It is beneficial to do this activity once every week, using a new object each Monday and revealing it Friday.

Dinosaur Shop (Dramatic Play)

Children set up and run a toy dinosaur shop. First, children sort and classify toy dinosaurs by color, size, type, or other criterion. Some children are customers; others are salespeople. "Orders" are put into boxes, which are then labeled with the corresponding numeral. Play money may be used to add realism and challenge to the activity. Thematic images—such as dinosaur, fossil, and appropriate habitat images—may be used to decorate the "shop."

Shape Walk

Go for a walk outside of the classroom to search for a specific shape.

Center Preview

Computer Center Building Blocks

Get your classroom Computer Center ready for Mystery Pictures 3: Match New Shapes and Memory Geometry 3: Shapes-A-Round from the ***Building Blocks*** software.

After you introduce Mystery Pictures 3 and Memory Geometry 3, each child should complete the activities individually as you (or an assistant) monitor and guide him or her periodically. Ideally, each child will have at least ten minutes of computer time at least twice during the week. Use children's center time as an opportunity for assessment.

Hands On Math Center

This week's Hands On Math Center activities are Shape Flip Book, Shape Pictures, Feely Box (Name), and Shape Step. Supply the center with these materials: Shape Flip Book, Shape Sets, pattern blocks, Feely Box, and large shapes on which to step, which can be made from construction paper.

Literature Connections

These books help develop shape recognition.

- *Building a House* by Byron Barton
- *The Wing on a Flea* by Ed Emberley
- *Shapes and Things* by Tana Hoban
- *So Many Circles, So Many Squares* by Tana Hoban
- *Seven Blind Mice* by Ed Young

(t)Matt Meadows; (b)Eclipse Studios

Weekly Planner

Learning Trajectories

Week 14 Objectives

- To identify and match shapes
- To find and describe the shape of objects in their environments
- To count objects to 10 and beyond

	Developmental Path	Instructional Activities
Shapes: Matching	*Shape Matcher, Sizes*	
	Shape Matcher, More Shapes	Mystery Pictures 3 Memory Geometry 3
	Shape Matcher, Sizes and Orientations	Mystery Pictures 3 Memory Geometry 3 Feely Box (Match)
	Shape Matcher, Combinations	
Shapes: Naming	*Shape Recognizer, All Rectangles*	
	Side Recognizer	Shape Step Feely Box (Match)
	Shape Recognizer, More Shapes	Mystery Pictures 3 Memory Geometry 3 Shape Step
	Shape Identifier	Shape Flip Book Trapezoids Shape Pictures Feely Box (Name) *Building Shapes*
	Parts of Shapes Identifier	
Counting	*Counter (10)*	
	Counter and Producer (10+)	Dinosaur Shop (Dramatic Play)
	Counter Backward from 10	Count and Move (Forward and Back)
	Counter from N (N +1, N − 1)	

Use this chart to plan for your specific class schedule. If you have your prekindergarteners for only three days, complete Monday, Tuesday, and Thursday of the week.

Pacing

Work Time

Whole Group	Small Group	Computer	Hands On	Program Resources
Shape Flip Book **Trapezoids** *Materials:* *Shape Sets *Pattern Blocks books with trapezoids (optional)		**Mystery Pictures 3**	**Shape Flip Book** **Shape Pictures** *Materials:* *Shape Sets *Pattern Blocks	***Teacher's Resource Guide*** Shape Flip Book Building Blocks ***Assessment*** Weekly Record Sheet
Count and Move (Forward and Back) **Feely Box (Match)** *Materials:* *Shape Sets	**Feely Box (Match and Name)** *Materials:* *Shape Sets	**Memory Geometry 3**	**Feely Box (Name)** *Materials:* *Shape Sets **Shape Step** *Materials:* large shapes to step on	Building Blocks ***Assessment*** Small Group Record Sheet
Building Shapes **Shape Step** *Materials:* large shapes to step on		• **Mystery Pictures 3** • **Memory Geometry 3**	**Shape Flip Book** **Shape Pictures** **Feely Box (Name)** **Shape Step**	***Big Book Building Shapes*** ***Teacher's Resource Guide*** Shape Sets Building Blocks ***Assessment*** Weekly Record Sheet
Count and Move (Forward and Back) **Feely Box (Name)** *Materials:* *Shape Sets	**Feely Box (Match and Name)** *Materials:* *Shape Sets	• **Mystery Pictures 3** • **Memory Geometry 3**	Shape Flip Book Shape Pictures Feely Box (Name) Shape Step	Building Blocks ***Assessment*** Small Group Record Sheet
Computer Show ***Building Shapes***		• **Mystery Pictures 3** • **Memory Geometry 3**	Shape Flip Book Shape Pictures Feely Box (Name) Shape Step	***Big Book Building Shapes*** Building Blocks ***Teacher's Resource Guide*** Family Letter Week 14 ***Assessment*** Weekly Record Sheet

*provided in Manipulative Kit

Monday Planner

Objectives

- To identify and match shapes
- To find and describe the shape of objects in their environments

Materials

- *Shape Sets
- *Pattern Blocks
- books with trapezoids (optional)

Vocabulary

Orientation is the position or direction relative to a framework, such as how much a shape is turned compared to another shape.

Math Throughout the Year

Review activity directions at the top of page 211, and complete each in class whenever appropriate.

Looking Ahead

If you have not already done so, prepare the Shape Flip Book from the ***Teacher's Resource Guide*** for today and the Feely Box for tomorrow. It would be helpful to have books with examples of trapezoids for today.

*provided in Manipulative Kit

Monday

1 Whole Group

15

Warm-Up: Shape Flip Book

- Review with children how to match the panels of the Shape Flip Book. The panels can be flipped separately: first panels show shapes, middle panels show pictures in which that shape is obvious, and last panels show pictures in which that shape is less obvious.
- Focus on less familiar shapes, such as trapezoid (a shape with one pair of parallel sides), rhombus (may also be called *diamond*—a shape with four straight sides the same length), hexagon (a shape with six straight sides), and kite (a shape with two pairs of adjacent sides the same length).
- Help children name as many shapes as time allows, and ask if they have anything with these shapes.

Trapezoids

- Using a Shape Set, show a rectangle, circle, square, and triangle. Ask children to name each.
- Show the rhombus; write and say its name (have children say it too). Ask children what details they notice. Encourage them to compare the rhombus to other objects in the classroom, as well as objects from where they live and places they go. Repeat the same "show and tell" for the trapezoid.
- Show pattern block shapes, one after another, having children name each one. Focus especially on the rhombus and trapezoid (if you have other books with trapezoid examples, such as age-appropriate architecture books, show them). Ask children what they could make with such shapes.
- Note: Both trapezoids and rhombuses are *polygons*, closed shapes made only of straight line segments. A trapezoid has one pair of parallel sides; a rhombus has two pairs of parallel sides all the same length. Children *might* notice the two parallel sides of a trapezoid, but we name it only because it is common to math manipulatives, not because we expect children to describe it with such sophistication. Furthermore, there is a special kind of *isosceles trapezoid*, in which the two nonparallel sides are the same length.

Eclipse Studios

Work Time

Computer Center

Introduce Mystery Pictures 3: Match New Shapes from the ***Building Blocks*** software, in which children make mystery pictures out of shapes that are revealed only one shape at a time. This level includes less familiar shapes, but the program continues to name each shape, building the child's comprehension and vocabulary. Each child should complete Mystery Pictures 3 this week.

Mystery Pictures 3

Hands On Math Center

These activities may be set up at the center all week.

Shape Flip Book

Children match the Shape Flip Book's panels and name the shapes.

Shape Pictures

Children make designs and pictures with Shape Sets and Pattern Blocks. Encourage them to name and discuss shapes, especially the trapezoids, rhombuses, and hexagons.

Monitoring Student Progress

If . . . children struggle during Shape Pictures,	**Then . . .** have them feel a shape's sides and corners to help with their descriptions.
If . . . children excel during Shape Pictures,	**Then . . .** have them tell why a shape is a trapezoid or a rhombus, for example, by describing their differences.

RESEARCH IN ACTION

Most early childhood math programs teach four familiar shapes (circle, square, triangle, and rectangle), and they typically show only standard examples of these shapes, such as an equilateral triangle with a horizontal base. Research shows, however, that such limited experience hurts children's development of geometric ideas. This program provides exposure to other interesting and important shapes, such as trapezoids and rhombuses as well as richer experiences with familiar shapes.

Reflect

Ask children:

- **How did you match new Mystery Pictures shapes?**

Children might say: I turned them in my head to see which one would fit.

Assess

Use the Weekly Record Sheet from ***Assessment*** to record children's progress. Use their time at the centers as an opportunity to complete your observations.

Tuesday Planner

Objectives

- To identify and match shapes
- To find and describe the shape of objects in their environments
- To count objects to 10 and beyond

Materials

- *Shape Sets
- large shapes to step on

Math Throughout the Year

Review activity directions at the top of page 211, and complete each in class whenever appropriate.

Looking Ahead

For today's Hands On Math Center, choose the shape(s) on which you would like to focus, and make large shapes on which children may step, for example, out of construction paper.

*provided in Manipulative Kit

Tuesday

Whole Group

Warm-Up: Count and Move (Forward and Back)

Everyone starts in a crouched position, and slowly rises to a standing position while counting aloud to 10. Then, counting backward from 10, slowly sink back down. Next, help children pretend to "climb" ten steps, counting each aloud, and then "climb" down, counting backward from 10.

Feely Box (Match)

- Secretly hide a trapezoid from a Shape Set in the Feely Box (a decorated box with a hole large enough to fit a child's hand but not so large that you can see into the box). Display five shapes from another Shape Set, including the one that exactly matches the one you hid.
- Have a child put his or her hand in the box to feel the shape; that child should then point to the matching shape on display. Encourage the child to name the shape and explain how he or she figured it out. Discuss the shape, emphasizing straightness of the sides and the number of sides and angles.

Work Time

20

Small Group

Feely Box (Match and Name)

- Start with the Match version of Feely Box, following the Whole Group directions. Be sure to repeat the activity with the trapezoid and rhombus. Children who do not get a chance today will get one during Thursday's Small Group.
- As soon as children are ready, switch to Feely Box (Name). Have a pair of children demonstrate how to play it during Work Time: Child One hides a Shape Set shape in the Feely Box, and Child Two puts his or her hand in the box to feel the shape and tell what it is. Encourage children to share how they figured out the shape, and discuss each shape, emphasizing its particular attributes.

Monitoring Student Progress

If . . . children struggle to identify shapes during Feely Box,	**Then . . .** use Pattern Blocks instead, or reduce the number of shapes, making sure they are clearly different.
If . . . children excel at shape identification during Feely Box,	**Then . . .** use more shapes, and have children describe the shape well enough without using its name for the rest of the class to guess.

 Computer Center Building Blocks

Introduce Memory Geometry 3: Shapes-A-Round from the ***Building Blocks*** software, in which children play a version of the traditional concentration game by matching shapes of the same orientation in Round 1 but of a different orientation in Round 2. Each child should complete the activity this week.

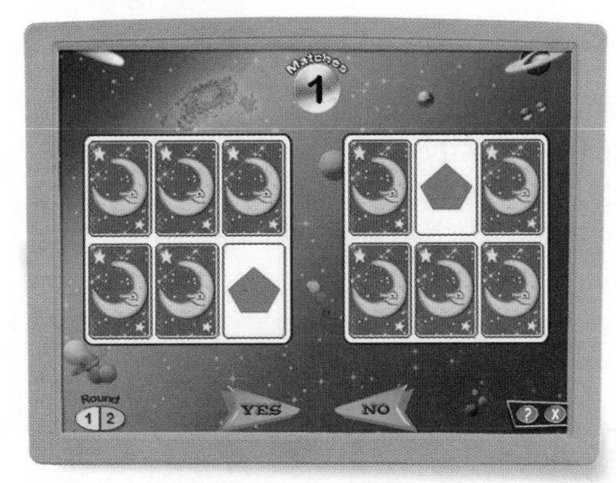

Memory Geometry 3

 Hands On Math Center

Children may do these and/or continue Monday's activities.

Feely Box (Name)

One child hides a Shape Set shape in the Feely Box, and a second child puts his or her hand in the box to guess the shape. Include rhombuses and trapezoids.

Shape Step

Evenly distribute large shapes to children, placing many on the floor. Tell each group of three or four children on which shape to step (especially trapezoids, rhombuses, or hexagons). As best you can, circulate to observe if target shapes are stepped on. Encourage children to discuss why the shape they stepped on was the correct shape and to tell each other which shapes to step on.

3 Reflect

5 

Ask children:

- **How did you know which shape is a trapezoid and which is a rhombus?**

Children might say: A trapezoid has two sides going the same. A rhombus has four sides with small or big corners.

4 Assess

During Small Group activities, use the Small Group Record Sheet from ***Assessment*** to observe and record children's progress.

Wednesday Planner

Objectives

- To identify and match shapes
- To find and describe the shape of objects in their environments

Materials

large shapes to step on

Math Throughout the Year

Review activity directions at the top of page 211, and complete each in class whenever appropriate.

Looking Ahead

If needed, preview ***Big Book Building Shapes*** for today's Warm-Up.

Wednesday

1 Whole Group

15

Warm-Up: *Building Shapes*

Show and read the book's cover to the class, and ask what they think the book is about. Then read it in its entirety. Discuss the book, reviewing the pages as children point out all the different shapes they see.

Shape Step

- If you do not already have large shapes on which children may step, do one of the following: tape (laminated) construction paper shapes to the floor; make chalk shapes outdoors; or, use a copy machine to enlarge shapes from the ***Teacher's Resource Guide's*** Shape Sets for laminating.
- Place large shapes on the floor. Show children trapezoids from the Shape Set (***Teacher's Resource Guide*** pages 174 and 176), and tell them that they will step only on trapezoids.
- Have a group of four or five children step on the trapezoids. Ask the rest of the class to watch carefully to make sure each group steps on all trapezoids. Whenever possible, ask children to explain why the shape they stepped on was the correct shape. Repeat with other groups of children.
- If children are ready, switch to the Sides and Angles version of this activity, in which you would ask children to step on shapes with three (triangles), four (quadrilaterals), or six (hexagons) sides.

Monitoring Student Progress

If . . .	Then . . .
If . . . children struggle during Shape Step,	**Then . . .** review a shape's attributes while children feel the shape and/or step on a target shape, asking children if it is the correct shape.
If . . . children excel during Shape Step,	**Then . . .** have them find examples of target shapes in unusual places and/or teach even less familiar shapes.

Eclipse Studios

2 Work Time

20

Computer Center

Continue to provide each child with a chance to complete Mystery Pictures 3 and Memory Geometry 3.

Hands On Math Center

Based on what children learn and benefit from most, choose from this week's Hands On Math Center activities: Shape Flip Book, Shape Pictures, Feely Box (Name), and Shape Step. Consult the Weekly Planner for corresponding materials and, if needed, Monday and Tuesday for activity directions.

3 Reflect

5

Ask children:

- **What shapes did you use on the computer?**

Children might say: Trapezoids, rhombuses, diamonds, hexagons, shapes with (insert number) of sides, and the like.

4 Assess

Use the Weekly Record Sheet from ***Assessment*** to record children's progress. Use their time at the centers as an opportunity to complete your observations.

Thursday Planner

Objectives
- To identify and match shapes
- To find and describe the shape of objects in their environments
- To count objects to 10 and beyond

Materials
*Shape Sets

Math Throughout the Year
Review activity directions at the top of page 211, and complete each in class whenever appropriate.

Looking Ahead
For tomorrow, make copies of Family Letter Week 14 from the *Teacher's Resource Guide*.

*provided in Manipulative Kit

Thursday

1 Whole Group

10

Warm-Up: Count and Move (Forward and Back)

Everyone starts in a crouched position, and slowly rises to a standing position while counting aloud to 10. Then, counting backward from 10, slowly sink back down. Next, help children pretend to "climb" ten steps, counting each aloud, and then "climb" down, counting backward from 10.

Feely Box (Name)

- Children take turns completing this activity in pairs within the Whole Group setting.
- Child One hides a Shape Set shape in the Feely Box, and Child Two puts his or her hand in the box to feel the shape and tell what it is. Encourage children to share how they figured out the shape, and discuss each shape, emphasizing its particular attributes.

2 Work Time

25

Small Group

Feely Box (Match and Name)

- Remember to start with any children who did not get an opportunity during Tuesday's Small Group.
- If needed, start with the Match version of Feely Box by secretly hiding a Shape Set trapezoid in the Feely Box, displaying five shapes from another Shape Set (including the one that exactly matches the one you hid), and having a child put his or her hand in the box to feel the shape and point to its match on display. Discuss with children how they determined shapes and review shape attributes.
- As soon as children are ready, switch to Feely Box (Name): Child One hides a Shape Set shape in the Feely Box, and Child Two puts his or her hand in the box to feel the shape and tell what it is. Children share how they figured out the shape, discussing each shape and its particular attributes.

Monitoring Student Progress

If . . . children struggle to identify shapes during Feely Box,

Then . . . use pattern blocks instead of Shape Sets or, with either collection, reduce the number of shapes to a few shapes that are clearly different, slowly increasing the number of shapes and their difficulty.

If . . . children excel at shape identification during Feely Box,

Then . . . use more shapes, and have children describe the shape well enough without using its name for rest of the class to guess.

Computer Center Building Blocks

Continue to provide each child with a chance to complete Mystery Pictures 3 and Memory Geometry 3.

Hands On Math Center

Based on what children learn and benefit from most, choose from this week's Hands On Math Center activities: Shape Flip Book, Shape Pictures, Feely Box (Name), and Shape Step. Consult the Weekly Planner for corresponding materials and, if needed, Monday and Tuesday for activity directions.

3 Reflect

5

Show children two rhombuses, and ask:

- **What is the same about these shapes? What is different?**

Children might say, depending on your examples: They are both diamonds, they are both rhombuses, one is bigger than the other, and the like.

4 Assess

During Small Group activities, use the Small Group Record Sheet from ***Assessment*** to observe and record children's progress.

Friday Planner

Objectives

- To identify and match shapes
- To find and describe the shape of objects in their environments

Materials

no new materials

Math Throughout the Year

Review activity directions at the top of page 211, and complete each in class whenever appropriate.

Looking Ahead

For next week, familiarize yourself with Mystery Pictures 4: Name New Shapes, Memory Geometry 4: Shapes of Things, and Memory Geometry 5: Shapes in the World from the ***Building Blocks*** software.

Friday

1 Whole Group 20

Warm-Up: Computer Show

- Have the class gather around the computer at the Computer Center. Ask a volunteer who you know saved a Mystery Picture during the week to log on the ***Building Blocks*** software and load the picture to show others (as an option, you may just make your own Mystery Picture to discuss). Encourage the volunteer to talk about what shapes they used and why.
- Summarize by asking the rest of the class what they learned about shapes this week on the computer.

Building Shapes

- Review the book with children, and return to the page with the rhombus, explaining that mathematicians call it a *rhombus* but that it is also sometimes called a diamond.
- Challenge children to find other target shapes in the book. Summarize that, if you look carefully, you can see shapes in a lot of things in the world and children can name such shapes.

2 Work Time 15

Computer Center

Continue to provide each child with a chance to complete Mystery Pictures 3 and Memory Geometry 3.

Monitoring Student Progress

If . . . a child struggles to complete this week's computer activities by week's end,	**Then . . .** partner him or her with an "expert" child and/or do not advance the child during the beginning of next week's activities.
If . . . a child has mastered this week's computer activities by week's end,	**Then . . .** challenge him or her to use an activity's Free Explore.

Hands On Math Center

Based on what children learn and benefit from most, choose from this week's Hands On Math Center activities: Shape Flip Book, Shape Pictures, Feely Box (Name), and Shape Step. Consult the Weekly Planner for corresponding materials and, if needed, Monday and Tuesday for activity directions.

3 Reflect

Briefly discuss with children what you have done in class this week, such as making pictures from shapes, and ask:

- **Why did you use the shapes you did to make your pictures?**

Children might say: I tried to make my picture look like one in the book you read; I wanted to make my favorite toy; I wanted to use a rhombus to make a top; or the like.

4 Assess

Use the Weekly Record Sheet from ***Assessment*** to record children's progress. Use their time at the centers as an opportunity to complete your observations.

RESEARCH IN ACTION

Children usually refer to *rhombus* as *diamond*. Accept that name all year if necessary. We encourage the mathematical term *rhombus* because *diamond* is often misunderstood. Many people think a diamond's orientation must always be straight up and down. We also want to encourage children to use the mathematical terms whenever possible.

Home Connection

From the ***Teacher's Resource Guide,*** distribute to children copies of Family Letter Week 14 to share with their family. Each letter has an area for children to show families an example of what they have been doing in class.

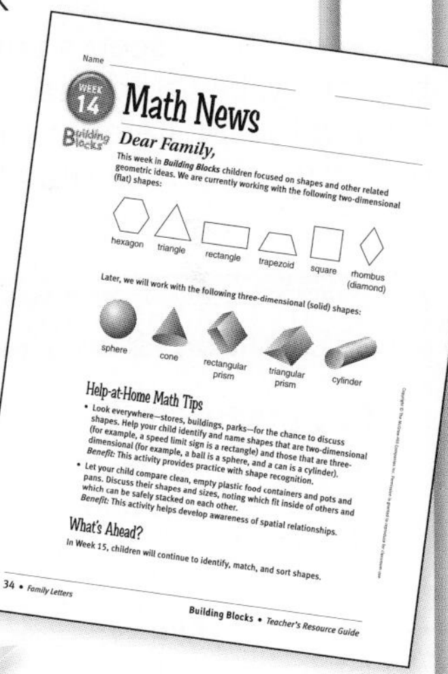

Name

WEEK 14

Math News

Building Blocks

Dear Family,

This week in ***Building Blocks*** children focused on shapes and other related geometric ideas. We are currently working with the following two-dimensional (flat) shapes:

hexagon, triangle, rectangle, trapezoid, square, rhombus (diamond)

Later, we will work with the following three-dimensional (solid) shapes:

sphere, cone, rectangular prism, triangular prism, cylinder

Help-at-Home Math Tips

- Look everywhere—stores, buildings, parks—for the chance to discuss shapes. Help your child identify and name shapes that are two-dimensional (for example, a speed limit sign is a rectangle) and those that are three-dimensional (for example, a ball is a sphere, and a can is a cylinder). *Benefit:* This activity provides practice with shape recognition.
- Let your child compare clean, empty plastic food containers and pots and pans. Discuss their shapes and sizes, noting which fit inside of others and which can be safely stacked on each other. *Benefit:* This activity helps develop awareness of spatial relationships.

What's Ahead?

In Week 15, children will continue to identify, match, and sort shapes.

34 • *Family Letters*   **Building Blocks** • *Teacher's Resource Guide*

Assess and Differentiate

A Gather Evidence

Review children's progress in mathematics by looking at the Weekly Record Sheets (Monday, Wednesday, Friday) and the Small Group Record Sheets (Tuesday, Thursday) from this past week.

B Summarize Findings

Using ***Online Assessment,*** summarize and analyze assessment data for each child based on your weekly observations and Record Sheets. Such information helps determine where each child is on the math trajectory for counting, comparing, and patterning. See ***Assessment*** for the print companion to each Learning Trajectory Record.

C Differentiate Instruction

Once you have seen a child exhibit specific levels of the trajectory, begin to encourage and work with that child toward the next level. Refer to Appendix A for individualized instruction opportunities, including Special Education concerns.

Shapes: Matching

If . . .	Then . . .
If . . . the child can match many shapes of the same size and orientation,	**Then . . .** *Shape Matcher, More Shapes* Matches a wider variety of shapes of same size and orientation.
If . . . the child can match shapes of varied sizes and orientations,	**Then . . .** *Shape Matcher, Sizes and Orientations* Matches a wider variety of shapes of different sizes and orientations. For example, matches the same triangle even if the examples are turned differently.

Shapes: Naming

If . . .	Then . . .
If . . . the child can recognize and point to the sides of shapes,	**Then . . .** *Side Recognizer* Parts: Identifies sides as distinct objects. For example, the child explains that a shape is a triangle because it has three sides and he or she runs his or her finger along the length of each side.
If . . . the child can recognize familiar shapes and some examples of less familiar shapes,	**Then . . .** *Shape Recognizer, More Shapes* Recognizes most familiar shapes and typical examples of other shapes, such as hexagon, rhombus (diamond), and trapezoid.
If . . . the child can name most shapes and recognize right angles,	**Then . . .** *Shape Identifier* Names most common shapes, including rhombuses, without making mistakes, such as calling ovals circles; implicitly recognizes right angles, thus can distinguish between rectangles and parallelograms without right angles.

Counting

If . . .	Then . . .
If . . . the child can accurately count to 10 and attempts to count beyond 10 in varied arrangements,	**Then . . .** *Counter and Producer (10+)* Accurately counts and produces objects to 10, then beyond to 30; has explicit understanding of cardinality (that numbers tell "how many?") and keeps track of objects that have and have not been counted even in different arrangements. Writes or draws to represent 1–10 (then 20 and 30), and gives next number to 20s or 30s; recognizes errors in others' counting and can eliminate most errors in own counting if asked to try hard.
If . . . the child can accurately and consistently count backward from 10,	**Then . . .** *Counter Backward from 10* Counts back verbally from 10 or when removing items from a group.

WEEK 15

Overview

Big Ideas

- shape matching
- shape identification
- adding and subtracting small numbers

Teaching for Understanding

Geometric shapes can be used to represent and understand objects in the world around us. Even preschool children can learn a lot about less common yet important geometric shapes; they can even learn to reason about the attributes of such shapes. This week we emphasize reasoning about geometric shapes in order to correctly identify and describe them.

Sorting Shapes

Some activities this week will require children to sort shapes and identify the rule for sorting. They will do this by examining the categories into which shapes have been sorted. Children will identify shapes with certain attributes, as well as reason about shapes by correcting mistakes in others' naming, classification, or description of shapes.

Number Activities

Children learn to use counting strategies to add and subtract small numbers. As we have been and continue to explain, even very young children are able to figure out how many items there *will be* in a small group when one more item is added to it. In such activities, adding or subtracting two items is acceptable as long as children continue to be successful at that level.

Meaningful Connections

We connect geometry and number by discussing with children the number of sides and angles (corners) shapes have. Another key focus this week is linking knowledge of math content to the development of mathematical processes, such as knowing triangles have three sides and applying that information to finding and naming a triangular example. Children learn specific processes including sorting, reasoning, and communicating. In this way, mathematics is also connected to general thinking and language skills.

What's Ahead?

In the weeks to come, children will focus on patterns that repeat. They will connect such patterns to number and geometric ideas.

Overview

How Children Learn about Shape and Number

Knowing where children are on the learning trajectory for geometry and number and what the next steps are helps facilitate their development. Children who easily surpass these trajectory levels might be challenged by less familiar shapes and larger quantities and encouraged to assist other children.

Shapes: Matching

What to Look For What types of shapes can the child match?

Shape Matcher, More Shapes Matches a wider variety of shapes of same size and orientation.

Shape Matcher, Sizes and Orientations Matches a wider variety of shapes of different sizes and orientations. For example, matches the same triangle even if the examples are turned differently.

Shapes: Naming

What to Look For What types of shapes and shape parts can the child identify?

Side Recognizer Parts: Identifies sides as distinct objects. For example, the child explains that a shape is a triangle because it has three sides and runs his or her finger along the length of each side.

Shape Recognizer, More Shapes Recognizes most familiar shapes and typical examples of other shapes, such as hexagon, rhombus (diamond), and trapezoid.

Shape Identifier Names most common shapes, including rhombuses, without making mistakes, such as calling ovals circles; implicitly recognizes right angles, thus can distinguish between rectangles and parallelograms without right angles.

Parts of Shape Identifier Identifies shapes in terms of their components. For example, no matter how "skinny" a triangle is, the child knows it is a triangle because it has three sides.

Counting

What to Look For Which numerals can the child count to and from accurately?

Counter Backward from 10 Counts back verbally from 10 or when removing items from a group.

Adding and Subtracting

What to Look For What small sums can the child determine?

Small Number (+/–) Finds sums for joining problems up to 3 + 2 by counting all objects. For example: There are two crayons. One more crayon is added. How many crayons are there in all? Child counts two, counts one more, and then counts "1, 2, 3...3!"

English Learner

Connect repeating visual and rhythmic patterns to English language patterns. Review the following: *repeat, How do you know?, and so on*, and *attributes*. Refer to page 91 of the ***Teacher's Resource Guide.***

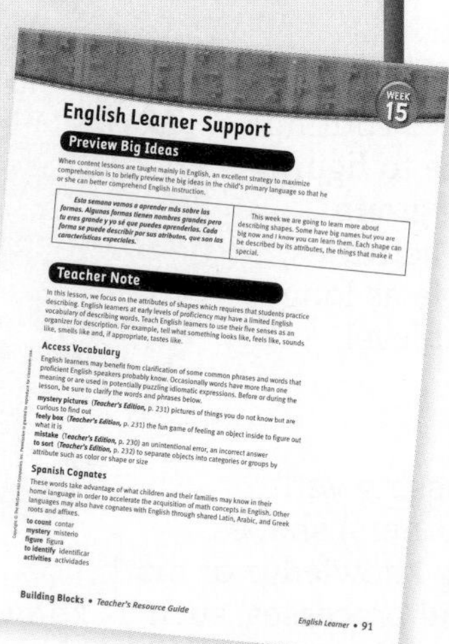

WEEK 15

English Learner Support

Preview Big Ideas

When content lessons are taught mainly in English, an excellent strategy to maximize comprehension is to briefly preview the big ideas in the child's primary language so that he or she can better comprehend English instruction.

Esta semana vamos a aprender más sobre las formas. Algunas formas tienen nombres grandes pero tu eres grande y yo sé que puedes aprenderlos. Cada forma se puede describir por sus atributos, que son las características especiales.	This week we are going to learn more about describing shapes. Some have big names but you are big now and I know you can learn them. Each shape can be described by its attributes, the things that make it special.

Teacher Note

In this lesson, we focus on the attributes of shapes which requires that students practice describing. English learners at early levels of proficiency may have a limited English vocabulary of describing words. Teach English learners to use their five senses as an organizer for description. For example, tell what something looks like, feels like, sounds like, smells like and, if appropriate, tastes like.

Access Vocabulary

English learners may benefit from clarification of some common phrases and words that proficient English speakers probably know. Occasionally words have more than one meaning or are used in potentially puzzling idiomatic expressions. Before or during the lesson, be sure to clarify the words and phrases below.

mystery pictures (***Teacher's Edition,*** p. 231) pictures of things you do not know but are curious to find out
feely box (***Teacher's Edition,*** p. 231) the fun game of feeling an object inside to figure out what it is
mistake (***Teacher's Edition,*** p. 230) an unintentional error, an incorrect answer
to sort (***Teacher's Edition,*** p. 232) to separate objects into categories or groups by attribute such as color or shape or size

Spanish Cognates

These words take advantage of what children and their families may know in their home language in order to accelerate the acquisition of math concepts in English. Other languages may also have cognates with English through shared Latin, Arabic, and Greek roots and affixes.

to count contar
mystery misterio
figure figura
to identify identificar
activities actividades

Building Blocks • *Teacher's Resource Guide*

English Learner • 91

Technology Project

Children may get additional shape practice using Mystery Pictures Free Explore. Encourage children to drag as many shapes as they would like to make their own pictures on the computer. If they click the play button, children can invite a classmate to complete their mystery picture, which is challenging because the correct shape in the correct orientation must be selected. As you visit children at the computer, help them describe what they are making, what shapes they are using, and why.

Math Throughout the Year

Math Throughout the Year activities are recommended to build on the mathematical skills highlighted in each week. Here are suggested activities for **Week 15.**

Counting Jar

Using the jar you have been all year, vary the sizes of the items you place in the jar each week. Children spill the jar's items to count them. This week is a good time to encourage children to record their counts; provide self-sticking notes for children to write the numeral that tells how many items are in the jar. Offer writing assistance as needed and/or allow children to make marks that represent an amount instead of writing numerals.

Shape Books

Provide the Shape Flip Book, magazines and newspapers (to be cut), crayons, scissors, construction paper, and paste for this activity. Review the Shape Flip Book with children, explaining that, together, you will make a class shape book. Discuss how shapes are all around us, in things, and look for some together. As children find shapes, help them label each—actually writing out shape names. Children cut shapes from magazines and newspapers, trace the item they found the shape in if applicable (children could draw the item instead), and keep their work for a page in the shape book. Bind books with staples, yarn, or use a three-pronged folder. Optionally, this activity may be done during Whole Group or the Hands On Math Center.

Shape Walk

Go for a walk outside of the classroom to search for a specific shape.

Center Preview

Computer Center Building Blocks

Get your classroom Computer Center ready for Mystery Pictures 4: Name New Shapes, Memory Geometry 4: Shapes of Things, and Memory Geometry 5: Shapes in the World from the ***Building Blocks*** software.

After you introduce Mystery Pictures 4 and Memory Geometry 4 and 5, each child should complete the activities individually as you (or an assistant) monitor and guide him or her periodically. Ideally, each child will have at least ten minutes of computer time at least twice during the week. Use children's center time as an opportunity for assessment.

Hands On Math Center

This week's Hands On Math Center activities are Shape Pictures, Feely Box (Name), Shape Flip Book, and Memory Geometry. Supply the center with these materials: Shape Sets, pattern blocks, Feely Box, Shape Flip Book, and Memory Geometry Card Sets C1 and C2.

Literature Connections

These books help develop shape recognition.

- *Up Goes the Skyscraper!* by Gail Gibbons
- *The Village of Round and Square Houses* by Ann Grifalconi
- *Shapes and Things* by Tana Hoban
- *So Many Circles, So Many Squares* by Tana Hoban
- *Manhattan Skyscrapers* by Norman McGrath and Eric Peter Nash

(t)Matt Meadows; (b)Eclipse Studios

WEEK 15 Weekly Planner

Learning Trajectories

Week 15 Objectives

- To identify and match shapes
- To find and describe the shape of objects in their environments
- To count forward to and backward from 10
- To add and subtract small numbers

	Developmental Path	Instructional Activities
Shapes: Matching	*Shape Matcher, Sizes*	
	Shape Matcher, More Shapes	Memory Geometry 4 and 5 Memory Geometry (print)
	Shape Matcher, Sizes and Orientations	Mystery Pictures 4
	Shape Matcher, Combinations	
Shapes: Naming	*Shape Recognizer, All Rectangles*	
	Side Recognizer	Shape Step
	Shape Recognizer, More Shapes	Shape Step Mystery Pictures 4 Memory Geometry 4 and 5 Memory Geometry (print) Guess My Rule
	Shape Identifier	Mr. Mixup (Shapes) Shape Pictures Feely Box (Name) Shape Flip Book
	Parts of Shapes Identifier	Guess My Rule
	Shape Class Identifier	
Counting	*Counter and Producer (10+)*	
	Counter Backward from 10	Count and Move (Forward and Back)
	Counter from N (N +1, N – 1)	
Adding and Subtracting	*Nonverbal (+/–)*	
	Small Number (+/–)	How Many Now?
	Find Result (+/–)	

Use this chart to plan for your specific class schedule. If you have your prekindergarteners for only three days, complete Monday, Tuesday, and Thursday of the week.

Work Time

Whole Group	Small Group	Computer	Hands On	Program Resources
Shape Step *Materials:* large tape shapes ***Mr. Mixup (Shapes)** *Materials:* *Shape Set		**Mystery Pictures 4**	**Shape Pictures** *Materials:* *Shape Sets *Pattern Blocks **Feely Box (Name)** *Materials:* *Shape Sets **Shape Flip Book**	Building Blocks ***Teacher's Resource Guide*** • Shape Set • Shape Flip Book ***Assessment*** Weekly Record Sheet
Count and Move (Forward and Back) **Guess My Rule** *Materials:* *Shape Sets	**Guess My Rule** *Materials:* *Shape Sets **Shape Step** *Materials:* large tape shapes	**Memory Geometry 4**	**Shape Pictures** **Feely Box (Name)** **Shape Flip Book** **Memory Geometry**	Building Blocks ***Teacher's Resource Guide*** • Shape Flip Book • Memory Geometry Card Sets C1 and C2 ***Assessment*** Small Group Record Sheet
Shape Step *Materials:* large tape shapes **How Many Now?** *Materials:* *counters dark cloth		**Memory Geometry 5**	**Shape Pictures** **Feely Box (Name)** **Shape Flip Book**	Building Blocks ***Teacher's Resource Guide*** • Shape Set • Shape Flip Book ***Assessment*** Weekly Record Sheet
Count and Move (Forward and Back) **Guess My Rule** *Materials:* *Shape Sets	**Guess My Rule** *Materials:* *Shape Sets **Shape Step** *Materials:* large tape shapes	• **Mystery Pictures 4** • **Memory Geometry 4** • **Memory Geometry 5**	**Shape Pictures** **Feely Box (Name)** **Shape Flip Book** **Memory Geometry**	Building Blocks ***Teacher's Resource Guide*** • Shape Flip Book • Memory Geometry Card Sets C1 and C2 ***Assessment*** Small Group Record Sheet
Discuss Shape Pictures ***Mr. Mixup (Shapes)** *Materials:* *Shape Set		• **Mystery Pictures 4** • **Memory Geometry 4** • **Memory Geometry 5**	**Shape Pictures** **Feely Box (Name)** **Shape Flip Book**	Building Blocks ***Teacher's Resource Guide*** • Family Letter Week 15 • Shape Flip Book ***Assessment*** Weekly Record Sheet

*provided in Manipulative Kit

Monday Planner

Objectives
- To identify and match shapes
- To find and describe the shape of objects in their environments

Materials
- large tape shapes
- *Mr. Mixup
- *Shape Sets
- *Pattern Blocks

Math Throughout the Year
Review activity directions at the top of page 227, and complete each in class whenever appropriate.

Looking Ahead
If you have not already done so, prepare the Feely Box and Shape Flip Book *(Teacher's Resource Guide)* for today's Hands On Math Center.

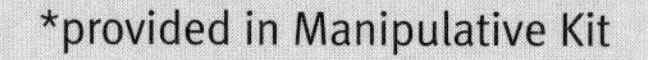
*provided in Manipulative Kit

Monday

1 Whole Group
15

Warm-Up: Shape Step

- Use masking or colored tape to make large shapes on the floor. Show children rhombuses from the Shape Set (*Teacher's Resource Guide* page 174), and tell them they will step only on rhombuses.
- Have a group of four or five children step on the rhombuses. Ask the rest of the class to watch carefully to make sure each group steps on all rhombuses. Whenever possible, ask children to explain why the shape they stepped on was the correct shape. Repeat with other groups of children.
- Repeat with new shapes, such as hexagons, engaging different groups of children each time.

Monitoring Student Progress

If . . . children struggle during Shape Step,	**Then . . .** review the target shape's attributes while children feel and/or trace the shape with their fingers. You might also step on a target shape, and then ask children whether it is the correct shape.
If . . . children excel during Shape Step,	**Then . . .** have them find examples of target shapes in unusual places and/or teach even less familiar shapes.

Mr. Mixup (Shapes)
- Ask children: Do you remember Mr. Mixup? Explain that they are going to help Mr. Mixup name shapes. Remind children to stop Mr. Mixup right when he makes a mistake to correct him. Use a silly voice, and have fun!
- Using Shape Set shapes, have Mr. Mixup start by confusing the names of a square and a rhombus. After children have identified the correct names, ask them to explain how their angles are different (squares must have all right angles; rhombuses may have different angles). Review that all rhombuses and squares, which are actually a special kind of rhombus with all right angles, have four straight sides of equal length.
- Repeat with a trapezoid, a hexagon, and any other shapes you would like children to practice.

Eclipse Studios

Work Time

20

Computer Center

Introduce Mystery Pictures 4: Name New Shapes from the ***Building Blocks*** software, in which children have to identify each shape to create the mystery pictures. Each child should complete the activity this week.

Mystery Pictures 4

Hands On Math Center

These recurring activities may be set up at the center all week.

Shape Pictures

- Children make designs and pictures with Shape Sets and Pattern Blocks. Encourage them to name and discuss shapes, especially the trapezoids, rhombuses, and hexagons.
- To simplify, have children feel a shape's sides and corners to help with their descriptions. For a challenge, have children tell exactly why a shape is a trapezoid or a rhombus, for example, by describing their differences.
- For Friday's Whole Group, sketch or photograph children's designs, especially those that show symmetry (mirror images) or linear patterns.

Feely Box (Name)

In pairs, one child hides a Shape Set shape in the Feely Box; the other child puts his or her hand in the box to feel the shape to tell what it is. Help children explain how he or she figured out the shape, emphasizing the straightness and number of sides.

Shape Flip Book

Children match the book's panels, naming each shape.

RESEARCH IN ACTION

Watch what children do with Pattern Blocks. Children are naturally attracted to and make symmetric designs. Showing and discussing their designs are wonderful starting points for learning about symmetry. Children's designs may have line symmetry (for example, if folded, shape halves fit on each other), rotational symmetry (when a shape can be turned and fit on itself, like a parallelogram), both types of symmetry, or linear patterns (sequences in a row with a repeated core unit, such as ABC).

Reflect

5

Ask children:

- **How do you know which shapes are rhombuses and which are trapezoids?**

Children might say: Rhombuses have four sides all the same, and trapezoids have different sides, or the like.

Assess

Use the Weekly Record Sheet from ***Assessment*** to record children's progress. Use their time at the centers as an opportunity to complete your observations.

Tuesday Planner

Objectives

- To identify and match shapes
- To find and describe the shape of objects in their environments
- To count forward to and backward from 10

Materials

- *Shape Sets
- large tape shapes

Math Throughout the Year

Review activity directions at the top of page 227, and complete each in class whenever appropriate.

Looking Ahead

If you prefer, make large shapes out of masking or colored tape on the floor ahead of time for today's Small Group.

*provided in Manipulative Kit

Tuesday

1 Whole Group

10

Warm-Up: Count and Move (Forward and Back)

Everyone starts in a crouched position, and slowly rises to a standing position while counting aloud to 10. Then, counting backward from 10, slowly sinks back down.

Guess My Rule

- Ask children to watch carefully as you sort Shape Set shapes into piles based on something that makes them alike. Tell children to silently guess your sorting rule, such as trapezoids versus triangles.
- Sort shapes one at a time, continuing until there are at least two shapes in each pile. Signal "shhh," and pick up another shape. Looking uncertain, ask all children to point silently to the pile in which the shape belongs, and then place the shape in its pile.
- Ask children to explain to a classmate what the sorting rule is. As they talk, help them consider the following: Why did you guess that rule? How did you know? Show me how the shapes match the rule. Repeat with other shapes and new rules, such as rectangles versus rhombuses.

2 Work Time

25

Small Group

Guess My Rule

Help children follow the Whole Group directions for this activity, using different rules, such as rectangles versus all other shapes, triangles versus rhombuses, trapezoids versus non-trapezoids, or hexagons versus trapezoids.

Monitoring Student Progress

If . . . children struggle during Guess My Rule,	**Then . . .** use simpler rules, such as squares versus circles.
If . . . children excel during Guess My Rule,	**Then . . .** have them think of a sorting rule for the class, telling only you at first.

Shape Step

- If you have not already done so, make large tape shapes on the floor for children to step on, and then show them to children.
- Tell children which shape to step on. As best you can, observe whether target shapes are stepped on. Encourage groups to discuss why the shape was the correct shape.

Computer Center Building Blocks

Introduce Memory Geometry 4: Shapes of Things from the ***Building Blocks*** software, in which children play a version of the traditional concentration game by matching shapes to common objects, such as an octagon to a stop sign. Each child should complete Memory Geometry 4 this week.

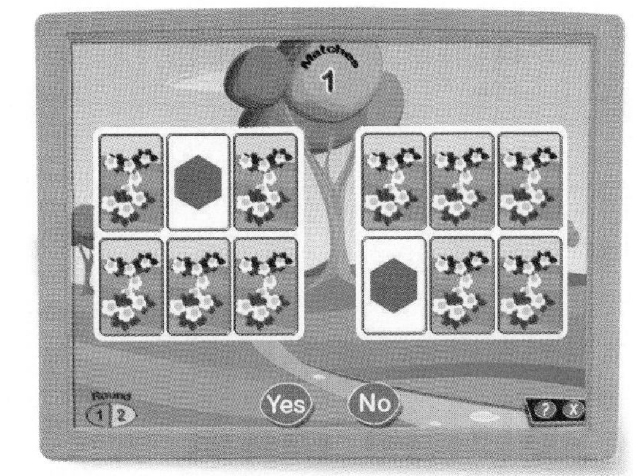

Memory Geometry 4

Hands On Math Center

Based on what children learn and benefit from most, choose from this week's Hands On Math Center activities: Shape Pictures, Feely Box (Name), and Shape Flip Book. Consult the Weekly Planner for corresponding materials and, if needed, yesterday for activity directions. Children might also benefit from a hands-on version of Memory Geometry in which they match shapes to common objects using Sets C1 and C2. You can find general directions for Memory Geometry on page 137.

3 Reflect

5

Ask children:

■ **What shape(s) did you find today?**

Children might say: We found mostly rectangles (insert corresponding target shape name); rectangles are everywhere like your desk! We also found circles like our clock.

4 Assess

During Small Group activities, use the Small Group Record Sheet from ***Assessment*** to observe and record children's progress.

Wednesday Planner

Objectives

- To identify and match shapes
- To find and describe the shape of objects in their environments
- To count forward to and backward from 10
- To add and subtract small numbers

Materials

- large tape shapes
- *counters
- dark cloth

Math Throughout the Year

Review activity directions at the top of page 227, and complete each in class whenever appropriate.

*provided in Manipulative Kit

Wednesday

1 Whole Group

25

Warm-Up: Shape Step

- Use masking or colored tape to make large shapes on the floor. Review rhombuses if needed, and tell children to step only on rhombuses.
- Have a group of four or five children step on the rhombuses. Ask the rest of the class to watch carefully to make sure each group steps on all rhombuses. Whenever possible, ask children to explain why the shape they stepped on was the correct shape. Repeat with other groups of children.
- Repeat with new shapes, such as hexagons, engaging different groups of children each time.
- If children are ready, switch to the Sides and Angles version of this activity, in which you would ask children to step on shapes with three (triangles), four (quadrilaterals), or six (hexagons) sides without saying the shape's name.

How Many Now?

- Show five counters to children, and, as a group, count and say how many counters there are.
- Add one counter, and ask children how many there are now. Count together to check.
- Repeat by adding and removing one counter and, eventually, do the same with two counters.
- Once children are able, play this activity's hidden version by following the same steps, but hide the counters under a dark cloth (paper plate is optional, as counters can simply be placed on a desk and covered). Keep counters hidden as you add or remove counters, as well as during children's response portion of the activity. For example, add two counters to the others under the cloth and, after children have answered, uncover the counters and count with the class to check.

Monitoring Student Progress

If . . . children need help during How Many Now?	**Then . . .** use fewer counters, or provide answer choices.
If . . . children need a challenge during How Many Now?	**Then . . .** use more counters.

Eclipse Studios

Work Time

Computer Center Building Blocks

Introduce Memory Geometry 5: Shapes in the World from the ***Building Blocks*** software, in which children play a version of the traditional concentration game by matching shapes to shapes from the "world," such as a rectangle to a cereal box. Each child should complete Memory Geometry 5 this week.

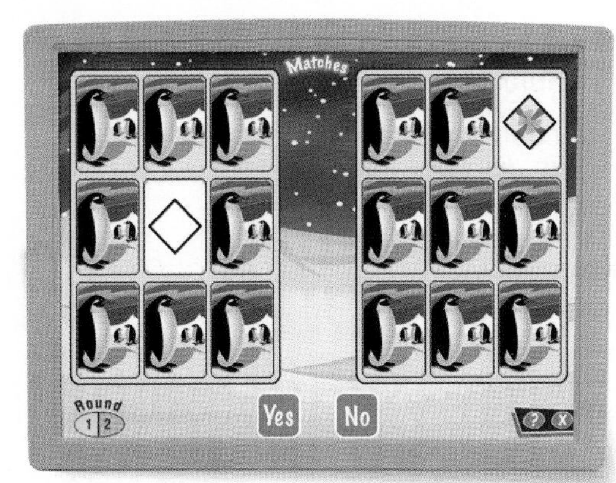

Memory Geometry 5

Hands On Math Center

Based on what children learn and benefit from most, choose from this week's Hands On Math Center activities: Shape Pictures, Feely Box (Name), and Shape Flip Book. Consult the Weekly Planner for corresponding materials and, if needed, Monday for activity directions.

Reflect

Ask children:

- **How did you know how many counters there were?**

Children might say: I counted; I know 6 is one more than 5; or the like.

Assess

Use the Weekly Record Sheet from ***Assessment*** to record children's progress. Use their time at the centers as an opportunity to complete your observations.

RESEARCH IN ACTION

What if a child steps on a square when asked to step on a rhombus? That is good! A square is a special kind of rhombus; it is a rhombus because it has four equal sides, and it is special because it has all right angles.

Thursday Planner

Objectives
- To identify and match shapes
- To find and describe the shape of objects in their environments
- To count forward to and backward from 10

Materials
- *Shape Sets
- large tape shapes

Math Throughout the Year
Review activity directions at the top of page 227, and complete each in class whenever appropriate.

Looking Ahead
For tomorrow, make copies of Family Letter Week 15 from the *Teacher's Resource Guide*.

*provided in Manipulative Kit

Thursday

1 Whole Group 10

Warm-Up: Count and Move (Forward and Back)
Everyone starts in a crouched position, and slowly rises to a standing position while counting aloud to 10. Then, counting backward from 10, slowly sinks back down. Repeat with other movements if you prefer.

Guess My Rule
- Ask children to watch carefully as you sort Shape Set shapes into piles based on something that makes them alike. Tell children to silently guess your sorting rule, such as hexagons versus all other shapes.
- Sort shapes one at a time, continuing until there are at least two shapes in each pile. Signal "shhh," and pick up another shape. Looking uncertain, ask all children to point silently to the pile in which the shape belongs, and then place the shape in its pile.
- Ask children to explain to a classmate what the sorting rule is. As they talk, help them consider the following: Why did you guess that rule? How did you know? Show me how the shapes match the rule. Repeat with other shapes and new rules (See the appendix for complete sorting rule suggestion list).

2 Work Time 25

Small Group

Guess My Rule
Help children follow the Whole Group directions for this activity, using different rules, such as circles versus ovals, triangles versus rhombuses, squares versus rectangles, or hexagons versus trapezoids.

Eclipse Studios

Monitoring Student Progress

If . . . children struggle during Guess My Rule,	**Then . . .** use simpler rules, such as squares versus triangles.
If . . . children excel during Guess My Rule,	**Then . . .** have them think of a sorting rule for the class, telling only you at first.

Shape Step

Using large tape shapes on the floor, tell children which shape to step on. As best you can, observe whether target shapes are stepped on. Encourage groups to discuss why the shape was the correct shape.

Many computer programs, such as ***Building Blocks*** activities, can help reinforce ideas and skills by letting children know whether or not they are correct, and most children do not mind being corrected.

Computer Center

Continue to provide each child with a chance to complete this week's computer activities: Mystery Pictures 4, Memory Geometry 4, and Memory Geometry 5.

Hands On Math Center

Based on what children learn and benefit from most, choose from this week's Hands On Math Center activities: Shape Pictures, Feely Box (Name), and Shape Flip Book. Consult the Weekly Planner for corresponding materials and, if needed, Monday for activity directions. Children might also benefit from a hands-on version of Memory Geometry in which they match shapes to common objects using Sets C1 and C2. You can find general directions for Memory Geometry on page 137.

3 Reflect

Ask children:

- **How did you know the sorting rule during Guess My Rule?**

Children might say: I look at all shapes to see what is the same about them; I count the sides first; I look at the piles and compare; or the like.

4 Assess

During Small Group activities, use the Small Group Record Sheet from *Assessment* to observe and record children's progress.

Friday Planner

Objectives
- To identify and match shapes
- To find and describe the shape of objects in their environments

Materials
- *Mr. Mixup
- *Shape Set

Math Throughout the Year
Review activity directions at the top of page 227, and complete each in class whenever appropriate.

Looking Ahead
For next week, familiarize yourself with Pattern Zoo 1: Recognize AB, Pattern Planes 1: Duplicate AB, and Marching Patterns 1: Extend AB from the ***Building Blocks*** software.

*provided in Manipulative Kit

Friday

1 Whole Group
20 

Warm-Up: Discuss Shape Pictures
- Display children's designs you copied from this week's Hands On Math Center Shape Pictures activity, and discuss each type of design, drawing special attention to those with line or rotational symmetry. Talk about how designs with line (mirror) symmetry are the same on one side as the other, and you may discuss rotational symmetry as being the same all around.
- Ask children what patterns they see. For example, square, triangle, square, triangle, and so on.

Mr. Mixup (Shapes)
- Remind children to stop Mr. Mixup right when he makes a shape mistake to correct him. Use a silly voice, and have fun!
- With Shape Set shapes, have Mr. Mixup first confuse the names of familiar shapes. After children have identified the correct names, ask them to compare and/or describe shapes, such as a square being a special kind of rhombus with all right angles.
- Repeat with a trapezoid, a hexagon, and any other shapes you would like children to practice.

Monitoring Student Progress

If . . . children struggle correcting Mr. Mixup,	**Then . . .** exaggerate Mr. Mixup's errors.
If . . . children excel at correcting Mr. Mixup,	**Then . . .** have Mr. Mixup make less obvious mistakes about less familiar shapes.

2 Work Time
15

Computer Center *Building Blocks*
Continue to provide each child with a chance to complete this week's computer activities: Mystery Pictures 4, Memory Geometry 4, and Memory Geometry 5.

Hands On Math Center

Based on what children learn and benefit from most, choose from this week's Hands On Math Center activities: Shape Pictures, Feely Box (Name), and Shape Flip Book. Consult the Weekly Planner for corresponding materials and, if needed, Monday for activity directions.

3 Reflect

5

Briefly discuss with children what you have done in class this week, such as guessing and naming shapes, and ask:

- **How do you know the names of shapes in the computer games?**

Children might say: I know by how many sides a shape has; I know because the sides are all the same; and the like.

4 Assess

Use the Weekly Record Sheet from ***Assessment*** to record children's progress. Use their time at the centers as an opportunity to complete your observations.

RESEARCH IN ACTION

We will continue to emphasize linear patterns next week. However, any complex designs children build may also contain wonderful, noteworthy patterns.

Home Connection

From the ***Teacher's Resource Guide,*** distribute to children copies of Family Letter Week 15 to share with their family. Each letter has an area for children to show families an example of what they have been doing in class.

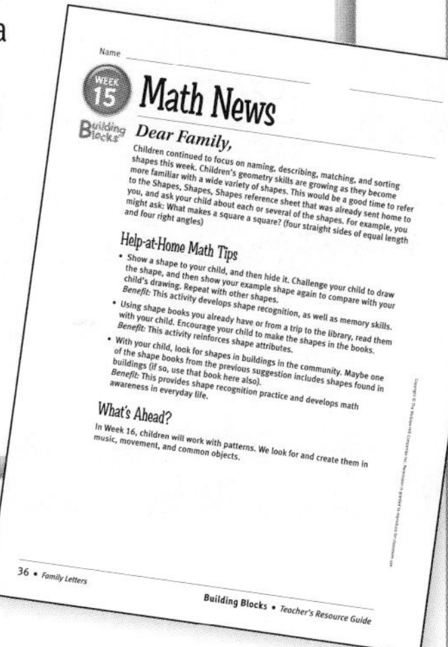

Name

WEEK 15 Math News

Dear Family,

Children continued to focus on naming, describing, matching, and sorting shapes this week. Children's geometry skills are growing as they become more familiar with a wide variety of shapes. This would be a good time to refer to the Shapes, Shapes, Shapes reference sheet that was already sent home to you, and ask your child about each or several of the shapes. For example, you might ask: What makes a square a square? (four straight sides of equal length and four right angles)

Help-at-Home Math Tips

- Show a shape to your child, and then hide it. Challenge your child to draw the shape, and then show your example shape again to compare with your child's drawing. Repeat with other shapes.
Benefit: This activity develops shape recognition, as well as memory skills.
- Using shape books you already have or from a trip to the library, read them with your child. Encourage your child to make the shapes in the books.
Benefit: This activity reinforces shape attributes.
- With your child, look for shapes in buildings in the community. Maybe one of the shape books from the previous suggestion includes shapes found in buildings (if so, use that book here also).
Benefit: This provides shape recognition practice and develops math awareness in everyday life.

What's Ahead?

In Week 16, children will work with patterns. We look for and create them in music, movement, and common objects.

36 • *Family Letters*

Building Blocks • *Teacher's Resource Guide*

Assess and Differentiate

A Gather Evidence

Review children's progress in mathematics by looking at the Weekly Record Sheets (Monday, Wednesday, Friday) and the Small Group Record Sheets (Tuesday, Thursday) from this past week.

B Summarize Findings

Using ***Online Assessment,*** summarize and analyze assessment data for each child based on your weekly observations and Record Sheets. Such information helps determine where each child is on the math trajectory for geometry and number. See ***Assessment*** for the print companion to each Learning Trajectory Record.

C Differentiate Instruction

Once you have seen a child exhibit specific levels of the trajectory, begin to encourage and work with that child toward the next level. Refer to Appendix A for individualized instruction opportunities, including Special Education concerns.

Shapes: Matching

If . . .	Then . . .
If . . . the child can match many shapes of the same size and orientation,	**Then . . .** *Shape Matcher, More Shapes* Matches a wider variety of shapes of same size and orientation.
If . . . the child can match shapes of varied sizes and orientations,	**Then . . .** *Shape Matcher, Sizes and Orientations* Matches a wider variety of shapes of different sizes and orientations. For example, matches the same triangle even if the examples are turned differently.

Shapes: Naming

If . . .	Then . . .
If . . . the child can recognize and point to the sides of shapes,	**Then . . .** *Side Recognizer* Parts: Identifies sides as distinct objects. For example, child explains that a shape is a triangle because it has three sides and runs finger along the length of each side.
If . . . the child can recognize familiar shapes and some examples of less familiar shapes,	**Then . . .** *Shape Recognizer, More Shapes* Recognizes most familiar shapes and typical examples of other shapes, such as hexagon, rhombus (diamond), and trapezoid.
If . . . the child can name most shapes and recognize right angles,	**Then . . .** *Shape Identifier* Names most common shapes, including rhombuses, without making mistakes, such as calling ovals circles; implicitly recognizes right angles, thus can distinguish between rectangles and parallelograms without right angles.
If . . . the child can name some shapes by their attributes,	**Then . . .** *Parts of Shapes Identifier* Identifies shapes in terms of their components. For example, no matter how "skinny" a triangle is, the child knows it is a triangle because it has three sides.

Counting

If . . .	Then . . .
If . . . the child can accurately and consistently count backward from 10,	**Then . . .** *Counter Backward from 10* Counts back verbally from 10 or when removing items from a group.

Adding and Subtracting

If . . .	Then . . .
If . . . the child can add or subtract small amounts by counting,	**Then . . .** *Small Number (+/–)* Finds sums for joining problems up to 3 +2 by counting all objects. For example: There are two crayons. One more crayon is added. How many crayons are there in all? Child counts two, counts one more, and then counts "1, 2, 3...3!"

Appendix

Appendix A

Appendix B

Appendix C

Using Games

When introducing a game to young children, the following strategy is usually successful. Start it several days before you want children to play the game independently. Gather the whole class around you. Each "step" below is a short and simple action, such as dealing out cards, picking up a card, and so forth.

Teacher	Children
Demonstrates (with one or more children) and describes step A Demonstrates (with one or more children) and describes step A	Describe step A
Demonstrates and describes step B	Describe steps A & B
Demonstrates and describes step C	Describe steps A, B, & C
...and so on	

Keep modeling until children can tell you all the steps two days in a row.

Each child should play each new game at least once during the week in which it is introduced. For some situations, it may be necessary for you or an aide to play with single children at first.

Guidelines for Games

Teachers should be involved with the games, observing, so children know they are important. Here are some answers to some common questions.

- *What do you do about losing control of the class?*

  Hold class meetings to decide how to solve their problems. Use "one person" and "the other person" rather than people's names.

- *Do you assign the games and partners?*
  - Children may be able to choose when they play the games, but recording sheets help keep track of whether they are choosing wisely. Assign children to play certain games as necessary (if they haven't "signed up" by themselves all week).
  - Choosing partners of about equal ability is usually beneficial. Choosing or helping children choose partners with whom they will interact is helpful.
- *Do you let children change game rules?*
  - If they ask other members of group, yes.
  - If they simplify a game too much, refer to the original rules.
- *Do you emphasize competition?*
  - We often just play so that everyone finishes.
  - If it comes up, treat "Who won?" as a trivial question.
  - When children seem to desire a winner, we often treat the first person who finishes ("wins") as "the first winner," the next person as "the second winner," and so on.
- *What should the teacher do while children are playing?*
  - Playing with individual children or a small group is most useful.
  - Assess children's level and thinking strategies.
  - Also, assess what specific knowledge (for example, counting skills or number combinations) they possess.

Guess My Rule Sorting List

Guess My Rule can be played with many rules. The following is a number of suggested rules, ordered from those that children find easier to those that are more challenging.

Use a chalk board to draw different shapes. This makes it easy to create different types of shapes, such as many shapes and sizes of triangles. Some suggestions of different shapes are below.

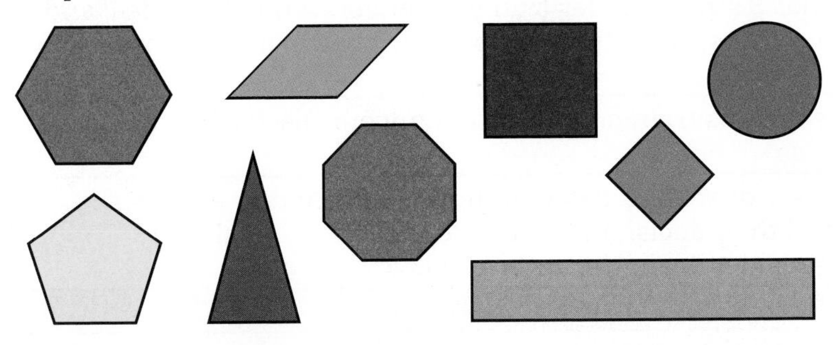

In addition, the higher levels include "foolers"—shapes that are close to a rectangle, for example, but are not rectangles. Here are some examples for rectangles and for triangles.

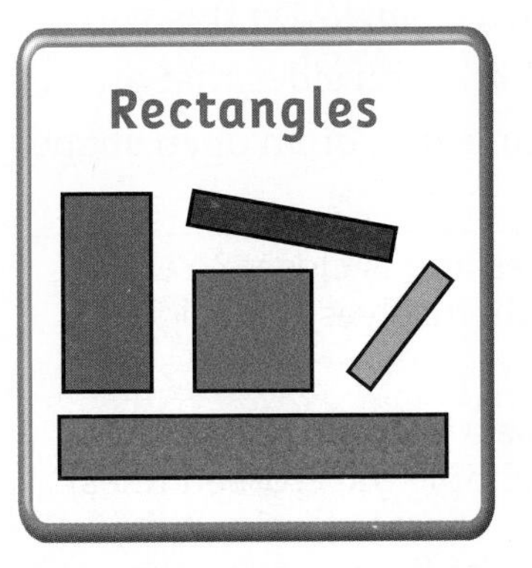

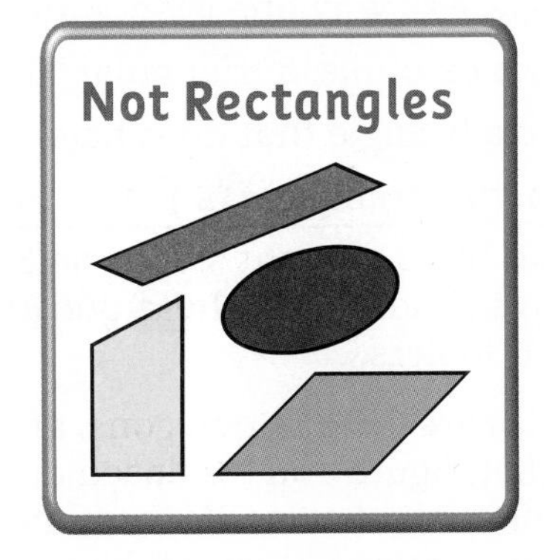

Here's the list of rules:

- circles versus squares (same orientation)
- circles versus triangles
- circles versus rectangles
- triangles versus squares
- triangles versus rectangles
- circles versus triangles versus squares (start having different orientations)
- triangles versus rhombuses
- trapezoids versus rhombuses
- trapezoids versus not trapezoids
- hexagons versus trapezoids
- triangles versus not triangles
- squares versus not squares (for example, all other shapes)
- rectangles versus not rectangles
- triangles versus not triangles
- trapezoids versus not trapezoids
- rhombuses versus not rhombuses.
- triangles versus diamonds
- polygons versus non-polygons (open shapes, or shapes with curved parts)
- closed figures versus non-closed figures
- numbers of sides/angle (for example, 3 versus 4)
- circles versus "foolers" (shapes very close in appearance to circles)
- squares versus "foolers"
- triangles versus "foolers"
- rectangles versus "foolers"
- symmetrical versus non-symmetrical
- regular polygons versus irregular polygons
- having one or more right angles versus no right angles

Direct Teaching of Concepts

Some children require a more structured approach to learning shapes and their attributes. Research shows this sequence is effective. We illustrate it with triangles, but it can be adapted for any shape or, indeed, any mathematical idea.

General Teaching Sequence	Illustration: Teaching "Triangle"
Session 1	
1. Describe the defining attributes of the shape and illustrate them with examples and non-examples.	Show shapes, including triangles and other shapes, such as four-sided shapes (quadrilaterals). Include shapes that look like triangles but are not. Say that a triangle has 3 straight sides (run your fingers along the sides) and 3 angles (touch the corners, or vertices as you count them) and is closed—no "gaps."
2. Draw each attribute (if necessary, but connecting figures outline with dots).	Have children trace various triangles with their fingers, discussing the 3 sides and the 3 angles.
3. Show shapes and have children indicate whether each has a particular defining attribute. Provide feedback immediately.	Again show the variety of shapes. Ask children to tell whether each is a triangle and defend their decision. Make sure they mention all the attributes—3 straight sides, 3 angles, and closed.
Session 2	
1. Review all session 1 work.	Review the above.
2. Children draw or build with manipulatives examples of the shape.	Have children draw or make triangles. Ask them how they know they are triangles.
3. Show paired shapes, one with a defining attribute and one without it. Have children identify which is an example of the shape and which is not, explaining their answer.	Show a "fooler" (shape that looks like a triangle but is not), such as a chevron, next to a triangle. Ask why one is and one is not a triangle. Do this with all the attributes (for example, a triangular "fooler" with sides that are not straight next to a triangle that does have straight sides; or an open shape next to an actual triangle).
4. Show figures and model asking whether all the defining attributes are present. Have students do the same for new figures.	"Is this a shape that has 3 straight sides and 3 angles and is closed? Yes, so it is a triangle!" Have children ask these questions themselves to decide on more triangles and "foolers." **Math Note:** Triangles are 3-sided polygons, and to be a polygon, a figure has to be a plane (flat) figure and be closed and be simple (no crossed lines) and be made of just straight sides. We assume these characteristics, but if they emerge in conversations, they can be added to the list of necessary attributes.
Repeat in session 2 as necessary	

Differentiating Instruction

Some students may need extra supports during lessons and activities described in the curriculum. These students may not have the background knowledge, motor skills, or cognitive skills necessary to complete the activities as described. Extra supports or modifications may be required for these students. Modifications may be made by providing information to students prior to lesson implementation, altering presentation of materials, changing questioning techniques, or changing the way the student shows his or her answer.

There are several reasons for modifying activities. You may need to provide extra support for the entire class by building in scaffolds in order for the group to grasp the concepts you are trying to teach. Or you may need to provide extra supports for one or two students within the context of the lesson.

Whole-class modifications may be things you do prior to the lesson with children who are having difficulties. This often entails previewing vocabulary so that children can focus on the mathematical concepts, rather than on novel vocabulary. Sometimes, the objective of a whole-group lesson is participation, rather than gaining the mathematical concept. A child who is having difficulty needs to participate in the lesson without holding up the other students.

Modifications for small group and everyday activities and off-computer centers should enable the student to learn the mathematical concept whenever possible. The most common ways of altering these activities are through presenting the materials in different ways, adding other materials and changing questioning techniques.

Whenever adding extra supports for students, keep these ideas in mind:

1. modify the activity only as much as necessary for the student to be successful,
2. before changing the objective of the lesson for a student, change presentation of materials or questioning techniques that will maintain the integrity of the lesson,
3. document the type of support given, and
4. fade supports as soon as possible to allow the child to complete the task in the same manner as his or her classmates.

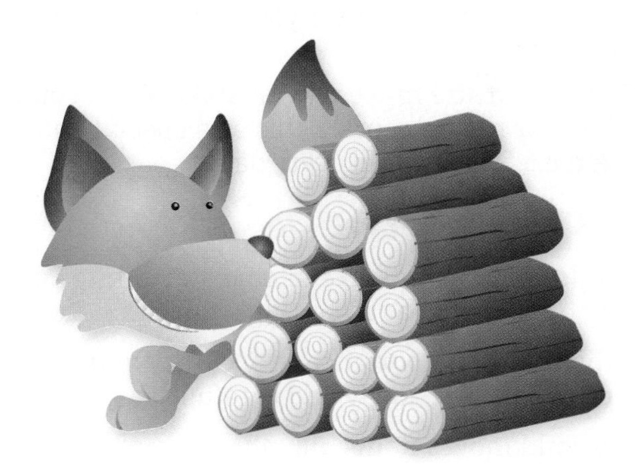

Differentiating Instruction: Working with Struggling Learners

Following are specific suggestions for working with struggling learners.

Verbal Counting Help

If the child makes a mistake in verbal counting, try the following:

- Emphasize the importance of accuracy and encourage the student to count slowly and carefully.
- Invite the child to count with you. Then ask them to do it (the same task again) alone.
- Mirror
  - Have the child mirror what you say, number by number. "Say each number after I say it. One"
  - Pause for response.
  - If no response, repeat "one" and then tell the child to say "one."
  - If the child says "two," then say "three" and continue, allow the child to mirror you or continue your counting.
  - If the child still makes the mistake when counting on his or her own, mark this as a special "warm-up" exercise for the child every day.

Object Counting Help

If the child makes a mistake in object counting such as in "passing out objects" as in setting a table, try the following:

- Emphasize the importance of accuracy and encourage the child to work slowly and carefully.
- If the child moves items out without matching the items, or if they pull out more than is needed, remind them of the task and say, "Each person only needs one object. We don't need these." Then repeat, "Give just one to each person."
- If children lose track, for instance, of where they started or stopped, model making a verbal plan, such as "Go from the top to the bottom. Give an object to every one person." Then enact that plan as a demonstration. Give the child a new problem.
- Note this is for matching/"passing out objects" only: If the child is counting, see the section on keeping track during counting.

One-to-One Errors (includes errors in keeping-track-of-what has been counted)

- Emphasize the importance of accuracy and encourage the child to count slowly and carefully so that each item is counted once and only once.
- Explain a keeping-track strategy. If moving objects is possible and desirable in the activity, suggest a strategy of moving items to a different pile or location. Otherwise, explain making a verbal plan, such as "Go from top to bottom. Start from the top and count every one."—then do so.
- If the mistake is counting "0"; that is, on a "race-type" game board with spaces, or a ruler, counting the zero (the space you are on) as "1," then "jump start" the counting by saying, "This is where you start. Here (highlight the 1) is 1."
- If the child returns and re-counts objects (e.g., in a circular arrangement):
  - 1st: Stop them and tell them they counted that item already. Suggest that they start on one they can remember (e.g., one at the "top" or "the corner" or "the blue one"—whatever makes sense in the activity; if there is no identifier, highlight an item in some way).
  - 2nd: Ask children to click on items as they count, providing highlighting to the object, marking it. If they click on a highlighted item, the character immediately says they counted that item already.
  - Go to "Guided Counting Sequence".

Cardinality ("The How Many Rules") Errors (Counter: object counting)

- Ask children to re-count.
- Demonstrate the cardinality rule on the collection. That is, count the collection, pointing to each item in turn, then gesture at them all, saying "Five in all!"
- Demonstrate the cardinality rule on a small (subitizable) collection in an easily recognizable arrangement (see the Snapshots activity).

Differentiating Instruction: Working with Struggling Learners

Cardinality Errors (Counter To, Production tasks—Knowing when to stop)

- Remind the child of the goal number and ask the child to re-count.
- If the child counts too few, count the existing collection quickly and ask the child to put on another object. Say, "And that makes [x + 1]." (Continue until the correct number is reached).
- If there are too many, ask the child to remove one or more items, and then recount (again, continue). The child is allowed to take off more than one object. Count the existing collection quickly and say, "There are too many. Erase some so that we have (z)."
- If there were too many yet again, count the existing collection, stopping at the correct number, and say "That's (z). That's all we need. Take away the others."
- Demonstration. Count the number of objects, highlighting one at a time. Complete the group for the child by adding objects or stopping when z is reached. Say, "We needed (z)."

Guided Counting Sequence

- Ask the children to count out loud as they point to each object. Suggest a keeping track strategy if necessary.
- If there are still errors after this remediation, say, "Count with me," and name the keeping track strategy you will model. Have the child point to each item and say the correct counting word, thus walking the child through the counting. For Counter (object counting) activities, demonstrate the cardinality rule—repeat the last counting number, gesture in a circular motion to all the items, and say "That's how many there are in all." For "Counter to" activities, emphasize the goal number (z), saying "Z! That's what we wanted!"

Counting On

- Tell the child, "Try again" and remind them of the goal number.
- Reset the group to have the starting number x. Say, "Start at x (gesture at the x items). What's one more?" If a child responds inappropriately, model the answer and try again. If correct, ask, "What's two more?" (follow the same strategy).
- Reset the group to x. "Count with me. Start at x (gesture at the x items). Count from there." Model either reaching the goal number or counting on (adding) the correct number (keeping track on fingers of how many have been added).

Skip Counting

- If the child is "counting out" objects to x people, and they initially placed y objects out in all, give them a problem in which x > y and remind them that "each person needs y objects."
- 1st: Tell the child, "Try again," and remind them of the goal z.
- 2nd: Say, "Count with me. Count by y's." [If you are counting items, move the appropriate amount with each count.]
- 3rd: "Count by y's like this: [counting verbally only: y, 2y, 3y] Count with me." If you are counting items, move the appropriate amount with each count.

Differentiating Instruction

Below are suggestions for differentiating instruction for everyday activities.

Modifying Math Throughout the Year Activities	
Cleanup, pick a number	Have children work in pairs. One picks up objects while the other counts. Be sure that each child has a turn to count. OR have the child pick up several objects first, then count them. Counting them all at once will be easier than counting as they are being picked up.
Counting Jar	Give choices rather than have children guess at random. Make sure that one of the choices is right. List choices on the jar and have children copy their choices onto stick notes.
Counting Wand	Children work in pairs. One taps the others on the shoulder (this child is working on 1:1 correspondence) while her partner counts the taps.
Dinosaur shop – dramatic play	Have shopping lists available as in Dinosaur shop—fill orders.
Dough Shapes	Flattening play dough is difficult for some children. Encourage them to try it, but have pre-flattened pieces ready to go on old cookie sheets. Then they can just cut the shapes.
Foam Puzzles	Some children may have trouble putting the pieces in the puzzle. Either have them take the pieces out (exposure to the shapes and materials is beneficial) or just place the pieces on top of the openings until they acquire the necessary fine motor skills to complete the activity.
I Spy	Name an object and have the children guess the shape.
I'm Thinking of a Number	Display numeral cards, in order, as a visual cue.
Name Faces of Blocks	Display a shape poster and if the child is unable to tell you the name of the face, ask him or her to point to the shape on the poster.

Differentiating Instruction

Modifying Math Throughout the Year Activities	
Numerals Every Day	Have children experiment with numerals using different materials. Examples: painting on the chalkboard with water, writing them in the sand, shaping them from play dough. Another good tactile experience is to take cardstock or other heavy paper and draw the numeral in glue. Then sprinkle sand over the glue. Children can use their fingers to feel the numerals. Ask them to sing the directions for writing the numeral rather than just saying them. Music often helps children to remember, even if it is a made-up song.
Numerals in Dough–A special activity	Rolling the play dough may be difficult for some children. Have pre-rolled pieces available for children to use to form numerals.
Set the Table	Assign only one task to the child. For example, instead of having him count straws and place one at each place setting, have him count them and give them to a friend to put on the table.
Simon Says Numbers	Be sure to model and count aloud.
Snack Time	Show a model.
Time–A special activity	For children who are really struggling with this, create a schedule box. Divide a shoe box into sections or use an egg carton. Create pictures that represent activities of the day. Place the pictures, in order, in the box. Review the schedule with the child in the morning. After each activity, the child moves a marker through the box to represent that the activity has happened. After the child becomes familiar with the materials, begin to point out when the day is half over, what comes next, what happened before, and so on.

Differentiating Instruction

Below are ideas for modifying Hands On Math Center activities for children who need extra help.

Hands On Center Modifications	
As Tall As Me	Finding objects exactly as long as children's arms may be challenging. Begin with finding objects that are shorter, then taller, working up to the same length as.
Build Stairs	Some children will have difficulty putting the cubes together. If that is the case, have steps made at the centers for students to take apart.
Circles and Cans	Cover the bottom of cans with different colors of paper. Trace circles that are the same color; then children can use the colored circles as cues.
Compare Game	Limit the range of numbers in the cards (use only 1, 2, or 3 dots).
Dinosaur Shop (Fill Orders)	Create shopping lists for children. Make cards that have pictures of items for the customer to buy (8 tyrannosaurus rexes, or 3 stegosauruses). The customer picks one card and gives it to the salesperson who then uses the card as a cue to fill the order.
Dinosaur Shop (Sort and Label)	Have children only sort or only count, not both.
Draw Numbers	Provide a model for the child to copy. Have him copy the same group number several times and get a friend to check it. Eventually have him self-check.
Feely Box	Trace the shapes onto a piece of poster board and display it in the center for children to use as a cue as they try to figure out what the shapes are.
Find Groups	Prepare two shoe boxes with slots cut in the top. Begin with one for group of 2 and one for group of 3. Tape a picture of a group of 2 objects to the "2" box and a group of 3 objects to the "3"box. Have an assortment of pictures of groups of 2 or 3 that the child can sort into the appropriate box. To make the activity self-checking: line the bottom of the inside of the box with colored paper. Place a small circle of the same colored paper on the back of the pictures that belong in that box.
Find the Number (pizza)	The child lifts one container. If this is not the matching pizza, then she leaves it uncovered and checks the next one.
Get Just Enough	Instead of having the child count the initial group of objects and then count the second group of objects, have her place each object from the second group with an item from the first group, one at a time. The objective becomes straight 1 to 1 correspondence rather than counting and comparing.
Goldilocks and the Three Bears	Photocopy pages of the book and cut out the 3 bears, 3 chairs, 3 bowls, and 3 beds. Children can use the cut out pictures to match to the pictures in the book.

Differentiating Instruction

Hands On Center Modifications	
Make Groups	Provide a model for the child to copy. Have him copy the same group number several times and get a friend to check it. Eventually have him self-check.
Match Shape Sets	Limit the number of shapes the child needs to match. Create several groups of shapes so that eventually, the student will match all of them, just not all at the same time.
Memory Geometry	Modify the rules so that the child flips one card over and then keeps flipping cards, leaving them face up, until she finds the one that matches.
Memory Number (dot cards to numeral cards)	Modify the rules so that the child flips one card over and then keeps flipping cards, leaving them face-up, until she finds the one that matches.
Number Pizzas (make)	Have plates prepared ahead of time that have dots to put the correct number of pizzas. The dots will serve as a visual cue for the correct number, but be certain that the child still counts the toppings as she places them on the dots.
Pattern Block Cutouts	Begin completing the picture for the child. Place the first one or two shapes. Use a "talk aloud" strategy –"Let me see, this spot needs a shape with four equal sides. Here is a shape with four equal sides. Let me see whether it fits." Once you have placed a few pieces, let the child take over with help as needed.
Pattern Block Puzzles	Begin completing the picture for the child. Place the first one or two shapes. Use a "talk aloud" strategy –"Let me see, this spot needs a shape with four equal sides. Here is a shape with four equal sides. Let me see whether it fits." Once you have placed a few pieces, let the child take over with help as needed.
Pattern Strips	Use only very familiar shapes such as the pattern block triangle and square for patterns. Give limited choices of which pattern block should go next, either verbally, or give two blocks and have the children decide which should be next. Or glue blocks together to make "units" and have the child use the units instead of individual blocks. To provide even more support, make 1 ABA unit and 1 BAB unit. As long as they place these units in a linear fashion, they have to create a pattern.
Pizza Game 1	Play as a team. One player rolls the die, the other counts the toppings. Both players recount the toppings to check if the number is correct. Once the pizza is decorated, the team wins.
Places Scenes	Make copies of Places Scenes background, and write the target numeral on top of the scene. Then place dots on the scene to represent that number of counters. Children can either use this as a model or place their counters directly on the model. Be sure they are counting as they place their counters on the scene.
Rectangles and Boxes	Cover the bottom of the boxes with different colors of paper. Trace rectangles that are the same color; and then children can use the colored rectangles as cues.
Road Blocks	Provide templates for the children to use in building their roads. On a piece of construction paper: write the numeral 5, draw a road that is five blocks long (be sure it is the exact size). The children can build their roads on top of or next to the model.

Differentiating Instruction

Hands On Center Modifications	
Set the Table	Make and laminate paper placemats on which you have traced each of the objects that needs to be on the mat.
Shape Step	Have children hold the shape from the shape set that corresponds to the shape that they are to step on. This serves as a visual cue. The shapes for the floor could be cut from felt.
Straw Shapes	Provide a model for the child to copy. Two options: (a) make a model, and glue it to a piece of paper so that it will stay together, or (b) draw the shape (exact size), and have the child place the straws over the drawing to make the shape.
Stringing Beads	Glue beads together to make "units" and have children build patterns with these. To provide even more support, make 1 ABA unit and 1 BAB unit. No matter how they string it, they have to create a pattern!!
What's the Missing Step?	Display cards with pictures of the steps paired with the numeral. Be sure that they are in the correct order. Draw steps made of cubes on index cards, one step on each card. On the bottom of the card, write the correct numeral. For example, draw a step with three cubes and write the numeral 3 on the bottom of that card. Students can refer to these cards when answering the questions. The child gets three chances to get it right
X-Ray Vision 1	Display another set of counting cards in order at the center. Students can use these as a visual cue to identify the card.
X-Ray Vision 2	Display another set of counting cards in order at the center. Students can use these as a visual cue to identify the card.

Differentiating Instruction

Following are ideas for modifying Small Group activities for students with special needs.

Small Group Modifications	
Build Cube Stairs	Instead of asking how many cubes should be next, show the choices and ask which one should be next, still focusing on the concept of "one more." Make "stair cards" that have pictures of steps with 1, 2, 3, 4, 5, and so on, blocks (one step on a card). Put them in order and display them so the child can use these as a visual aid in answering the questions.
Compare Game	Limit the range of numbers in the cards (use only 1, 2, or 3 dots).
Demonstrate Counting	Rather than hiding the counters in your hand, hold them for the child to see. Then say you need to count to see how many there are. Count and move them from your hand to the child's hand. Then have the child count them from her hand to yours, with help as needed.
Dinosaur Shop (Fill Orders)	Create a shopping list for the children. Make cards that have pictures of items for the customer to buy (8 tyrannosaurus rexes, or 3 stegosauruses). You are the customer and pick one card. The group then uses the card as a cue to fill the order.
Dinosaur Shop (Sort and Label)	Be sure to have the child sort first and finish sorting before beginning counting.
Feely Box (Shapes)	Reduce the number of choices from the second shape set. Or show the child a shape from the second shape set and ask him to show you "thumbs-up" if he thinks that is the shape in the feely box and "thumbs-down" if he doesn't. Be certain to ask him for justifications.
Find Groups	Give children groups to choose from, rather than ask them to tell you how many are in a group. Show a group of three and a group of four, and ask them to point to the group of four.
Get Just Enough	Instead of having the child count the initial group of objects and then count the second group of objects, have her place each object from the second group with an item from the first group, one at a time. The objective becomes straight 1 to 1 correspondence rather than counting and comparing.
Guess My Rule	Give choices of the rule (has 2 sides or has 3 sides) or state rules as choices and have children show "thumbs up" or "thumbs down" to show agreement or disagreement.
How Many Now?	Use cards that have numerals on one side and dots on the other. The children can use the side with the dots as a visual aid.
How Many Now? Hidden version	Show children how to count on their fingers. Before hiding 3 chips, have the child put out three fingers, then one more when you add one more chip. Or show the numeral card representing the number of hidden chips and let children use that to figure out how many chips there are altogether.
Is It or Not?	As you name a shape, display that shape and keep it displayed (on the chalkboard or somewhere nearby) as you show other shapes and ask "Is it or not?" The displayed shape serves as a visual cue for children.
Length Riddles	Present an array of objects that you will make up riddles about (3–5 objects). As you say the riddle, give the child a measuring device made of the correct number of cubes. The child uses the cube tower to find the object.

Differentiating Instruction

Small Group Modifications

Activity	Modification
Make Groups	Be certain to provide a model for each group you ask the child to create. Continue to work on counting.
Make Number Pizzas	Have plates prepared ahead of time that have dots to put the correct number of pizzas. The dots will serve as a visual cue for the correct number, but be certain that the child still counts the toppings as she places them on the dots.
Match and Name Shapes	Continue to ask children to name a shape; that is the ultimate objective. However, if a child does not know the shape or is struggling with the task, give him verbal choices of shape names. If the child is still struggling, present two different shapes, and ask him to show you to a specific shape. For example, display a square and a trapezoid, and ask him to pick up the square. Then have him feel the shape and describe it to you.
Memory Number or Geometry	Flip over one card and then place it in front of the child. The child flips over another card; if it matches, he or she keeps the cards. If it doesn't, leave it faceup and flip another card. Keep flipping cards until he or she finds a match.
Mr. Mixup (Counting)	Give the child cards to use as visual cues in catching Mr. Mixup's mistakes. As Mr. Mixup is counting, have the child point to the number cards. When Mr. Mixup makes a mistake, show how the number that Mr. Mixup said is not the numeral to which the child is pointing.
Pattern Strips	Give limited choices of which pattern block should go next, either verbally, or give two blocks and have the children decide which should be next. Be sure to "warm up" using auditory patterns, which are often easier for children with processing difficulties to grasp.
Pizza Game 1	Play as a team. One player rolls the die, the other counts the toppings. Both players recount the toppings to check if the number is correct. Once the pizza is decorated, the team wins.
Shape Step	Have children hold the shape from the shape set they are to step on. This serves as a visual cue.
Straw Shapes	Provide a model for the child to copy. Two options: (a) make a model and glue it to a piece of paper so that it will stay together, or (b) draw the shape (exact size) and have the child place the straws over the drawing to make the shape.
What's the Missing Card?	Provide another set of cards for the child to use to help her determine which card is missing from the original set.
What's the Missing Step?	Display cards with pictures of the steps paired with the numeral. Be sure that they are in the correct order. Draw steps made of cubes on index cards, one step on each card. On the bottom of the card, write the correct numeral. For example, draw a step with three cubes and write the numeral 3 on the bottom of that card. Children can refer to these cards when answering the questions.
X-Ray Vision 1	Display another set of counting cards on the chalkboard or other nearby surface. Children can use these as a visual cue to identify the card.

Differentiating Instruction

Following are ideas for modifying Whole Group activities for struggling students.

Whole Group Modifications

Activity	Modification
As Long As My Arm	Finding objects exactly as long as children's arms may be challenging. Begin with finding objects that are shorter, then taller, working up to "as long as my arm."
Compare Number Pizzas	Be certain that the toppings are displayed in an easily recognizable array. You may want to glue the toppings onto the plates so that you can hold them up for the class to see without the toppings falling off.
Compare Snapshots	When you uncover the hidden plate, have the child put out that number of fingers. Once he has the correct number of fingers out, then cover the plate and ask him to compare.
Dancing Patterns	At first, stick with A B patterns (clap-kick, clap-kick or jump-turn, jump-turn).
Demonstrate Counting	Rather than hiding the counters in your hand, hold them for the child to see. Then say you need to count to see how many there are. Count and move them from your hand to the child's hand. Then have the child count them from her hand to yours, with help as needed.
Feely Box (Shapes)	Reduce the number of choices from the second shape set. OR show the child a shape from the second shape set and ask him to show you "thumbs-up" if he thinks that is the shape in the feely box and "thumbs-down" if he doesn't. Be certain to ask him for justifications.
Find Groups	To solicit active engagement in the lesson from children who have not yet grasped the concept of group — prior to the lesson, spend a minute individually introducing the term "group" and provide several examples of groups in the room. The child can then point out that group during the lesson. Be sure to have this child take her turn early, so that nobody "steals" the group she was going to point out.
Goldilocks and the Three Bears	Photocopy pages of the book and cut out the 3 bears, 3 chairs, 3 bowls, and 3 beds. Children can use the cut out pictures to match to the pictures in the book.
Guess My Rule	Give choices of the rule (has 2 sides or has 3 sides) or state rules as choices and have children show "thumbs-up" or "thumbs-down" to show agreement or disagreement.
How Many Now?	Use cards that have numerals on one side and dots on the other.
How Many Now? (Hidden Version)	Show children how to count on their fingers. Before hiding 3 chips, have the child put out three fingers, then one more when you add one more chip. OR show the numeral card representing the number of hidden chips and let children use that to figure out how many chips there are altogether.
I'm Thinking of a Number (Length)	Display numeral cards as a visual cue.
Is It or Not?	As you name a shape, display that shape, and keep it displayed (on the board or somewhere nearby) as you show other shapes, and ask "Is it or not?" The displayed shape serves as a visual cue for children.
Listen and Copy	First, have children copy simple patterns, without counting. Only work on them listening and copying the pattern.
Listen and Count	Have children hold up a finger each time you drop a marble, and then they count their fingers to determine the number of marbles dropped.
Make Groups	Have children work with a partner. Or show groups for the children to choose from. Point out a group of two trees, and ask whether it is a group of two or a group of three.

Differentiating Instruction

Whole Group Modifications	
Make Number Pizzas	Have plates prepared ahead of time that have dots to put the correct number of pizzas. The dots will serve as a visual cue for the correct number, but be certain that the child still counts the toppings as he or she places them on the dots.
Measuring Length	Mark a ruler at the inch marks with a heavy marker, making it easier to see these markings. Put a bright red line at the zero point of the measuring device to prompt children where to place the device.
Mr. Mixup (Comparing)	Begin by showing the two towers next to each other, asking which has more blocks. Then move one to the chair and one to the floor. Ask again which has more blocks, and then bring the towers back together.
Mr. Mixup (Counting)	Display numeral cards in order for children to use as visual cues in catching Mr. Mixup's mistakes.
Mr. Mixup (Counting 2)	Use numeral cards as visual cues to remind children what number Mr. Mixup is supposed to stop at or which number comes next.
Mr. Mixup (Shapes)	Begin by having Mr. Mixup say that shapes are the same when they are different. Once children are more firm on the vocabulary of shapes, do the activity as directed.
Mr. Mixup's Measuring Mess	Tell the children how long the rope is supposed to be. "I know this rope is 5 cubes long, but I can't seem to figure out how to prove it."
Number Jump	Have children work in partners, one jumps, the other counts. And then switch. This way, the child only needs to think about doing one thing at a time (jumping or counting, but not both).
Number Me	Be certain to model the action for the student. When you say "show me your 2 hands," show your 2 hands. Also be sure that the student knows his age before asking in a group setting.
Numerals	Have children experiment with numerals using different materials. Examples: painting on the chalkboard with water, writing them in the sand, shaping them from play dough. Another good tactile experience is to take cardstock or other heavy paper and draw the numeral in glue. Then sprinkle sand over the glue. Children can use their fingers to feel the numerals. Ask them to sing the directions for writing the numeral rather than just saying them. Music often helps children to remember, even if it is a made-up song.
Order Cards	Give choices of what should come next, either verbally or by showing Numeral Cards.
Pattern Strips	Give limited choices of which pattern block should go next, either verbally, or give two blocks, and have the children decide which should be next.
Places Scenes	Make copies of places scenes background, and write the target numeral on top of the scene. Then place dots on the scene to represent that number of counters. Children can either use this as a model or place their counters directly on the model. Be sure they are counting as they place their counters on the scene.
Rectangles and Boxes	Outline each face of the box with a large marker. This should make it easier for children to distinguish the shape.
Shape Flip Book	Initially remove some pages from the book, gradually adding them back as children become more familiar with the activity.
Shape Show	Use numeral cards to label the sides as children count, then they can use the visual cue. Sing the properties of the shapes, rather than saying them. Many children find things easier to remember when put to music. Preview vocabulary terms that may be new to the child (shape names).
Shape Step	Have children hold the shape from the shape set that they are to step on. This serves as a visual cue.

Differentiating Instruction

Whole Group Modifications	
Simon Says Numbers	Use numeral cards to display the number of times the children are to do the activity. For example, "Jump 4 times." Display the numeral card 4 as you jump with the children, counting aloud. The child can then focus on counting and jumping, rather than on trying to remember the number 4.
Snapshots	Prior to the lesson — Begin with 1 counter, and repeat several times until the student is confident on what the expectation is for the activity. Be sure to keep the counters in an easily recognizable array (a straight line for up to 3, dice patterns for 4–6, and so on). You may also ask the child to copy your plate if she is not able to say the number words. Give her a plate and counters (even better, the right number of counters) and have her replicate the array you have shown her.
Snapshots 2	Give children answers to choose from, either with numeral cards, or verbally. Or make plates ahead of time by glueing counters to plates and having the child use the pre-made plates to choose which one looks like the plate you uncovered.
The Very Hungry Caterpillar	Instead of asking "how many ______?" ask "the caterpillar ate three _____?" The child still has to count, but not retrieve the vocabulary "three."
Trapezoid	Preview the vocabulary with the student. Draw a "zoid" (a small furry creature, or anything else you can draw) "trapped" instead – hence the word trapezoid.
What's the Missing Step?	Display numeral cards for children to use as a visual cue.
What's This Step?	Add counting cards to label each step.
Where's My Number?	Two options: (a) keep your hands open so the child can count the counters; (b) instead of numeral cards, use dot cards so the child can match without having to know what the numeral represents.
X-Ray Vision 1	Display another set of counting cards on the chalkboard or other nearby surface. Children can use these as a visual cue to identify the card.
X-Ray Vision 2	Display another set of counting cards on the chalkboard or other nearby surface. Children can use these as a visual cue to identify the card.

Reading and Using Books

Research tells us that the more we read books to children, and the more we talk to them about the books and about mathematics using complex language, the more likely they will become good readers themselves.

- Encourage children to look carefully at the book itself: cover, the end papers, and the title page.
- Be sure to note the dedication page and the names of the author and illustrator.
- The first reading of a book should be done without interjecting questions or comments. This helps children develop a sense of narrative.
- Ask prediction or "what if?" questions before or during the second reading.
- Feel free to interrupt stories with questions and comments, as long as they are relevant to the story. Research shows that they help children maintain attention and understand and learn from the story.
- Hearing stories repeatedly helps children understand them more deeply and learn how to relate them to their lives.
- Occasionally point out key features of print (left-to-right/top-to-bottom progression, spaces between words, punctuation) as you read to children.
- Vary the size of the group when you read to children.
- Sit so that all children are comfortable and can easily see the illustrations. If you have extra copies of the book, share these with the group.
- Read slowly and smoothly. Be dramatic! Be silly! Play with different voices, make funny or scary sounds, shout, whisper, pause dramatically or speed up when there's a lot of action, and slow down at the end of a book.
- Take time to introduce a new book to children. Ask children to predict who or what the book will be about based on the cover or title page. Begin with a movement or finger play exercise related to the story. You could also take a picture walk through the book before reading the words.
- Help children make connections between stories and their own experiences.
- Ask open-ended questions to keep listeners involved with the book: What do you notice about this illustration? How did that page make you feel? Why do you think she or he did that?
- Build vocabulary by discussing one or two words that might be new to children.
- At the end of a book, thank children for being good listeners.
- Get to know books before sharing them with children.

Math Throughout the Year

Classrooms are filled with opportunities to include mathematics. Use children's spontaneous questions and observations as catalysts for mathematical exploration. By encouraging children's ideas and sharing of math stories, you can support their efforts to solve problems in different ways.

Throughout the days, weeks, months and year there are simple ways of incorporating math ideas and concepts. These become the basis of our program's Math Throughout the Year activities.

Some routines can be set up early in the year and continue throughout the year. Some of these will be done nearly every day. Others are done when they "fit" but not as a separate unit of instruction. By incorporating and emphasizing the math in these activities, you can develop and reinforce concepts. By connecting daily events to mathematical concepts, children may come to see math as an integral part of life and the classroom environment.

Managing Centers

Consider a reasonable limit to the number of children in each learning center. Hang a numeral, at first with dots on it as well, that indicates the number of children at each center. If you like, hang hooks so that children can hang their nametags when they are working at the center. Then they can talk about how many are there and how many could still join in the center.

Attendance

Attendance routines also present opportunities to show children the need for developing strategies for counting, organizing, and recording data.

Beginning of the Year:

With all the children present, begin counting everyone. Children will hear you say each number as you touch or point to each child. They may begin chiming in as you count. This provides a way of informally assessing counting and number language skills. Do not be surprised if most children cannot count the entire class with you. Do this for several days to establish the idea of what attendance is and the number you need to account for. If children are missing, acknowledge this and add them to the count.

Concrete Representation:

Give each child a connecting cube. (The cubes can be in a container, on top of name or picture cards, or passed out at Whole Group.) Explain that you have been counting "how many" children are present and now want to see this number. Have the children put their cubes together to create a long train/stick, counting each cube as it is added. By putting the cubes together, the group will be able to count and see how many children are present. If a child is absent, you can say, " Today there are 19 here but Juanita is not here, so we'll add a cube for her. Now we have 20."

Sorting and Graphing:

Ask children to group themselves by a particular attribute. Clothing is good for this. For example, children with jeans could sit near one adult while those without jeans could sit by another adult. Each group can be counted and the amounts compared.

Create a simple graph

to record the results. Later, create a picture question on a board, such as a zipper and no–zipper. As the children arrive, they put their name cards under the correct answer. This helps children see math as connected to the problems they are interested in.

Nametags:

Using nametags (with or without pictures) for attendance not only helps children with pre-reading skills, but is also an easy way for children to see how many and which children are present. Children can find their tag and put into a container. These can be laid out and counted like the cubes or put on a bulletin board. The class counts the number of names in the "present" section. Knowing the class total, they can figure out how many are absent. You could help them with this problem by asking who is missing, adding their nametags to the "absent" side .

Meals

Setting the table:

Table Helpers can get the tables and/or snacks ready for the class. Using the attendance count to set tables helps connect the quantity with a purpose. Counting will occur as they decide how many of each item is needed. The matching of milk cartons to children and straws to cartons provides opportunities for the development of the concept of one-to-one correspondence.

Counting servings

As children serve, or are served, quantities can be described or counted. Comparison language, such as more or less, can be useful at meal times. Some foods are easily counted. Some children will count each item, "One, two, three pretzels," while others will begin to visually recognize quantities and put the correct amount at their place.

Mealtime chats:

Snacks and meals provide opportunities to converse with small groups of children. Discussing the food can lead to integrating mathematical concepts. Children can describe the food using color, temperature, shape, texture, and food groups. Using attributes helps children see how items are similar or different and becomes a basis for sorting. These attributes, as well as likes/dislikes, can be graphed.

Clean Up

- Pick a number: Each child picks up that many objects.
- Use locations: Request everyone put away something that belongs "down," "up," "high," "low", etc.
- Shape clean-up: "Everyone look for circles to put away."
- Rolling blocks: In the block area, challenge the children to move the blocks without carrying them. Someone usually discovers that circles (cylinders) can roll.
- Clean and Sort: Children can sort materials by what area of the room they belong in and locations on shelves, in cabinets, etc.

Transitioning

Lining Up:

Moving a group of children between activities or points is challenging. Safety concerns often call for children to move together, sometimes in a line. Using a variety of game-like activities can help emphasize language and math concepts while creating fun in an often-tedious task. Some ideas for moving children to another activity, or for lining up, include:

- Use attributes such as clothing color, feature (stripes), age, birthdays. As the children become acquainted with this idea, use more than one descriptive attribute. For example, "Children with brown hair and black on their clothes can line up next."
- Lining up by height provides a visual seriation exercise. This method usually requires adult assistance. If done in groups of five, it provides a visual representation of seriation.
- Touch a shape. Describe a specific shape. As children line up (five at a time), they find that shape, touch one, and line up.
- Create a pattern. An adult calls children in a particular order. For example, girl, girl, boy. Talk about the pattern as it's being created. Children will begin to "see" the pattern and put themselves into one.

Counting to Check:

Teachers often count the number of children in line as a safety check. As children begin to gain confidence with numbers, let a helper count using a counting wand. The children can also "count off." This requires them to listen to the number before them and say the next number. These situations also allow assessment opportunities. Listen for how high a child can count, who is able to assist someone who is stuck, and who can answer "how many?"

When children are lining up, discussions about their positions in line can be the basis for the development of ordinal numbers. Occasionally count with ordinal numbers, "First, second, etc."

Exit games:

Before exiting the classroom, many teachers play transition type games to focus the children's attention. Simon Says type games are often used. These provide a quick assessment of vocabulary and listening skills. They can also reinforce mathematical concepts and provide informal assessment opportunities. Some examples include:

- Counting movements or motions. For example, "Jump five times" or "Clap six times." Early in the year, count the action with the children. Later, do the movements but model and explain how to count silently. Children who understand the number of motions will stop but others will continue doing the motion.
- Use spatial relations. Have the children put their hands "on top of," "behind your back," "under your knee," "next to." These terms work well with touching body parts.
- Use descriptions of lines. Children can make their arms horizontal, vertical, slanted, and parallel with their hands, or if the room allows, other parts of their bodies.

Physical Activity

Large muscle activities are an important part of young children's development. Programs generally include some sort of large muscle activity, either organized or free choice, indoors or out. These types of activities are a natural place to integrate mathematical concepts.

Counting motions

If children are engaging in a type of movement, help them count how many steps or do a certain number. If children are using a play structure as a rabbit home, encourage them to hop seven times.

Spatial relations:

Use comparison and spatial relation terms to describe what you observe children doing and to provide directions.

Games:

Encourage games such as hopscotch that provide opportunities to work with numbers and patterning.

Voting and Graphing

Pre-K classrooms provide many opportunities for visual representations of information. Creating a visual representation provides opportunities to organize and interpret information. It also provides a way of conveying information about your class to parents and others.

Visual representations can be a variety of things. Teachers are good at creating bulletin boards with information about each child or each child's interpretation of a question or idea. Think of opportunities for extending these representations into graphs.

People Graphs:

Ask a question such as, "Does your coat have buttons?" Write "yes" and "no" on two pieces of construction paper. The children then line up behind the correct sign. (See more ideas and explanation in the attendance section.)

Food Graphs:

Depicting the children's like or dislike of a food item is easy to do with young children. One teacher uses nametag stickers and writes each child's name or has the child write it. Two sheets of construction paper are used, one with a sad face and one with a happy face. As the children are done tasting the food item, they find the sticker with their name and put it on the appropriate sheet. The class can then discuss the graphs.

Gingerbread Graphs:

Use faces or gingerbread style people to graph how many people are in each child's family. Other ideas are to include labels from food such as different kinds of soups or cereal boxes. Children could add their name or even a plastic spoon to indicate their choice.

...and everywhere, all the time!

- Math is all around us. Thinking about it everyday helps children make connections in their lives. Stories and songs, which are key components of early childhood programs, are excellent ways to incorporate and develop mathematical ideas. Integrating math should occur throughout the day, week, and year as with any subject.
- Engage children in mathematics deeply. As you work with this program and set up your classroom environment, think about encouraging children's questioning and problem solving skills. Don't always provide the answers. Children can solve many problems that occur in a classroom. Let them make suggestions. You will be surprised at what they may think of.

*Children follow natural developmental progressions in learning. Curriculum research has revealed sequences of activities that are effective in guiding children through these levels of thinking. These developmental paths are the basis for **Building Blocks** learning trajectories.*

Learning Trajectories for Primary Grades Mathematics

Learning trajectories have three parts: a mathematical goal, a developmental path along which children develop to reach that goal, and a set of activities matched to each of the levels of thinking in that path that help children develop the next higher level of thinking. The **Building Blocks** learning trajectories give simple labels, descriptions, and examples of each level. Complete learning trajectories describe the goals of learning, the thinking and learning processes of children at various levels, and the learning activities in which they might engage. This document provides only the developmental levels.

The following provides the developmental levels from the first signs of development in different strands of mathematics through approximately age 8. Research shows that when teachers understand how children develop mathematics understanding, they are more effective in questioning, analyzing, and providing activities that further children's development than teachers who are unaware of the development process. Consequently, children have a much richer and more successful math experience in the primary grades.

Each of the following tables, such as "Counting," represents a main developmental progression that underlies the learning trajectory for that topic.

For some topics, there are "subtrajectories"—strands within the topic. In most cases, the names make this clear. For example, in Comparing and Ordering, some levels are "Composer" levels and others are building a "Mental Number Line." Similarly, the related subtrajectories of "Composition" and "Decomposition" are easy to distinguish. Sometimes, for clarification, subtrajectories are indicated with a note in italics after the title. For example, *Parts* and *Representing* are subtrajectories within the Shape Trajectory.

Frequently Asked Questions (FAQ)

1. *Why use learning trajectories?* Learning trajectories allow teachers to build the mathematics of children— the thinking of children as it develops naturally. So, we know that all the goals and activities are within the developmental capacities of children. We know that each level provides a natural developmental building block to the next level. Finally, we know that the activities provide the mathematical **Building Blocks** for school success.
2. *When are children "at" a level?* Children are at a certain level when most of their behaviors reflect the thinking—ideas and skills—of that level. Often, they show a few behaviors from the next (and previous) levels as they learn. Most levels are levels of thinking. However, some are merely "levels of attainment" and indicate a child has gained knowledge. For example, children must learn to name or write more numerals, but knowing more numerals does not require deeper or more complex thinking.
3. *Can children work at more than one level at the same time?* Yes, although most children work mainly at one level or in transition between two levels (naturally, if they are tired or distracted, they may operate at a much lower level). Levels are not "absolute stages." They are "benchmarks" of complex growth that represent distinct ways of thinking.
4. *Can children jump ahead?* Yes, especially if there are separate "sub-topics." For example, we have combined many counting competencies into one "Counting" sequence with sub-topics, such as verbal counting skills. Some children learn to count to 100 at age 6 after learning to count objects to 10 or more, some may learn that verbal skill earlier. The sub-topic of verbal counting skills would still be followed.
5. *How do these developmental levels support teaching and learning?* The levels help teachers, as well as curriculum developers, assess, teach, and sequence activities. Through planned teaching and also encouraging informal, incidental mathematics, teachers help children learn at an appropriate and deep level.
6. *Should I plan to help children develop just the levels that correspond to my children's ages?* No! The ages in the table are typical ages children develop these ideas. But these are rough guides only—children differ widely. Furthermore, the ages below are lower bounds on what children achieve without instruction. So, these are "starting levels" not goals. We have found that children who are provided high-quality mathematics experiences are capable of developing to levels one or more years beyond their peers.

Developmental Levels for Counting

The ability to count with confidence develops over the course of several years. Beginning in infancy, children show signs of understanding numbers. With instruction and number experience, most children can count fluently by age 8, with much progress in counting occurring in kindergarten and first grade. Most children follow a natural developmental progression in learning to count with recognizable stages or levels. This developmental path can be described as part of a learning trajectory.

Age Range	Level Name	Level	Description
1–2	Pre-Counter (Verbal)	1	At the earliest level a child shows no verbal counting. The child may name some number words with no sequence.
1–2	Chanter (Verbal)	2	At this level, a child may sing-song or chant indistinguishable number words.
2	Reciter (Verbal)	3	At this level, the child may verbally count with separate words, but not necessarily in the correct order.
3	Reciter (10)	4	A child at this level may verbally count to 10 with some correspondence with objects. He or she may point to objects to count a few items, but then lose track.
3	Corresponder	5	At this level, a child may keep one-to-one correspondence between counting words and objects—at least for small groups of objects laid in a line. A corresponder may answer "how many" by recounting the objects, starting over with one each time.
4	Counter (Small Numbers)	6	At around 4 years of age, the child may begin to count meaningfully. He or she may accurately count objects in a line to 5 and answer the "how many" question with the last number counted. When objects are visible, and especially with small numbers, the child begins to understand cardinality (that numbers tell how many).
4	Producer (Small Numbers)	7	The next level after counting small numbers is to count out objects to 5. When asked to show four of something, for example, this child may give four objects.
4	Counter (10)	8	This child may count structured arrangements of objects to 10. He or she may be able to write or draw to represent 1–10. A child at this level may be able to tell the number just after or just before another number, but only by counting up from 1.
5	Counter and Producer—Counter to (10+)	9	Around 5 years of age, a child may begin to count out objects accurately to 10 and then beyond to 30. He or she has explicit understanding of cardinality (that numbers tell how many). The child may keep track of objects that have and have not been counted, even in different arrangements. He or she may write or draw to represent 1 to 10 and then 20 and 30, and may give the next number to 20 or 30. The child also begins to recognize errors in others' counting and is able to eliminate most errors in his or her own counting.
5	Counter Backward from 10	10	Another milestone at about age 5 is being able to count backward from 10 to 1, verbally, or when removing objects from a group.
6	Counter from N (N+1, N−1)	11	Around 6 years of age, the child may begin to count on, counting verbally and with objects from numbers other than 1. Another noticeable accomplishment is that a child may determine the number immediately before or after another number without having to start back at 1.
6	Skip Counting by 10s to 100	12	A child at this level may count by 10s to 100 or beyond with understanding.
6	Counter to 100	13	A child at this level may count by 1s to 100. He or she can make decade transitions (for example, from 29 to 30) starting at any number.
6	Counter On Using Patterns	14	At this level, a child may keep track of a few counting acts by using numerical patterns (spatial, auditory, or rhythmic).
6	Skip Counter	15	At this level, the child can count by 5s and 2s with understanding.
6	Counter of Imagined Items	16	At this level, a child may count mental images of hidden objects to answer, for example, "how many" when 5 objects are visible and 3 are hidden.
6	Counter On Keeping Track	17	A child at this level may keep track of counting acts numerically, first with objects, then by counting counts. He or she counts up one to four more from a given number.
6	Counter of Quantitative Units	18	At this level, a child can count unusual units, such as "wholes" when shown combinations of wholes and parts. For example, when shown three whole plastic eggs and four halves, a child at this level will say there are five whole eggs.
6	Counter to 200	19	At this level, a child may count accurately to 200 and beyond, recognizing the patterns of ones, tens, and hundreds.
7	Number Conserver	20	A major milestone around age 7 is the ability to conserve number. A child who conserves number understands that a number is unchanged even if a group of objects is rearranged. For example, if there is a row of ten buttons, the child understands there are still ten without recounting, even if they are rearranged in a long row or a circle.
7	Counter Forward and Back	21	A child at this level may count in either direction and recognize that sequence of decades mirrors single-digit sequence.

Developmental Levels for Comparing and Ordering Numbers

Comparing and ordering sets is a critical skill for children as they determine whether one set is larger than another in order to make sure sets are equal and "fair." Prekindergartners can learn to use matching to compare collections or to create equivalent collections. Finding out how many more or fewer in one collection is more demanding than simply comparing two collections. The ability to compare and order sets with fluency develops over the course of several years. With instruction and number experience, most children develop foundational understanding of number relationships and place value at ages four and five. Most children follow a natural developmental progression in learning to compare and order numbers with recognizable stages or levels. This developmental path can be described as part of a learning trajectory.

Age Range	Level Name	Level	Description
2	Object Corresponder	1	At this early level, a child puts objects into one-to-one correspondence, but may not fully understand that this creates equal groups. For example, a child may know that each carton has a straw, but does not necessarily know there are the same numbers of straws and cartons.
2	Perceptual Comparer	2	At this level, a child can compare collections that are quite different in size (for example, one is at least twice the other) and know that one has more than the other. If the collections are similar, the child can compare very small collections.
2–3	First-Second Ordinal Counter	3	At this level the child can identify the "first" and often "second" object in a sequence.
3	Nonverbal Comparer of Similar Items	4	At this level, a child can identify that different organizations of the same number are equal and different from other sets (1–4 items). For example, a child can identify ••• and •˙• as equal and different from •• or ˙•.
3	Nonverbal Comparer of Dissimilar Items	5	At this level, a child can match small, equal collections of dissimilar items, such as shells and dots, and show that they are the same number.
4	Matching Comparer	6	As children progress, they begin to compare groups of 1–6 by matching. For example, a child gives one toy bone to every dog and says there are the same number of dogs and bones.
4	Knows-to-Count Comparer	7	A significant step occurs when the child begins to count collections to compare. At the early levels, children are not always accurate when a larger collection's objects are smaller in size than the objects in the smaller collection. For example, a child at this level may accurately count two equal collections, but when asked, says the collection of larger blocks has more.
4	Counting Comparer (Same Size)	8	At this level, children make accurate comparisons via counting, but only when objects are about the same size and groups are small (about 1–5 items).
5	Counting Comparer (5)	9	As children develop their ability to compare sets, they compare accurately by counting, even when a larger collection's objects are smaller. A child at this level can figure out how many more or less.

Age Range	Level Name	Level	Description
5	Ordinal Counter	10	At this level, a child identifies and uses ordinal numbers from "first" to "tenth." For example, the child can identify who is "third in line."
6	Counting Comparer (10)	11	This level can be observed when the child compares sets by counting, even when a larger collection's objects are smaller, up to 10. A child at this level can accurately count two collections of 9 items each, and says they have the same number, even if one collection has larger blocks.
6	Mental Number Line to 10	12	As children move into this level, they begin to use mental images and knowledge of number relationships to determine relative size and position. For example, a child at this level can answer which number is closer to 6, 4 or 9 without counting physical objects.
6	Serial Orderer to 6+	13	At this level, the child orders lengths marked into units (1–6, then beyond). For example, given towers of cubes, this child can put them in order, 1 to 6.
7	Place Value Comparer	14	Further development is made when a child begins to compare numbers with place value understanding. For example, a child at this level can explain that "63 is more than 59 because six tens is more than five tens, even if there are more than three ones."
7	Mental Number Line to 100	15	Children demonstrate the next level when they can use mental images and knowledge of number relationships, including ones embedded in tens, to determine relative size and position. For example, when asked, "Which is closer to 45, 30 or 50?" a child at this level may say "45 is right next to 50, but 30 isn't."
8+	Mental Number Line to 1,000s	16	At about age 8, children may begin to use mental images of numbers up to 1,000 and knowledge of number relationships, including place value, to determine relative size and position. For example, when asked, "Which is closer to 3,500—2,000 or 7,000?"a child at this level may say "70 is double 35, but 20 is only fifteen from 35, so twenty hundreds, 2,000, is closer."

Learning Trajectories

Developmental Levels for Recognizing Number and Subitizing (Instantly Recognizing)

The ability to recognize number values develops over the course of several years and is a foundational part of number sense. Beginning at about age two, children begin to name groups of objects. The ability to instantly know how many are in a group, called *subitizing,* begins at about age three. By age eight, with instruction and number experience, most children can identify groups of items and use place values and multiplication skills to count them. Most children follow a natural developmental progression in learning to count with recognizable stages or levels. This developmental path can be described as part of a learning trajectory.

Age Range	Level Name	Level	Description
2	Small Collection Namer	1	The first sign occurs when the child can name groups of 1 to 2, sometimes 3. For example, when shown a pair of shoes, this young child says, "two shoes."
3	Maker of Small Collections	2	At this level, a child can nonverbally make a small collection (no more than 4, usually 1 to 3) with the same number as another collection. For example, when shown a collection of 3, the child makes another collection of 3.
4	Perceptual Subitizer to 4	3	Progress is made when a child instantly recognizes collections up to 4 and verbally names the number of items. For example, when shown 4 objects briefly, the child says "4."
5	Perceptual Subitizer to 5	4	This level is the ability to instantly recognize collections up to 5 and verbally name the number of items. For example, when shown 5 objects briefly, the child says "5."
5	Conceptual Subitizer to 5+	5	At this level, the child can verbally label all arrangements to about 5, when shown only briefly. For example, a child at this level might say, "I saw 2 and 2, and so I saw 4."
5	Conceptual Subitizer to 10	6	This step is when the child can verbally label most arrangements to 6 shown briefly, then up to 10, using groups. For example, a child at this level might say, "In my mind, I made 2 groups of 3 and 1 more, so 7."
6	Conceptual Subitizer to 20	7	Next, a child can verbally label structured arrangements up to 20 shown briefly, using groups. For example, the child may say, "I saw 3 fives, so 5, 10, 15."
7	Conceptual Subitizer with Place Value and Skip Counting	8	At this level, a child is able to use groups, skip counting, and place value to verbally label structured arrangements shown briefly. For example, the child may say, "I saw groups of tens and twos, so 10, 20, 30, 40, 42, 44, 46...46!"
8+	Conceptual Subitizer with Place Value and Multiplication	9	As children develop their ability to subitize, they use groups, multiplication, and place value to verbally label structured arrangements shown briefly. At this level, a child may say, "I saw groups of tens and threes, so I thought, 5 tens is 50 and 4 threes is 12, so 62 in all."

Developmental Levels for Numerals

Age Range	Level Name	Level	Description
3	Quantity Representer	1	Represents and recalls sets with pictographic, iconic, representations of quantity. However, they may not incorporate written symbols into their own acting and thinking.
4	Numeral Representer	2	Can match small sets (1-5) with the corresponding numbers and represent and recall the size of sets using those numerals.
4–5	Functional Numeral User	3	Can use numerals to represent and communicate quantity. For example, can use numerals to remember results of counting or to compare quantities
6	Teen/Ten + Recognizer	4	Understand that a teen number is composed of a ten and one, two, three, ..., seven, eight or nine ones.
6–7	Decade Number Identifier	5	Understands decade words (e.g., sixty = 6 tens).
7	Digit Identifier	6	Understand that the two digits of a two-digit number represent amounts of tens and ones. In 29, for example, the 2 represents two tens and the 9 represents nine ones.

Developmental Levels for Composing (Knowing Combinations of Numbers)

Composing and decomposing are combining and separating operations that allow children to build concepts of "parts" and "wholes." Most prekindergartners can "see" that two items and one item make three items. Later, children learn to separate a group into parts in various ways and then to count to produce all of the number "partners" of a given number. Eventually children think of a number and know the different addition facts that make that number. Most children follow a natural developmental progression in learning to compose and decompose numbers with recognizable stages or levels. This developmental path can be described as part of a learning trajectory.

Age Range	Level Name	Level	Description
4	Pre-Part-Whole Recognizer	1	At the earliest levels of composing, a child only nonverbally recognizes parts and wholes. For example, when shown 4 red blocks and 2 blue blocks, a young child may intuitively appreciate that "all the blocks" includes the red and blue blocks, but when asked how many there are in all, the child may name a small number, such as 1.
5	Inexact Part-Whole Recognizer	2	A sign of development is that the child knows a whole is bigger than parts, but does not accurately quantify. For example, when shown 4 red blocks and 2 blue blocks and asked how many there are in all, the child may name a "large number," such as 5 or 10.
5	Composer to 4, then 5	3	At this level, a child knows number combinations. A child at this level quickly names parts of any whole, or the whole given the parts. For example, when shown 4, then 1 is secretly hidden, and then shown the 3 remaining, the child may quickly say "1" is hidden.

Age Range	Level Name	Level	Description
6	Composer to 7	4	The next sign of development is when a child knows number combinations to totals of 7. A child at this level quickly names parts of any whole, or the whole when given parts, and can double numbers to 10. For example, when shown 6, then 4 are secretly hidden, and then shown the 2 remaining, the child may quickly say "4" are hidden.
6	Composer to 10	5	This level is when a child knows number combinations to totals of 10. A child at this level may quickly name parts of any whole, or the whole when given parts, and can double numbers to 20. For example, this child would be able to say "9 and 9 is 18."
7	Composer with Tens and Ones	6	At this level, the child understands two-digit numbers as tens and ones, can count with dimes and pennies, and can perform two-digit addition with regrouping. For example, a child at this level may explain, "17 and 36 is like 17 and 3, which is 20, and 33, which is 53."
7–8	+/− Fact Fluency to 20	7	Quickly produces combinations (addends to 1–10)

Developmental Levels for Adding and Subtracting

Single-digit addition and subtraction are generally characterized as "math facts." It is assumed children must memorize these facts, yet research has shown that addition and subtraction have their roots in counting, counting on, number sense, the ability to compose and decompose numbers, and place value. Research has also shown that learning methods for addition and subtraction with understanding is much more effective than rote memorization of seemingly isolated facts. Most children follow an observable developmental progression in learning to add and subtract numbers with recognizable stages or levels. This developmental path can be described as part of a learning trajectory.

Age Range	Level Name	Level	Description
1	Pre +/−	1	At the earliest level, a child shows no sign of being able to add or subtract.
3	Nonverbal +/−	2	The first sign is when a child can add and subtract very small collections nonverbally. For example, when shown 2 objects, then 1 object being hidden under a napkin, the child identifies or makes a set of 3 objects to "match."
4	Small Number +/−	3	This level is when a child can find sums for joining problems up to 3 + 2 by counting with objects. For example, when asked, "You have 2 balls and get 1 more. How many in all?" the child may count out 2, then count out 1 more, then count all 3: "1, 2, 3, 3!"
5	Find Result +/−	4	**Addition** Evidence of this level in addition is when a child can find sums for joining (you had 3 apples and get 3 more; how many do you have in all?) and part-part-whole (there are 6 girls and 5 boys on the playground; how many children were there in all?) problems by direct modeling, counting all, with objects. For example, when asked, "You have 2 red balls and 3 blue balls. How many in all?" the child may count out 2 red, then count out 3 blue, then count all 5. **Subtraction** In subtraction, a child can also solve take-away problems by separating with objects. For example, when asked, "You have 5 balls and give 2 to Tom. How many do you have left?" the child may count out 5 balls, then take away 2, and then count the remaining 3.
5	Find Change +/−	5	**Addition** At this level, a child can find the missing addend (5 + _ = 7) by adding on objects. For example, when asked, "You have 5 balls and then get some more. Now you have 7 in all. How many did you get?" The child may count out 5, then count those 5 again starting at 1, then add more, counting "6, 7," then count the balls added to find the answer, 2. **Subtraction** A child can compare by matching in simple situations. For example, when asked, "Here are 6 dogs and 4 balls. If we give a ball to each dog, how many dogs will not get a ball?" a child at this level may count out 6 dogs, match 4 balls to 4 of them, then count the 2 dogs that have no ball.
5	Make It +/−	6	A significant advancement occurs when a child is able to count on. This child can add on objects to make one number into another without counting from 1. For example, when told, "This puppet has 4 balls, but she should have 6. Make it 6," the child may put up 4 fingers on one hand, immediately count up from 4 while putting up 2 fingers on the other hand, saying, "5, 6," and then count or recognize the 2 fingers.

Age Range	Level Name	Level	Description
6	Counting Strategies +/−	7	This level occurs when a child can find sums for joining (you had 8 apples and get 3 more...) and part-part-whole (6 girls and 5 boys...) problems with finger patterns or by adding on objects or counting on. For example, when asked "How much is 4 and 3 more?" the child may answer "4...5, 6, 7. 7!" Children at this level can also solve missing addend (3 + _ = 7) or compare problems by counting on. When asked, for example, "You have 6 balls. How many more would you need to have 8?" the child may say, "6, 7 [puts up first finger], 8 [puts up second finger]. 2!"
6	Part-Whole +/−	8	Further development has occurred when the child has part-whole understanding. This child can solve problems using flexible strategies and some derived facts (for example, "5 + 5 is 10, so 5 + 6 is 11"), can sometimes do start- unknown problems (_ + 6 = 11), but only by trial and error. When asked, "You had some balls. Then you get 6 more. Now you have 11 balls. How many did you start with?" this child may lay out 6, then 3, count, and get 9. The child may put 1 more, say 10, then put 1 more. The child may count up from 6 to 11, then recount the group added, and say, "5!"
6	Numbers-in-Numbers +/−	9	Evidence of this level is when a child recognizes that a number is part of a whole and can solve problems when the start is unknown (_ + 4 = 9) with counting strategies. For example, when asked, "You have some balls, then you get 4 more balls, now you have 9. How many did you have to start with?" this child may count, putting up fingers, "5, 6, 7, 8, 9." The child may then look at his or her fingers and say, "5!"
7	Deriver +/−	10	At this level, a child can use flexible strategies and derived combinations (for example, "7 + 7 is 14, so 7 + 8 is 15") to solve all types of problems. For example, when asked, "What's 7 plus 8?" this child thinks: 7 + 8 = 7 + [7 + 1] = [7 + 7] + 1 = 14 + 1 = 15. The child can also solve multidigit problems by incrementing or combining 10s and 1s. For example, when asked "What's 28 + 35?" this child may think: 20 + 30 = 50; + 8 = 58; 2 more is 60, and 3 more is 63. He or she can also combine 10s and 1s: 20 + 30 = 50. 8 + 5 is like 8 plus 2 and 3 more, so it is 13. 50 and 13 is 63.
8+	Problem Solver +/−	11	As children develop their addition and subtraction abilities, they can solve by using flexible strategies and many known combinations. For example, when asked, "If I have 13 and you have 9, how could we have the same number?" this child may say, "9 and 1 is 10, then 3 more makes 13. 1 and 3 is 4. I need 4 more!"
8+	Multidigit +/−	12	Further development is shown when children can use composition of 10s and all previous strategies to solve multidigit +/− problems. For example, when asked, "What's 37 − 18?" this child may say, "Take 1 ten off the 3 tens; that's 2 tens. Take 7 off the 7. That's 2 tens and 0...20. I have one more to take off. That's 19." Or, when asked, "What's 28 + 35?" this child may think, 30 + 35 would be 65. But it's 28, so it's 2 less...63.

Developmental Levels for Multiplying and Dividing

Multiplication and division builds on addition and subtraction understanding and is dependent upon counting and place-value concepts. As children begin to learn to multiply, they make equal groups and count them all. They then learn skip counting and derive related products from products they know. Finding and using patterns aids in learning multiplication and division facts with understanding. Children typically follow an observable developmental progression in learning to multiply and divide numbers with recognizable stages or levels. This developmental path can be described as part of a learning trajectory.

Age Range	Level Name	Level	Description
2	Non-quantitative Sharer "Dumper"	1	Multiplication and division concepts begin very early with the problem of sharing. Early evidence of these concepts can be observed when a child dumps out blocks and gives some (not an equal number) to each person.
3	Beginning Grouper and Distributive Sharer	2	Progression to this level can be observed when a child is able to make small groups (fewer than 5). This child can share by "dealing out," but often only between 2 people, although he or she may not appreciate the numerical result. For example, to share 4 blocks, this child may give each person a block, check that each person has one, and repeat this.
4	Grouper and Distributive Sharer	3	The next level occurs when a child makes small equal groups (fewer than 6). This child can deal out equally between 2 or more recipients, but may not understand that equal quantities are produced. For example, the child may share 6 blocks by dealing out blocks to herself and a friend one at a time.
5	Concrete Modeler ×/÷	4	As children develop, they are able to solve small-number multiplying problems by grouping—making each group and counting all. At this level, a child can solve division/sharing problems with informal strategies, using concrete objects—up to 20 objects and 2 to 5 people—although the child may not understand equivalence of groups. For example, the child may distribute 20 objects by dealing out 2 blocks to each of 5 people, then 1 to each, until the blocks are gone.
6	Parts and Wholes ×/÷	5	A new level is evidenced when the child understands the inverse relation between divisor and quotient. For example, this child may understand "If you share with more people, each person gets fewer."

Age Range	Level Name	Level	Description
7	Skip Counter ×/÷	6	As children develop understanding in multiplication and division, they begin to use skip counting for multiplication and for measurement division (finding out how many groups). For example, given 20 blocks, 4 to each person, and asked how many people, the children may skip count by 4, holding up 1 finger for each count of 4. A child at this level may also use trial and error for partitive division (finding out how many in each group). For example, given 20 blocks, 5 people, and asked how many each should get, this child may give 3 to each, and then 1 more.
8+	Deriver ×/÷	7	At this level, children use strategies and derived combinations to solve multidigit problems by operating on tens and ones separately. For example, a child at this level may explain "7 × 6, five 7s is 35, so 7 more is 42."
8+	Array Quantifier	8	Further development can be observed when a child begins to work with arrays. For example, given 7 × 4 with most of 5 × 4 covered, a child at this level may say, "There are 8 in these 2 rows, and 5 rows of 4 is 20, so 28 in all."
8+	Partitive Divisor	9	This level can be observed when a child is able to figure out how many are in each group. For example, given 20 blocks, 5 people, and asked how many each should get, a child at this level may say, "4, because 5 groups of 4 is 20."
8+	Multidigit ×/÷	10	As children progress, they begin to use multiple strategies for multiplication and division, from compensating to paper-and-pencil procedures. For example, a child becoming fluent in multiplication might explain that "19 times 5 is 95, because 20 fives is 100, and 1 less five is 95."

Developmental Levels for Measuring

Measurement is one of the main real-world applications of mathematics. Counting is a type of measurement which determines how many items are in a collection. Measurement also involves assigning a number to attributes of length, area, and weight. Prekindergarten children know that mass, weight, and length exist, but they do not know how to reason about these or to accurately measure them. As children develop their understanding of measurement, they begin to use tools to measure and understand the need for standard units of measure. Children typically follow an observable developmental progression in learning to measure with recognizable stages or levels. This developmental path can be described as part of a learning trajectory.

Age Range	Level Name	Level	Description
3	Length Quantity Recognizer	1	At the earliest level, children can identify length as an attribute. For example, they might say, "I'm tall, see?"
4	Length Direct Comparer	2	In this level, children can physically align 2 objects to determine which is longer or if they are the same length. For example, they can stand 2 sticks up next to each other on a table and say, "This one's bigger."
5	Indirect Length Comparer	3	A sign of further development is when a child can compare the length of 2 objects by representing them with a third object. For example, a child might compare the length of 2 objects with a piece of string. Additional evidence of this level is that when asked to measure, the child may assign a length by guessing or moving along a length while counting (without equal-length units). For example, the child may move a finger along a line segment, saying 10, 20, 30, 31, 32.
6	Serial Orderer to 6+	4	At this level, a child can order lengths, marked in 1 to 6 units. For example, given towers of cubes, a child at this level may put them in order, 1 to 6.
6	End-to-End Length Measurer	5	At this level, the child can lay units end-to-end, although he or she may not see the need for equal-length units. For example, a child might lay 9-inch cubes in a line beside a book to measure how long it is.

Age Range	Level Name	Level	Description
7	Length Unit Relater and Repeater	6	At this level, a child can relate size and number of units. For example, the child may explain, "If you measure with centimeters instead of inches, you'll need more of them because each one is smaller."
8+	Length Measurer	7	As a child develops measurement ability, they begin to measure, knowing the need for identical units, the relationships between different units, partitions of unit, and the zero point on rulers. At this level, the child also begins to estimate. The children may explain, "I used a meterstick 3 times, then there was a little left over. So, I lined it up from 0 and found 14 centimeters. So, it's 3 meters, 14 centimeters in all."
8+	Conceptual Ruler Measurer	8	Further development in measurement is evidenced when a child possesses an "internal" measurement tool. At this level, the child mentally moves along an object, segmenting it, and counting the segments. This child also uses arithmetic to measure and estimates with accuracy. For example, a child at this level may explain, "I imagine one meterstick after another along the edge of the room. That's how I estimated the room's length to be 9 meters."

Developmental Levels for Recognizing Geometric Shapes

Geometric shapes can be used to represent and understand objects. Analyzing, comparing, and classifying shapes helps create new knowledge of shapes and their relationships. Shapes can be decomposed or composed into other shapes. Through their everyday activities, children build both intuitive and explicit knowledge of geometric figures. Most children can recognize and name basic two-dimensional shapes at four years of age. However, young children can learn richer concepts about shape if they have varied examples and nonexamples of shape, discussions about shapes and their characteristics, a wide variety of shape classes, and interesting tasks. Children typically follow an observable developmental progression in learning about shapes with recognizable stages or levels. This developmental path can be described as part of a learning trajectory.

Age Range	Level Name	Level	Description
2	Shape Matcher—Identical Shape Matcher—Sizes Shape Matcher—Orientations	1	The earliest sign of understanding shape is when a child can match basic shapes (circle, square, typical triangle) with the same size and orientation. A sign of development is when a child can match basic shapes with different sizes. This level of development is when a child can match basic shapes with different orientations.
3	Shape Recognizer—Typical	2	A sign of development is when a child can recognize and name a prototypical circle, square, and, less often, a typical triangle. For example, the child names this a square. Some children may name different sizes, shapes, and orientations of rectangles, but also accept some shapes that look rectangular but are not rectangles. Children name these shapes "rectangles" (including the nonrectangular parallelogram).
3	Shape Matcher—More Shapes Shape Matcher—Sizes and Orientations Shape Matcher—Combinations	3	As children develop understanding of shape, they can match a wider variety of shapes with the same size and orientation. The child matches a wider variety of shapes with different sizes and orientations. The child matches combinations of shapes to each other.
4	Shape Recognizer—Circles, Squares, and Triangles	4	This sign of development is when a child can recognize some nonprototypical squares and triangles and may recognize some rectangles, but usually not rhombi (diamonds). Often, the child does not differentiate sides/corners. The child at this level may name these as triangles.
4	Constructor of Shapes from Parts—Looks Like *Representing*	5	A significant sign of development is when a child represents a shape by making a shape "look like" a goal shape. For example, when asked to make a triangle with sticks, the child may create the following: △.
5	Shape Recognizer—All Rectangles	6	As children develop understanding of shape, they recognize more rectangle sizes, shapes, and orientations of rectangles. For example, a child at this level may correctly name these shapes "rectangles."
5	Side Recognizer *Parts*	7	A sign of development is when a child recognizes parts of shapes and identifies sides as distinct geometric objects. For example, when asked what this shape is, the child may say it is a quadrilateral (or has 4 sides) after counting and running a finger along the length of each side.
5	Angle (Corner) Recognizer *Parts*	8	At this level, a child can recognize angles as separate geometric objects. For example, when asked, "Why is this a triangle," the child may say, "It has three angles" and count them, pointing clearly to each vertex (point at the corner).
5	Shape Recognizer—More Shapes	9	As children develop, they are able to recognize most basic shapes and prototypical examples of other shapes, such as hexagon, rhombus (diamond), and trapezoid. For example, a child can correctly identify and name all the following shapes:
6	Shape Identifier	10	At this level, the child can name most common shapes, including rhombi, without making mistakes such as calling ovals circles. A child at this level implicitly recognizes right angles, so distinguishes between a rectangle and a parallelogram without right angles. A child may correctly name all the following shapes:
6	Angle Matcher *Parts*	11	A sign of development is when the child can match angles concretely. For example, given several triangles, the child may find two with the same angles by laying the angles on top of one another.

Learning Trajectories

Age Range	Level Name	Level	Description
7	Parts of Shapes Identifier	12	At this level, the child can identify shapes in terms of their components. For example, the child may say, "No matter how skinny it looks, that's a triangle because it has 3 sides and 3 angles."
7	Constructor of Shapes from Parts—Exact *Representing*	13	A significant step is when the child can represent a shape with completely correct construction, based on knowledge of components and relationships. For example, when asked to make a triangle with sticks, the child may create the following:
8	Shape Class Identifier	14	As children develop, they begin to use class membership (for example, to sort) not explicitly based on properties. For example, a child at this level may say, "I put the triangles over here, and the quadrilaterals, including squares, rectangles, rhombi, and trapezoids, over there."
8	Shape Property Identifier	15	At this level, a child can use properties explicitly. For example, a child may say, "I put the shapes with opposite sides that are parallel over here, and those with 4 sides but not both pairs of sides parallel over there."

Age Range	Level Name	Level	Description
8	Angle Size Comparer	16	The next sign of development is when a child can separate and compare angle sizes. For example, the child may say, "I put all the shapes that have right angles here, and all the ones that have bigger or smaller angles over there."
8	Angle Measurer	17	A significant step in development is when a child can use a protractor to measure angles.
8	Property Class Identifier	18	The next sign of development is when a child can use class membership for shapes (for example, to sort or consider shapes "similar") explicitly based on properties, including angle measure. For example, the child may say, "I put the equilateral triangles over here, and the right triangles over here."
8	Angle Synthesizer	19	As children develop understanding of shape, they can combine various meanings of angle (turn, corner, slant). For example, a child at this level could explain, "This ramp is at a 45° angle to the ground."

Developmental Levels for Composing Geometric Shapes

Children move through levels in the composition and decomposition of two-dimensional figures. Very young children cannot compose shapes but then gain ability to combine shapes into pictures, synthesize combinations of shapes into new shapes, and eventually substitute and build different kinds of shapes. Children typically follow an observable developmental progression in learning to compose shapes with recognizable stages or levels. This developmental path can be described as part of a learning trajectory.

Age Range	Level Name	Level	Description
2	Pre-Composer	1	The earliest sign of development is when a child can manipulate shapes as individuals, but is unable to combine them to compose a larger shape.
3	Pre-Decomposer	2	At this level, a child can decompose shapes, but only by trial and error. For example, given only a hexagon, the child can break it apart to make a simple picture by trial and error.
4	Piece Assembler	3	Around age 4, a child can begin to make pictures in which each shape represents a unique role (for example, one shape for each body part) and shapes touch. A child at this level can fill simple outline puzzles using trial and error.
5	Picture Maker	4	As children develop, they are able to put several shapes together to make one part of a picture (for example, 2 shapes for 1 arm). A child at this level uses trial and error and does not anticipate creation of the new geometric shape. The children can choose shapes using "general shape" or side length, and fill "easy" outline puzzles that suggest the placement of each shape (but note that the child is trying to put a square in the puzzle where its right angles will not fit).
5	Simple Decomposer	5	A significant step occurs when the child is able to decompose ("take apart" into smaller shapes) simple shapes that have obvious clues as to their decomposition.
5	Shape Composer	6	A sign of development is when a child composes shapes with anticipation ("I know what will fit!"). A child at this level chooses shapes using angles as well as side lengths. Rotation and flipping are used intentionally to select and place shapes. For example, in this puzzle, all angles are correct, and patterning is evident.
6	Substitution Composer	7	A sign of development is when a child is able to make new shapes out of smaller shapes and uses trial and error to substitute groups of shapes for other shapes in order to create new shapes in different ways. For example, the child can substitute shapes to fill outline puzzles in different ways.
6	Shape Decomposer (with Help)	8	As children develop, they can decompose shapes by using imagery that is suggested and supported by the task or environment. For example, given hexagons, the child can break them apart to make this shape.
7	Shape Composite Repeater	9	This level is demonstrated when the child can construct and duplicate units of units (shapes made from other shapes) intentionally, and understands each as being both multiple, small shapes and one larger shape. For example, the child may continue a pattern of shapes that leads to tiling.
7	Shape Decomposer with Imagery	10	A significant sign of development is when a child is able to decompose shapes flexibly by using independently generated imagery. For example, the child can break hexagons apart into shapes such as these.
8	Shape Composer—Units of Units	11	Children demonstrate further understanding when they are able to build and apply units of units (shapes made from other shapes). For example, in constructing spatial patterns, the child can extend patterning activity to create a tiling with a new unit shape—a unit of unit shapes that he or she recognizes and consciously constructs. For example, the child may build *T*s out of 4 squares, use 4 *T*s to build squares, and use squares to tile a rectangle.
8	Shape Decomposer — Units of Units	12	As children develop understanding of shape, they can decompose shapes flexibly by using independently generated imagery and planned decompositions of shapes that themselves are decompositions.

Learning Trajectories

Developmental Levels for Comparing Geometric Shapes

As early as four years of age, children can create and use strategies, such as moving shapes to compare their parts or to place one on top of the other for judging whether two figures are the same shape. From PreK to Grade 2, they can develop sophisticated and accurate mathematical procedures for comparing geometric shapes. Children typically follow an observable developmental progression in learning about how shapes are the same and different with recognizable stages or levels. This developmental path can be described as part of a learning trajectory.

Age Range	Level Name	Level	Description
3	"Same Thing" Comparer	1	The first sign of understanding is when the child can compare real-world objects. For example, the children may say two pictures of houses are the same or different.
4	"Similar" Comparer	2	This sign of development occurs when the child judges two shapes to be the same if they are more visually similar than different. For example, the child may say, "These are the same. They are pointy at the top."
4	Part Comparer	3	At this level, a child can say that two shapes are the same after matching one side on each. For example, a child may say, "These are the same" (matching the two sides).
4	Some Attributes Comparer	4	As children develop, they look for differences in attributes, but may examine only part of a shape. For example, a child at this level may say, "These are the same" (indicating the top halves of the shapes are similar by laying them on top of each other).
5	Most Attributes Comparer	5	At this level, the child looks for differences in attributes, examining full shapes, but may ignore some spatial relationships. For example, a child may say, "These are the same."
7	Congruence Determiner	6	A sign of development is when a child determines congruence by comparing all attributes and all spatial relationships. For example, a child at this level may say that two shapes are the same shape and the same size after comparing every one of their sides and angles.
7	Congruence Superposer	7	As children develop understanding, they can move and place objects on top of each other to determine congruence. For example, a child at this level may say that two shapes are the same shape and the same size after laying them on top of each other.
8+	Congruence Representer	8	Continued development is evidenced as children refer to geometric properties and explain with transformations. For example, a child at this level may say, "These must be congruent because they have equal sides, all square corners, and I can move them on top of each other exactly."

Developmental Levels for Spatial Sense and Motions

Infants and toddlers spend a great deal of time learning about the properties and relations of objects in space. Very young children know and use the shape of their environment in navigation activities. With guidance they can learn to "mathematize" this knowledge. They can learn about direction, perspective, distance, symbolization, location, and coordinates. Children typically follow an observable developmental progression in developing spatial sense with recognizable stages or levels. This developmental path can be described as part of a learning trajectory.

Age Range	Level Name	Level	Description
4	Simple Turner	1	An early sign of spatial sense is when a child mentally turns an object to perform easy tasks. For example, given a shape with the top marked with color, the child may correctly identify which of three shapes it would look like if it were turned "like this" (90 degree turn demonstrated), before physically moving the shape.
5	Beginning Slider, Flipper, Turner	2	This sign of development occurs when a child can use the correct motions, but is not always accurate in direction and amount. For example, a child at this level may know a shape has to be flipped to match another shape, but flips it in the wrong direction.
6	Slider, Flipper, Turner	3	As children develop spatial sense, they can perform slides and flips, often only horizontal and vertical, by using manipulatives. For example, a child at this level may perform turns of 45, 90, and 180 degrees. For example, a child knows a shape must be turned 90 degrees to the right to fit into a puzzle.
7	Diagonal Mover	4	A sign of development is when a child can perform diagonal slides and flips. For example, a children at this level may know a shape must be turned or flipped over an oblique line (45 degree orientation) to fit into a puzzle.
8	Mental Mover	5	Further signs of development occur when a child can predict results of moving shapes using mental images. A child at this level may say, "If you turned this 120 degrees, it would be just like this one."

Developmental Levels for Patterning and Early Algebra

Algebra begins with a search for patterns. Identifying patterns helps bring order, cohesion, and predictability to seemingly unorganized situations and allows one to make generalizations beyond the information directly available. The recognition and analysis of patterns are important components of the young children's intellectual development because they provide a foundation for the development of algebraic thinking. Although prekindergarten children engage in pattern-related activities and recognize patterns in their everyday environment, research has revealed that an abstract understanding of patterns develops gradually during the early childhood years. Children typically follow an observable developmental progression in learning about patterns with recognizable stages or levels. This developmental path can be described as part of a learning trajectory.

Age Range	Level Name	Level	Description
2	Pre-Explicit Patterner	1	A child at the earliest level does not recognize patterns. For example, a child may name a striped shirt with no repeating unit a "pattern."
3	Pattern Recognizer	2	At this level, the child can recognize a simple pattern. For example, a child at this level may say, "I'm wearing a pattern" about a shirt with black, white, black, white stripes.
4	Pattern Fixer	3	At this level the child fills in missing elements of a pattern, first with ABABAB patterns. When given items in a row with an item missing, such as ABAB_BAB, the child identifies and fills in the missing element (A).
4	Pattern Duplicator AB	4	A sign of development is when the child can duplicate an ABABAB pattern, although the children may have to work alongside the model pattern. For example, given objects in a row, ABABAB, the child may make his or her own ABABAB row in a different location.
4	Pattern Extender AB	5	At this level the child extends AB repeating patterns. For example, given items in a row—ABABAB—the child adds ABAB to the end of the row.
4	Pattern Duplicator	6	At this level, the child is able to duplicate simple patterns (not just alongside the model pattern). For example, given objects in a row, ABBABBABB, the child may make his or her own ABBABBABB row in a different location.
5	Pattern Extender	7	A sign of development is when the child can extend simple patterns. For example, given objects in a row, ABBABBABB, he or she may add ABBABB to the end of the row.
6	Pattern Unit Recognizer	8	At this level, a child can identify the smallest unit of a pattern. For example, given objects in a ABBABBABB pattern, the child identifies the core unit of the pattern as ABB.
7	Numeric Patterner	9	Describes a pattern numerically, can translate between geometric and numeric representation of a series. For example, given objects in a geometric pattern, child describes the numeric progression.

Age Range	Level Name	Level	Description
7	Beginning Arithmetic Patterner	10	Recognizes and uses relatively transparent arithmetic patterns with perceptual or pedagogical support, first that involve properties of zero. Accepts number sentences not in the form of $a + b = c$ (e.g., $c = a + b$, or $a + b = c + d$); this represent a move from Pre-Equivalence, "equals-as-ananswer" to Equal Numbers Relater. For example, child recognizes and uses patterns (e.g., that can be symbolized as $a + b - b = a$)
8	Relational Thinker +/−	11	Recognizes and uses patterns that involve addition and subtraction and is an Equals Relater who can compare two sides of a number sentence with reasoning without actually carrying out the computations. For example, child recognizes $3 + 6 - 3 = 6$ as a true statement without performing computations.
8	Relational Thinker—Symbolic +/−	12	Recognizes and uses patterns that involve addition and subtraction and is an Equals Relater who can compare two sides of a number sentence with reasoning without actually carrying out the computations. For example, child recognizes $a + b = b + a$ (presented that way, symbolically) as a true statement in all cases.
9	Relational Thinker with Multiplication	13	Recognizes and uses patterns that involve multiplication as repeated addition and the use of the distributive property to partition number facts. For example, child recognizes $3 \times 6 + 3 = 4 \times 6$ as a true statement without performing all computations.

Developmental Levels for Classifying and Analyzing Data

Data analysis contains one big idea: classifying, organizing, representing, and using information to ask and answer questions. The developmental continuum for data analysis includes growth in classifying and counting to sort objects and quantify their groups. Children eventually become capable of simultaneously classifying and counting, for example, counting the number of colors in a group of objects. Children typically follow an observable developmental progression in learning about patterns with recognizable stages or levels. This developmental path can be described as part of a learning trajectory.

Age Range	Level Name	Level	Description
2	Similarity Recognizer	1	The first sign that a child can classify is when he or she recognizes, intuitively, two or more objects as "similar" in some way. For example, "that's another doggie."
2	Informal Sorter	2	A sign of development is when a child places objects that are alike in some attribute together, but switches criteria and may use functional relationships as the basis for sorting. A child at this level might stack blocks of the same shape or put a cup with its saucer.
3	Attribute Identifier	3	The next level is when the child names attributes of objects and places objects together with a given attribute, but cannot then move to sorting by a new rule. For example, the child may say, "These are both red."
4	Attribute Sorter	4	At the next level the child sorts objects according to given attributes, forming categories, but may switch attributes during the sorting. A child at this stage can switch rules for sorting if guided. For example, the child might start putting red beads on a string, but switches to spheres of different colors.
5	Consistent Sorter	5	A sign of development is when the child can sort consistently by a given attribute. For example, the child might put several identical blocks together.
6	Exhaustive Sorter	6	At the next level, the child can sort consistently and exhaustively by an attribute, given or created. This child can use terms "some" and "all" meaningfully. For example, a child at this stage would be able to find all the attribute blocks of a certain size and color.
6	Multiple Attribute Sorter	7	A sign of development is when the child can sort consistently and exhaustively by more than one attribute, sequentially. For example, a child at this level can put all the attribute blocks together by color, then by shape.
7	Classifier and Counter	8	At the next level, the child is capable of simultaneously classifying and counting. For example, the child counts the number of colors in a group of objects.
7	List Grapher	9	In the early stage of graphing, the child graphs by simply listing all cases. For example, the child may list each child in the class and each child's response to a question.
8+	Multiple Attribute Classifier	10	A sign of development is when the child can intentionally sort according to multiple attributes, naming and relating the attributes. This child understands that objects could belong to more than one group. For example, the child can complete a two-dimensional classification matrix or form subgroups within groups.

Age Range	Level Name	Level	Description
8+	Classifying Grapher	11	At the next level the child can graph by classifying data (e.g., responses) and represent it according to categories. For example, the child can take a survey, classify the responses, and graph the result.
8+	Classifier	12	A sign of development is when the child creates complete, conscious classifications logically connected to a specific property. For example, a child at this level gives a definition of a class in terms of a more general class and one or more specific differences and begins to understand the inclusion relation.
8+	Hierarchical Classifier	13	At the next level, the child can perform hierarchical classifications. For example, the child recognizes that all squares are rectangles, but not all rectangles are squares.
8+	Data Representer	14	Signs of development are when the child organizes and displays data through both simple numerical summaries such as counts, tables, and tallies, and graphical displays, including picture graphs, line plots, and bar graphs. At this level the child creates graphs and tables, compares parts of the data, makes statements about the data as a whole, and determines whether the graphs answer the questions posed initially.

Building Blocks Software Activities

Building Blocks software provides computer math activities that address specific developmental levels of the math learning trajectories. ***Building Blocks*** software is critical to ***Building Blocks PreK*** and provides support activities for specific concepts typically taught in grades K–6.

Some ***Building Blocks*** activities have different levels of difficulty indicated by ranges in the Activity Names below. The list provides an overview of all of the ***Building Blocks*** activities along with the domains, descriptions, and appropriate age ranges.

Domain: Trajectory	Activity Name	Description	Age Range
Geometry: Composition/ Decomposition	Create a Scene	Students explore shapes by moving and manipulating them to make pictures.	4–12
Geometry: Composition/ Decomposition	Piece Puzzler 1–5, Piece Puzzler Free Explore, and Super Shape 1–7	Students complete puzzles using pattern or tangram shapes.	4–12
Geometry: Imagery	Geometry Snapshots 1–8	Students match configurations of a variety of shapes (e.g., line segments in different arrangements, 3–6 tiled shapes, embedded shapes) to corresponding configurations, given only a brief view of the goal shapes.	5–12
Geometry: Shapes (Identifying)	Memory Geometry 1–5	Students match familiar geometric shapes (shapes in same or similar sizes, same orientation) within the framework of a Concentration card game.	3–5
Geometry: Shapes (Matching)	Mystery Pictures 1–4 and Mystery Pictures Free Explore	Students construct predefined pictures by selecting shapes that match a series of target shapes.	3–8
Geometry: Shapes (Parts)	Shape Parts 1–7	Students build or fix some real-world object, exploring shape and properties of shapes.	5–12
Geometry: Shapes (Properties)	Legends of the Lost Shape	Students identify target shapes using textual clues provided.	8–12
Geometry: Shapes (Properties)	Shape Shop 1–3	Students identify a wide range of shapes given their names, with more difficult distractors.	8–12
Measurement: Length	Comparisons	Students are shown pictures of two objects and are asked to click on the one that fits the prompt (longer, shorter, heavier, and so on).	4–8
Measurement: Length	Deep Sea Compare	Students compare the length of two objects by representing them with a third object.	5–7
Measurement: Length	Reptile Ruler	Students learn about linear measurement by using a ruler to determine the length of various reptiles.	7–10
Measurement: Length	Workin' on the Railroad	Students identify the length (in nonstandard units) of railroad trestles they built to span a gully.	6–9
Multiplication/Division	Arrays in Area	Students build arrays and then determine the area of those arrays.	8–11
Multiplication/Division	Comic Book Shop	Students use skip counting to produce products that are multiples of 10s, 5s, 2s, and 3s. The task is to identify the product, given a number and bundles.	7–9
Multiplication/Division	Egg-stremely Equal	Students divide large sets of eggs into several equal parts.	4–8
Multiplication/Division	Field Trip	Students solve multidigit multiplication problems in a field-trip environment (e.g., equal number of students on each bus; number of tickets needed for all students).	8–11
Multiplication/Division	Snack Time	Students use direct modeling to solve multiplication problems.	6–8
Multiplication/Division	Word Problems with Tools 5–6, 10	Students use number tools to solve single and multidigit multiplication and division problems.	8–11
Multiplication/Division	Clean the Plates	Students use skip counting to produce products that are multiples of 10s, 5s, 2s, and 3s.	7–9
Numbers: Adding and Subtracting	Barkley's Bones 1–10 and 1–20	Students determine the missing addend in $X + \_\_ = Z$ problems to feed bone treats to a dog ($Z = 10$ or less).	5–8

Software Activities

Domain: Trajectory	Activity Name	Description	Age Range
Number: Adding and Subtracting	Double Compare 1–10 and 1–20	Students compare sums of cards (to 10 or 20) to determine which sum is greater.	5–8
Number: Adding and Subtracting	Word Problems with Tools 1–4, 7–9, 11–12	Students use number tools to solve single and multidigit addition and subtraction problems.	8–12
Number: Adding and Subtracting and Counting	Counting Activities (Road Race Counting Game, Numeral Train Game, et. al.)	Students identify numerals or dot amounts (totals to 20) and move forward a corresponding number of spaces on a game board.	3–9
Number: Adding and Subtracting and Multiplying and Dividing	Function Machine 1–4	Students provide inputs to a function and examine the resulting outputs to determine the definition of that function. Functions include either addition, subtraction, multiplication, or division.	6–12
Number: Comparing	Ordinal Construction Company	Students learn ordinal positions (1st through 10th) by moving objects between floors of a building.	5–7
Number: Comparing	Rocket Blast 1–3	Given a number line with only initial and final endpoints labeled and a location on that line, students determine the number label for that location.	6–12
Number: Comparing and Counting	Party Time 1–3 and Party Time Free Explore	Students use party utensils to practice one-to-one correspondence, identify numerals that represent target amounts, and match object amounts to target numerals.	4–6
Number: Comparing and Multiplication and Division	Number Compare 1–5	Students compare two cards and choose the one with the greater value.	4–11
Number: Comparing, Counting, Adding, and Subtracting	Pizza Pizzazz 1–5 and Pizza Pizzazz Free Explore	Students count items, match target amounts, and explore missing addends related to toppings on pizzas.	3–8
Number: Counting (Object)	Countdown Crazy	Students click digits in sequence to count down from 10 to 0.	5–7
Number: Counting (Object)	Memory Number 1–3	Students match displays containing both numerals and collections to matching displays within the framework of a Concentration card game.	4–6
Number: Counting (Object) and Adding and Subtracting	Dinosaur Shop 1–4 and Dinosaur Shop Free Explore	Students use toy dinosaurs to identify numerals representing target amounts, match object amounts to target numerals, add groups of objects, and find missing addends.	4–7
Number: Counting (Objects)	Book Stacks	Students fill an order by counting up from a two-digit number through the next decade. Students count on (through at least one decade) from a given number as they load books onto a cart.	6–8
Number: Counting (Objects)	School Supply Shop	Students count school supplies bundled in groups of ten to reach a target number up to 100.	6-8
Number: Counting (Objects)	Tire Recycling	Students use skip counting by 2s and 5s to count tires as the tires are moved.	6–8
Number: Counting (Strategies)	Build Stairs 1–3, and Build Stairs Free Explore	Students practice counting, sequencing, and ordering by building staircases.	4–7
Number: Counting (Strategies)	Math-O-Scope	Students identify the numbers that surround a given number in the context of a 100s Table.	7–9
Number: Counting (Strategies)	Tidal Tally	Students identify missing addends (hidden objects) by counting on from given addends (visible objects) to reach a numerical total.	6–9
Number: Counting (Verbal)	Count and Race	Students count up to 50 by adding cars to a racetrack one at a time.	3–6
Number: Counting (Verbal)	Before and After Math	Students identify and select numbers that come either just before or right after a target number.	4–7
Number: Counting (Verbal)	Kitchen Counter	Students click on objects one at a time while the numbers from 1 to 10 are counted aloud.	3–6
Number: Subitizing	Number Snapshots 1–10	Students match numerals or dot collections to corresponding numerals or collections given only a brief view of the goal collections.	3–12
Patterning	Marching Patterns 1–3	Students extend a linear pattern of marchers by one full repetition of an entire unit (AB, AAB, ABB, and ABC patterns).	5–7
Patterning	Pattern Planes 1–3	Students duplicate a linear pattern of flags based on an outline that serves as a guide (AB, AAB, ABB, and ABC patterns).	4–6
Patterning	Free Explore	Students explore patterning by creating rhythmic patterns of their own.	3–6

Number Formation Guide

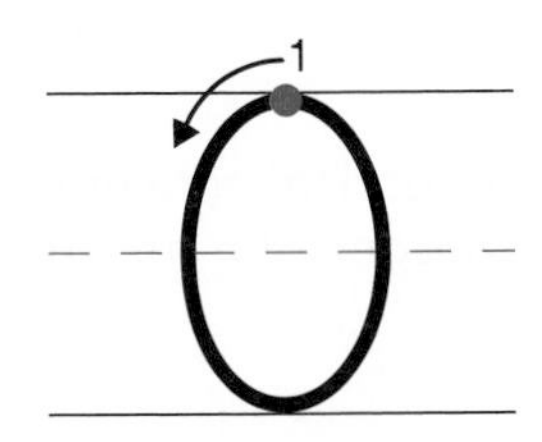

0 Starting point, curving left all the way around to starting point: 0

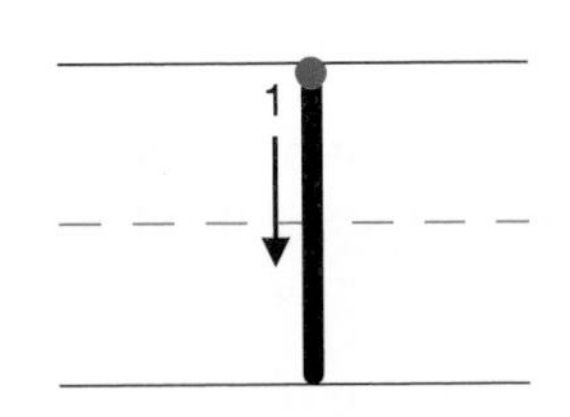

1 Starting point, straight down: 1

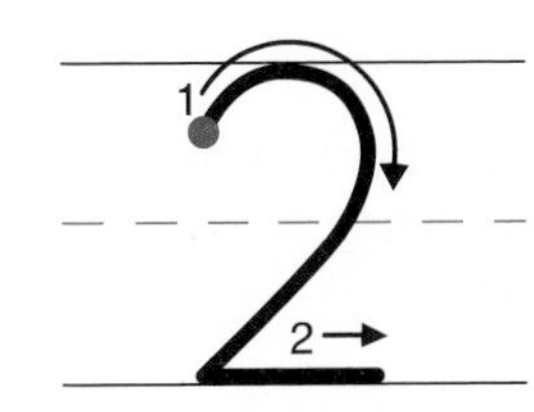

2 Starting point, around right, slanting left and straight across right: 2

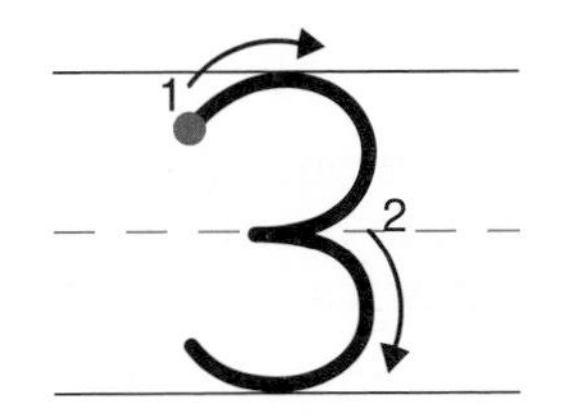

3 Starting point, around right, in at the middle, around right: 3

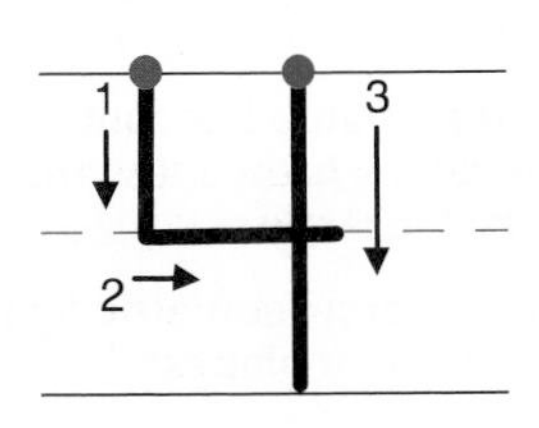

4 Starting point, straight down
Straight across right
Starting point, straight down, crossing line: 4

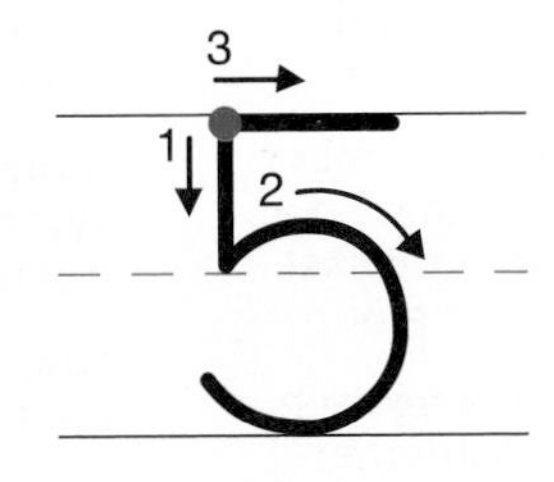

5 Starting point, straight down, curving around right and up
Starting point, straight across right: 5

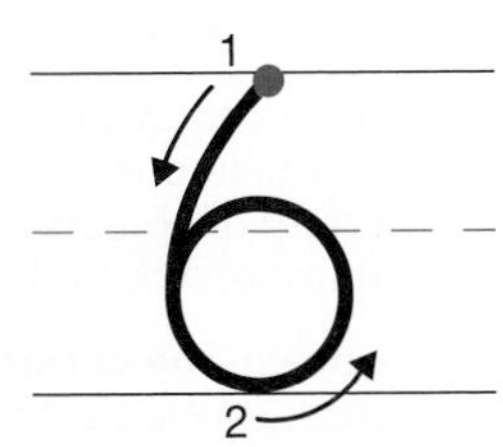

6 Starting point, slanting left, around the bottom curving up, around right and into the curve: 6

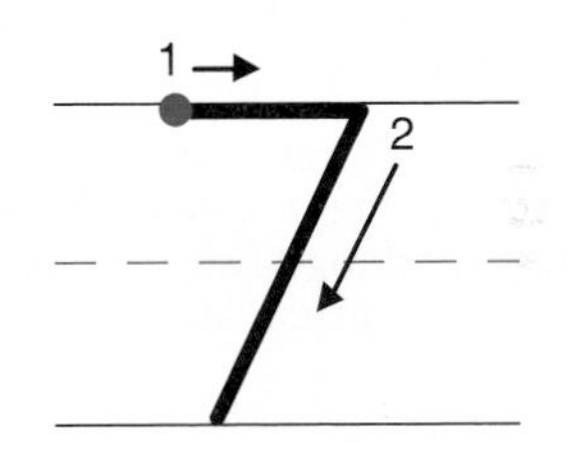

7 Starting point, straight across right, slanting down left: 7

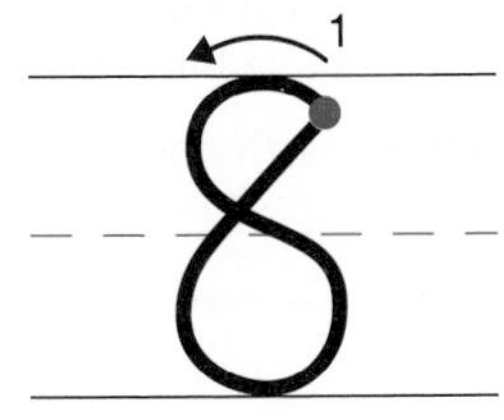

8 Starting point, curving left, curving down and around right, slanting up right to starting point: 8

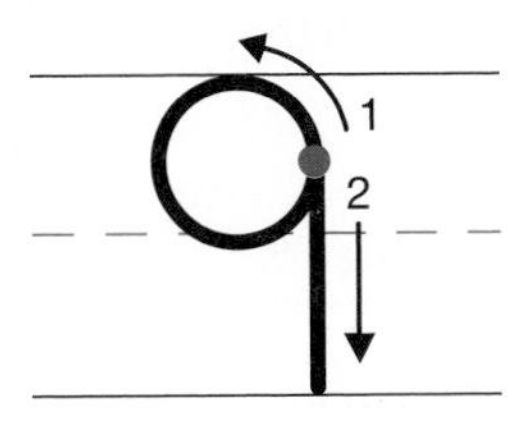

9 Starting point, curving around left all the way, straight down: 9

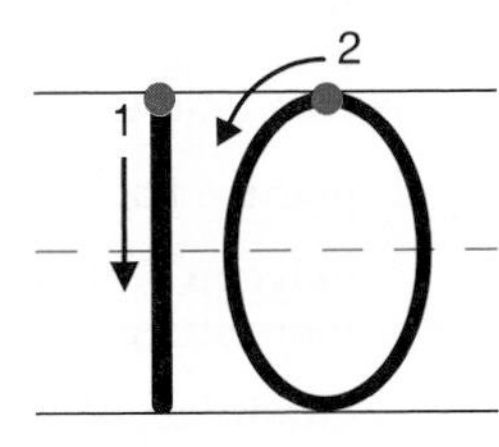

10 Starting point, straight down
Starting point, curving left all the way around to starting point: 10

Glossary

These definitions are intended to help teachers understand preschooler's development of specific mathematical concepts, and to talk to them about these concepts. They are not formal math definitions but rather simple descriptions of a mixture of mathematics and everyday vocabulary.

angle Two lines that meet to make a corner.

area An amount of surface contained within a boundary, that can be measured.

capacity The amount of material a container like a cup can hold. See also **volume.**

cardinal number A number used to represent how many in a group; if items in a group are counted, "one, two, three," then the cardinal number of that group is three.

closed A two-dimensional figure is closed when it is made up of several line segments that are joined together, exactly two sides meet at every vertex, and no sides cross each other.

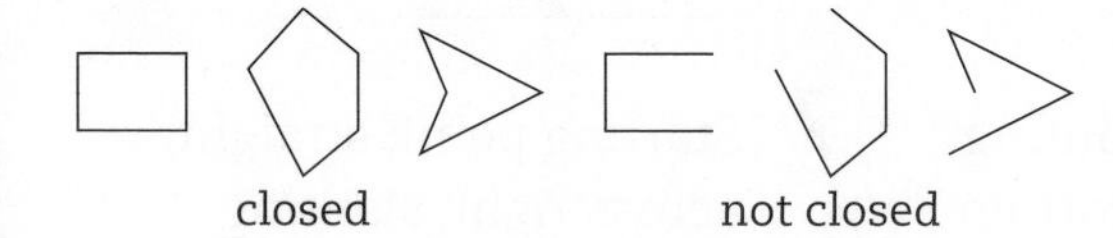

congruent Exactly alike in shape and size.

core unit The basic repeated element of a sequential pattern. In an ABCABCABC... pattern, ABC is the core unit.

distance The shortest path between two points. See also **length.**

group or collection One or more items considered together.

hexagon A closed shape with six straight sides.

horizontal Heading side-to-side, like the horizon, and at a right angle to up and down.

kite A four-sided figure with two pairs of adjacent sides that are the same length.

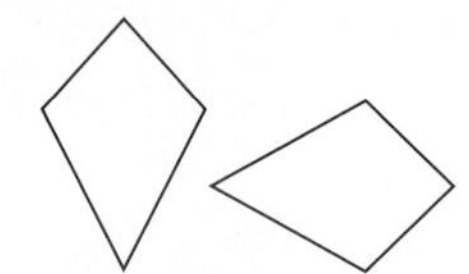

length How far it is from one end of a figure or object to the other end.

line symmetry Plane figures have line, or mirror, symmetry when their shape is reversed on opposite sides of a line, like R|Я. If the plane is folded at the line, the figures will fit together.

match To find things that are the same in some well-defined way, such as items that are exactly the same size and shape.

measurement Assigning a number to continuous quantities. Measuring consists of two aspects: identifying a unit of measure (like an inch or a liter) and counting the number of those units in the given quantity (e.g., iterating the unit, or placing the unit end-to-end, along an object).

object counting The ability to recite number words in one-to-one correspondence with a group of objects and recognize that the last number word tells how many in the group.

one-to-one correspondence Matching each item in one set to a single item in another set. For example, a child might match cups to plates to determine if there is exactly one cup for every plate. In **object counting,** one number word (e.g., "one, two...") is matched to each object counted.

orientation How a figure is turned compared to a reference line.

parallel lines Lines that remain the same distance apart like railroad tracks.

pattern A regular relationship in quantity, number or form. For example, ABABABAB is a repeating, sequential pattern in which the core unit AB repeats.

plane A flat surface.

polygon A plane figure bounded by three or more straight sides.

quadrilateral A polygon with four straight sides.

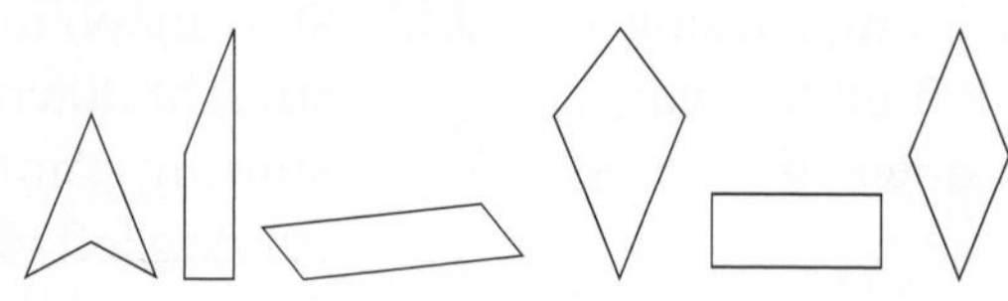

rectangle A polygon with four straight sides (i.e., a quadrilateral) and four right angles. A rectangle's opposite sides are parallel and the same length.

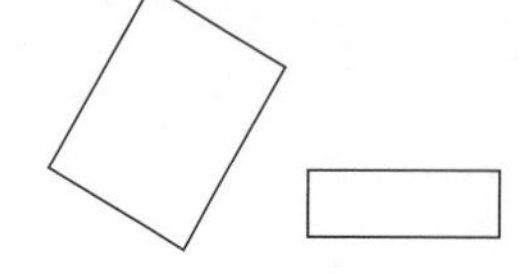

rhombus A plane figure with four straight sides all the same length.

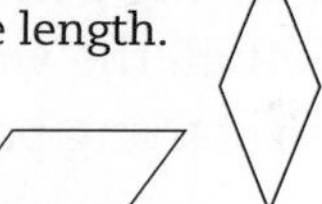

right angle Two lines that meet like a corner of a typical doorway. Often informally called "square corner," right angles measure 90 degrees. Lines intersecting at a right angle are perpendicular.

rotational symmetry A figure has rotational symmetry when it can be turned less than a full turn to fit on itself exactly.

shape Informal name for a geometric figure made up of points, lines, or planes.

square A polygon that has four equal straight sides and all right angles. Note that a square is both a special kind of rectangle and a special kind of rhombus.

subitize The ability to quickly recognize the number of objects in a small group without counting.

transitive reasoning A relationship among three items when what holds true between 1 and 2 as well as 2 and 3 must apply to 1 and 3. For example, if A is longer than B and B is longer than C, then A must be longer than C.

trapezoid A quadrilateral with one pair of parallel sides.

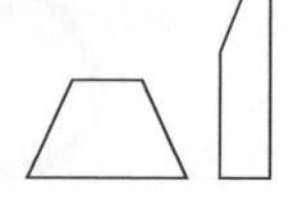

triangle A polygon with three sides.

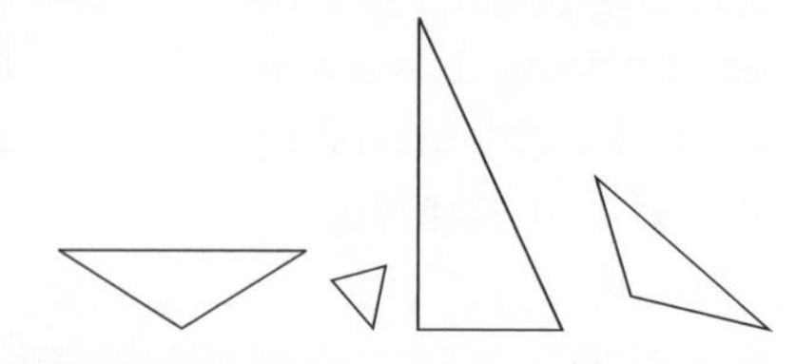

verbal counting Reciting number words in order.

vertical Heading up and down, at a right angle to the horizon.

volume The measure in units of capacity.

weight The measure of the heaviness of an object.

Hierarchies of Shapes

Quadrilaterals

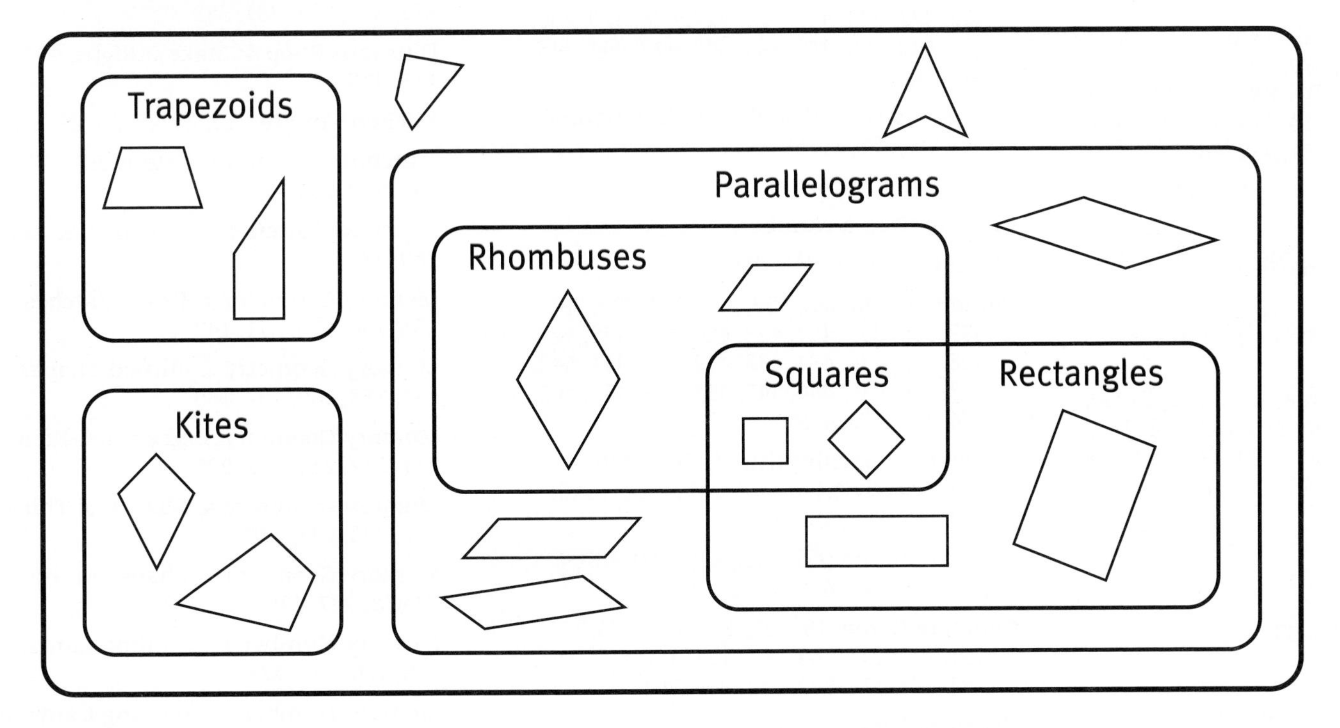

Triangles

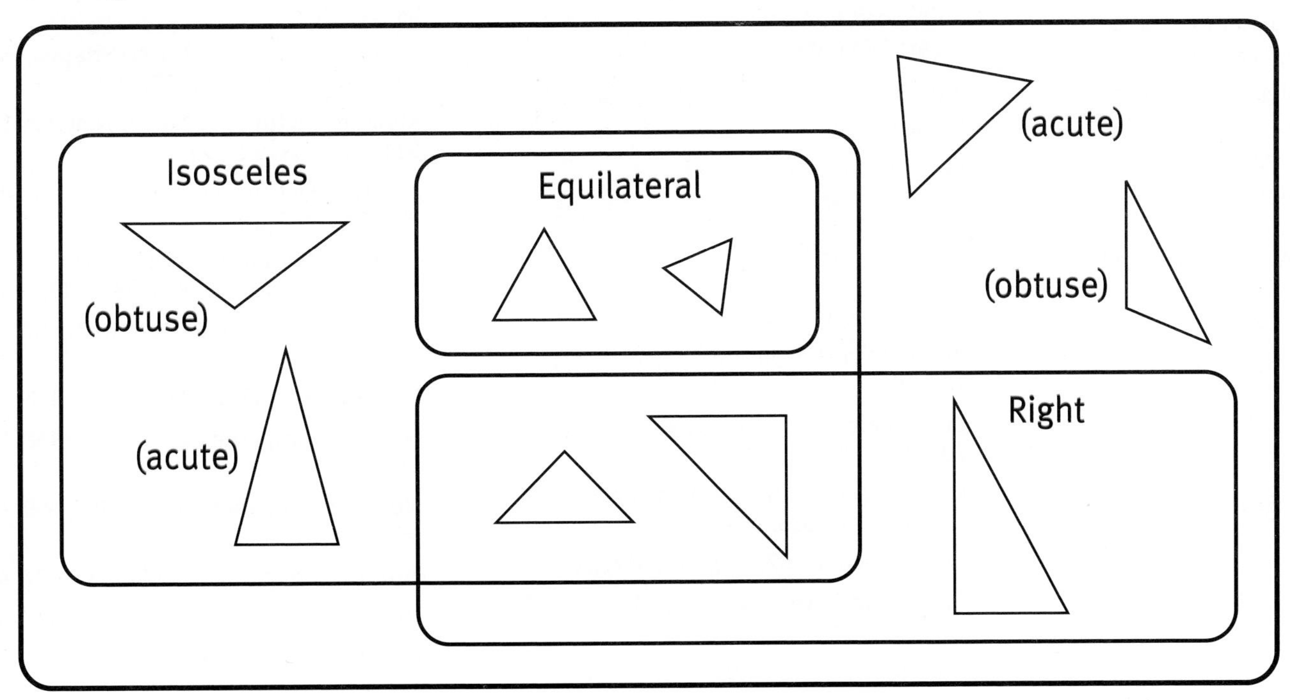

Index

D

E

Index

F

G

H

I

K

L

M

Index

N

O

P

Q

R

S

T

V

W

X